T0172753

Effective Project Management Through Applied Cost and Schedule Control

COST ENGINEERING

A Series of Reference Books and Textbooks

Editor

KENNETH K. HUMPHREYS, Ph.D.

Consulting Engineer
Granite Falls, North Carolina

Additional Volumes in Preparation

Effective Project Management Through Applied Cost and Schedule Control

edited by

James A. Bent
James Bent Associates, Inc.
Bountiful, Utah

Kenneth K. Humphreys
Consulting Engineer
Granite Falls, North Carolina

CRC Press
Taylor & Francis Group
Boca Raton London New York

CRC Press is an imprint of the
Taylor & Francis Group, an **Informa** business
A TAYLOR & FRANCIS BOOK

CRC Press
Taylor & Francis Group
6000 Broken Sound Parkway NW, Suite 300
Boca Raton, FL 33487-2742

First issued in paperback 2019

© 1996 by Taylor & Francis Group, LLC
CRC Press is an imprint of Taylor & Francis Group, an Informa business

No claim to original U.S. Government works

ISBN-13: 978-0-8247-9715-7 (hbk)
ISBN-13: 978-0-367-40134-4 (pbk)

This book contains information obtained from authentic and highly regarded sources. Reasonable efforts have been made to publish reliable data and information, but the author and publisher cannot assume responsibility for the validity of all materials or the consequences of their use. The authors and publishers have attempted to trace the copyright holders of all material reproduced in this publication and apologize to copyright holders if permission to publish in this form has not been obtained. If any copyright material has not been acknowledged please write and let us know so we may rectify in any future reprint.

Except as permitted under U.S. Copyright Law, no part of this book may be reprinted, reproduced, transmitted, or utilized in any form by any electronic, mechanical, or other means, now known or hereafter invented, including photocopying, microfilming, and recording, or in any information storage or retrieval system, without written permission from the publishers.

For permission to photocopy or use material electronically from this work, please access www.copyright.com (http://www. copyright.com/) or contact the Copyright Clearance Center, Inc. (CCC), 222 Rosewood Drive, Danvers, MA 01923, 978-750- 8400. CCC is a not-for-profit organization that provides licenses and registration for a variety of users. For organizations that have been granted a photocopy license by the CCC, a separate system of payment has been arranged.

Trademark Notice: Product or corporate names may be trademarks or registered trademarks, and are used only for identifition and explanation without intent to infringe.

Library of Congress catalog number: 96-13999

Library of Congress Cataloging-in-Publication Data

Catalog record is available from the Library of Congress

Visit the Taylor & Francis Web site at
http://www.taylorandfrancis.com

and the CRC Press Web site at
http://www.crcpress.com

Preface: A Search for Project Excellence

I recently carried out project management benchmarking evaluations of two large overseas projects in the $300 to $400 million range. In both cases, the evaluations identified major problems in both program techniques and personnel skills, resulting in cost impacts of more than $30 million for each project. Ratings were below average for both projects.

I reviewed the evaluations, in detail, with each company's project groups and senior management. The results, even though poor, were properly received by each company's project groups, but the reception by each company's senior management was very different. One company expressed complete acceptance by the senior executive and with this acceptance responded with an extensive development program to rectify the situation. The second company's senior executive rejected the major findings, even though the findings were fully supported by the project groups. With this rejection came no support for the remedial action plan.

It appeared that the rejection was due to the fact that the senior executive had personally instituted elements of the project program, and these elements were the major categories contributing to the poor benchmarking rating. The situation illustrated the frequent case that *personal pride can override professional competence.*

And I learned, once again, that where there is serious lack of project understanding by senior management, it is but a short step to there being little or no *real support* from the company for the project function. This, then, means that project managers potentially face an enormous obstacle to project success.

This book is written for all categories of project personnel—department managers and executives who plan, schedule, control, organize, staff, and manage engineering and construction projects. It is also a valuable update for people who have been in the field for a while and need to reinforce their techniques. It is especially beneficial to engineering and construction management who operate in today's demanding multiproject environment, who are responsible for setting policies and standards in that environment, and who want to keep their professional skills on the leading edge.

This volume has two major objectives: (1) to provide a reference of an advanced, state-of-the-art project control program and (2) to develop and enhance the project skills of all who use this book.

Throughout the book, I have interwoven information on the three major overlapping parts of the total project program, namely:

- *programs*—defined as techniques, procedures, and methods
- *people skills*—defined as experience, application, and analytical ability
- *culture of project groups*—defined as project commitment and working togetherness

The focus is on advanced application skills that directly correlate to all phases of a project, at both owner and contractor levels. When properly utilized, these skills can lead directly to cost savings or, at the very least, to risk mitigation, schedule stability, and cost containment. With poor skills/programs, cost increase is certain.

This book features, in great detail, a state-of-the-art project control and trending program, with more than 200 exhibits of reports, techniques, and data. Such a program requires *real* business skills, analytical ability, and effective project control techniques. When properly applied, the program results in the successful evaluation of project changes and their impact on costs, schedules, and contract conditions.

Many thanks and appreciation go to my friends and colleagues who have made contributions to this book: to Ken (my Welsh colleague) and Betsy Humphreys for their word processing and translation of my Welsh English to American English; to Ken for his contributions to the material, for his technical editing, and for his many suggestions and improvements for so effectively managing the scheduling and coordination of all materials and authors; to Judith Bart for her word processing/editing of my initial manuscript; to Klane Forsgren for his chapters on environmental and economic matters; to Jim Neill for his chapter on value engineering; to Michael and Kevin Curran for their chapter on range estimating; and to Graham Garratt for persuading and motivating me to undertake this work.

James A. Bent

Contents

PART 4 PROJECT MANAGEMENT KEYS AND INTERFACE

Introduction and Mission Statement

I. MISSION STATEMENT

The focus of this book is on advanced application skills that directly correlate to all phases of a project, at both owner and contractor levels. When properly utilized, these skills can lead directly to cost savings or, at the very least, to risk mitigation, schedule stability and cost containment. This book features, in great detail, a state-of-the-art project control and trending program, with more than 200 exhibits of reports, techniques, and data. Such a program requires *real* business skills, analytical ability, and effective project control techniques. This, then, results in the constant evaluation of project changes, cause and effect, their impact on contract conditions and the cost–schedule impact.

II. TOTAL PROJECT EXECUTION PROGRAM

Throughout the book, the authors have interwoven information on the three major overlapping parts of the *total project program*, namely:

- Programs—defined as techniques, procedures, and methods
- People skills—defined as experience, application, and analytical ability
- Culture of project groups—defined as project commitment and working togetherness

Figure FM.1 is a flowchart, showing the above three parts and their constituent subcategories. As the first two parts cover the same categories, the subcategories are then combined into two overall categories:

- What to do—skills and techniques

THREE MAJOR OVERLAPPING PARTS

Figure FM.1   Total project execution program.

- How to do it—working togetherness

It is our experience and judgement that *how to do it* is the more important, by a small margin only.

III.   CURRENT INDUSTRY SKILL LEVELS

Construction Industry Institute (CII) project management studies, confirmed by our direct experience of and data from thousands of individuals and hundreds of companies, show that today's project skill levels are low and inadequate.

A detailed evaluation of current skill levels and a definition of the best skills are provided in the following chapters:

- Chapter 1, "Benchmarking The Technical Core of Total Quality Management,"
- Chapter 2, "The Impact of Personnel Skills on Company Culture and Bottom Line."

As shown in these chapters, data for the period 1991–1994 shows an overall worldwide skill level of 58%. It is our judgment that the skill level needs to be at 80% for successful execution of projects.

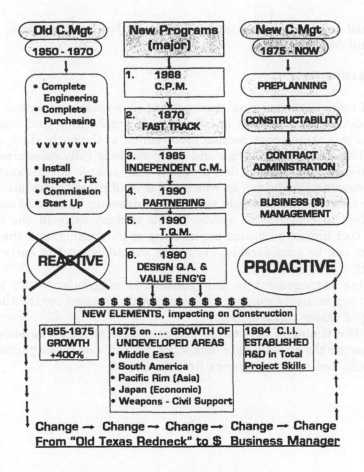

Figure FM.2 Changing role of managing construction.

The rest of the book then covers, in great detail, the skills outlined in these two chapters.

IV. CHANGING ROLE OF MANAGING CONSTRUCTION

Figure FM.2 is a flowchart that highlights the major changes in construction management from 1950 to the present day.

A. Old Versus New

The old program of sequential completion of the individual phases of engineering, procurement, and construction, in which construction management rarely got involved until four weeks prior to opening the site, has been replaced by the very challenging—but very efficient—*fast-track program*. This program and the *CPM scheduling program* are the two greatest advances in methodology that have occurred in the past 30 years; they followed the explosion of work that took place after World War II. Today's computer programs, while not of the same caliber, greatly assist in the collection and collation of data that, in turn, is developed into *essential information*.

B. A Proactive Role

The best of today's construction management now takes a *proactive* role through the newly developed programs of construction preplanning and constructability. This results in a strong construction involvement at the early stage of the project to ensure that engineering design and early planning fully recognize the requirements of an economic construction program. An example of construction preplanning is *backwards scheduling*, in which the overall project schedule is structured around the construction schedule with design drawing issues and material deliveries matched to construction needs. If this is done early in the design stage, there is no cost impact on design engineering or purchasing, and the construction cost savings can be considerable, even with the added cost of early construction involvement.

Business management is now considered of greater value than the old standard of aggressively pushing the work, with cost and contractual considerations being of lesser consequence.

Note: The timing of new major programs and elements as shown in Figure FM.2 represents the approximate date when the individual categories were widely used (proven), not when they were first developed.

Effective Project Management Through Applied Cost and Schedule Control

1

Benchmarking—The Technical Core of Total Quality Management

I. INTRODUCTION

Benchmarking is the technical core of the Total Quality Management (TQM) process. It identifies the quality of current personnel skill levels and company procedures/methods, and then compares this quality with the latest state-of-the-art in the industry. The resulting difference, or gap, identifies:

- the need for change,
- dimensions of the needed change, and
- the result that can be achieved by effecting the change.

The need for change is evident only when the latest state-of-the-art techniques are recognized, and when a company realizes and accepts that its current program is out of date. Further, the company must acknowledge that personnel skill levels need to be and can be upgraded. It should be noted that many companies perceive themselves to be better-run than they actually are, which makes recognition of their actual performance a difficult experience.

In the *capital projects* side of the business, an effective benchmarking study of a company's project program is essential if a company is to meet the demands of today's competitive, low-cost business environment. To be successful, an operating or contracting company must ensure that its project engineering group executes its capital projects program with an efficient program, skilled personnel, and with a positive, contributing company culture. *Only benchmarking can first identify and then implement these dimensions.*

II.   AN OVERVIEW OF TOTAL QUALITY MANAGEMENT

A.   The Process

Total Quality Management is a systematic process for continuous improvement in all facets and at all levels of company operations. The focus is on both the efficiencies of organizational structure and the skills of company personnel.

B.   The Self-Examination Process

The first step in TQM is conducting an in-depth self-examination by assigning top (i.e., high-quality) personnel to individual teams or task forces. Typical areas of focus include:

- personnel leadership (at all levels),
- employee involvement in company policy,
- benchmarking methods and procedures,
- employee empowerment,
- management inspection process (measuring progress), and
- the manager as a role model.

As is evident, effectiveness of the TQM process is rooted in the initial assignment of high-quality personnel as team leaders, and in the continuing assignment of high-quality personnel as the program is implemented. With regard to the benchmarking aspects of TQM, an outside expert is essential if the examination is to be unbiased, independent, and of the requisite quality. Conversely, if a company does not assign high-quality personnel or fails to use an outside expert for benchmarking, the self-examination and subsequent program will be of little value.

C.   Reasons Why Total Quality Management Fails

The TQM process, first begun in the manufacturing industry in the 1980s, is now sweeping the construction industry. Operators, contractors, subcontractors, and suppliers are embracing and implementing it with enthusiasm. In fact, many companies are making it a condition for doing business that their contractors/suppliers have a qualified TQM program in place.

Many of these firms are ecstatic with their TQM results, and report significant improvement in customer satisfaction, employee morale/commitment, contractor/supplier relationships, return on investments, and company culture. Other firms find themselves disillusioned, finding employee dissatisfaction/ poor morale, failing company relationships, unhappy clients, poor company culture, large training/organizational costs, and little or no return on investment. Such failures are unfortunately common, and in 1992, the managing director of one of the world's largest oil companies was fired for just such a failure; his firing came in conjunction with the first reported financial loss in the company's history. Current experience has identified several common reasons for TQM failures.

1.   Lack of Senior Management Commitment

The TQM process requires tremendous cultural change, particularly in the areas of self-examination and employee empowerment. This fact is often not understood or fully accepted by senior management, who may profess acceptance but are lukewarm in commitment and enthusiasm. These individuals are usually the autocratic dinosaurs of yesteryear and, fortunately, are a rapidly dying breed. Beyond this, however, many existing company policies (such as purchasing for an initial low price rather than the lowest evaluated cost) may need to be changed to accommodate TQM. See Chapter 2 for a further discussion of outdated policies and the need for change.

Another commitment problem arises when TQM is viewed as an event rather than a process. While management typically includes some funding for *quality training* in the annual budget and may even appoint a quality manager, such actions are rarely signs of actual commitment. To be successful, TQM must be regarded and implemented as an ongoing, consistent program rather a training exercise or the management fad of the month. Total Quality Management is a *top-down* program whose success requires serious understanding, acceptance, and commitment from senior management.

2.   Lack of an Implementation Plan

Successfully implementing TQM generally takes years, not months. For large companies in particular, the total time period can be six years or more since every employee, contractor, client, and supplier is affected. Such an undertaking demands a detailed plan and implementation program. Still, many companies go straight from the senior management acceptance phase to the implementation phase, omitting planning altogether. These companies usually find their efforts fragmented and unsuccessful.

3.   A Poor Training Program

Quality training for all levels of management and technical personnel is essential to a successful TQM program. The curriculum must cover interface management, interpersonal skills, and technical methods. Holding interactive-type workshops is also essential. Training alone will not be sufficient, and new training materials, standards, policies, and methods must be created if the training is to be effective. This may be a challenge in itself—since TQM was first developed for the manufacturing industry, converting the terminology and examples to the construction industry requires substantial experience and skill.

4.   Lack of Experienced Facilitators/Team Leaders

The initial self-examination and ongoing program require the development of small employee teams that will identify, evaluate, and recommend improvement opportunities. Although these teams need strong leadership if they are to be productive, such leaders are rarely available. Consequently, having an appropriate leadership training program is essential to TQM success.

5.   Lack of Measurement

When and if the problems of management commitment, planning, training, and experience are solved, the challenge of effective measurement must be met. This involves monitoring the newly implemented policies, standards, and work processes for cost-effectiveness and team-building value. Questions such as the following need to be posed:

- Are employee skills improving?
- Is the organization actually more effective?
- Are old matrix interface conflicts disappearing?
- Is an improved company culture developing?
- Is work being performed in less time, at less cost, and at increased levels of productivity?

The deliverables for these items have to be readily determinable and measurable, so employees can see that TQM is more than a theoretical exercise.

6. Impatience

As noted earlier, implementing TQM is a matter of years rather than months. For many people, particularly those who may be results-motivated, this is too long to wait. Impatience is common in TQM implementation situations, and soon creates frustration with the drawn-out process. In response, management sometimes gives in to the temptation of a quick fix with a preliminary and shallow program. Failure always results.

The problem of impatience can be solved, however, by formulating *quick start* teams simultaneously with appointment of the self-examination teams. The quick start teams' output can be put into practice in an area of company operations where implementation is relatively easy and the probability of success is high. The resulting success can be used to motivate the rest of the organization, where it will mitigate impatience and generate enthusiasm.

D. TQM Failure Impacts on the Bottom Line

Without question, TQM failure will result in increased cost and lower return on revenue or even in a loss. It will also result in low skill levels, poor company culture, matrix interface conflicts, low morale, loss of resources, and poor public image.

III. THE BENCHMARKING PROCESS

A. Hiring the Benchmarking Expert

Many companies assume that if a management consulting company is working in the construction industry, it is a benchmarking expert. The author, having seen such situations, knows this assumption to be false. In addition to knowledge of the construction industry, the benchmarking expert needs:

- a great many years of hands-on project experience,
- proven technical skills backed by reputation and credibility,
- significant interviewing and people skills,
- established measurement criteria, and
- an industry-wide level of comparison (state-of-the-art).

B. The Measurement Process

The company's level of technical expertise is determined by interviewing its technical personnel and by evaluating the quality of the company's formal, written procedures and techniques. The interviews should be conducted with all levels of personnel and with key interfacing personnel from supporting service groups.

C. Project Control Benchmarking (Deliverables)

Project control organizations usually consist of the following groups:

- cost estimating,
- cost control,
- planning and scheduling, and
- cost accounting.

In some instances, cost accounting is part of the company's general accounting department rather than part of the project control organization. When conducting interviews, two three-hour sessions are held with a mix of junior-to-senior personnel. Since all of the project controls disciplines are dependent on input information, it is proper and even requisite to conduct interviews with the organizations that provide the input as well as with those in the project control organization who use the input. Key interfacing organizations are:

- project management,
- discipline engineering,
- operations/maintenance,
- procurement,
- contracts,
- construction,
- sales/proposal, and
- quality assurance/quality control/safety.

The deliverables are an assessment of:

- personnel technical experience,
- personnel technical skills,
- personnel supervisory skills, and
- company culture (both in- and out-group).

The assessments are then compared with the current industry state-of-the-art, and, if possible, to those of similar companies. They could also be compared with a chart of skill levels as described below. A *deliverables report* illustrates all these issues and also provides a *remedial action plan* for reinforcing the strengths and weaknesses.

D. 1991–1994 Chart of Group Performance and Skill Levels

The chart contained in Figure 1.1 shows skill levels of participants in senior, professional project courses for the period 1991–1994. The participating companies are all leading companies from USA, Europe, UK, Australia, South Africa, Middle East, and Indonesia and therefore represent a good international mix. Even though these are many of the world's leading companies, the skills level shows an individual group range of 40–71% with an overall average of 58% over the four years. The author has a second, independent benchmarking database that has an overall average level of 55%. The correlation of these two independent

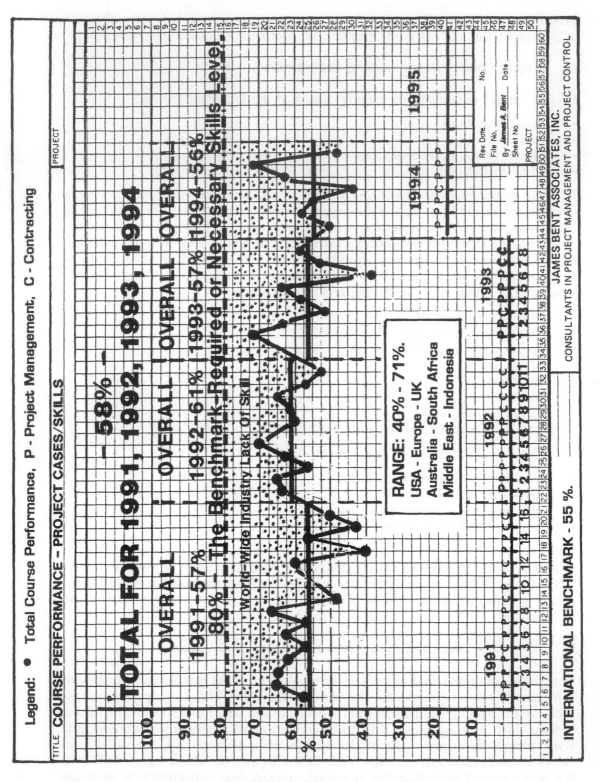

Figure 1.1 Course performance—project cases/skills.

databases is remarkably close. The skill levels of this chart are determined by how well course work-groups resolved a wide-ranging set of case histories in the areas of project estimating, cost control, scheduling, contractual risk, contract strategy, project organization, project planning and business analysis. These results confirm many current studies that point to low project management skills in present-day industry. It is the author's position that skill levels must exceed 80% if the cost savings illustrated in Chapter 2 are to be realized.

IV. SUMMARY

The title of this book features *Applied and Effective* and is meant to emphasize and illustrate the skills and techniques essential to a high-quality project control program. The benchmarking deliverables described in this chapter are, then, a summary of the skills needed to achieve effective cost and schedule control on engineering/construction projects. Moreover, there is a distinct relationship between skill level and project cost, a subject that is discussed more fully in Chapter 2, "The Impact of Personnel Skills on Company Culture and Bottom Line."

The overall assessment by the Construction Industry Institute (CII) that proper use of these techniques can lead to the U.S. construction industry saving $15 billion per year is challenging, to say the very least.

It must be emphasized that failure of TQM can impose a significant burden on any project program. Project personnel must ensure that the impacts of this burden are properly reflected in their project planning.

2

The Impact of Personnel Skills on Company Culture and Bottom Line

I. INTRODUCTION

One of the key considerations in managing projects effectively is the question of company commitment and support of the project function. Is the support real or not? Many studies and the author's extensive experience show that a direct correlation exists between people effectiveness, technical skills, and the working environment or company culture.

Individuals may be extremely skilled, but their effectiveness can be severely limited in a company with a poor project culture. When initiative is stifled, decision making is constantly influenced by political considerations, advancement is by patronage, and departmental loyalty is more important than the needs of the project. Even in contracting companies, where execution of projects is the major product, poor and inefficient cultures may be found.

Such negative cultures are unfortunately all too common. They were brought into existence when the matrix organization was destroyed by poor people skills, individual jealousies, company politics, little or no desire to excel, and ultimately, by management neglect.

II. MATRIX THEORY

Most projects, being small ones, are handled in a matrix organization environment: they are executed by many departments carrying out the work. This is usually done with a project manager who does not have adequate decision-making authority. This authority is shared between the departmental managers who actually do the work and the project manager who coordinates the work.

The fundamental of matrix theory in a project engineering environment requires that the project objectives and schedule/critical path(s) be clearly defined so that all working groups will then accept, commit to, and work to those objectives and schedules. Theoretically, there will be unanimous support from everyone, so that all will be working to the same set of priorities and objectives.

With a strong project management culture and effective management leadership, the matrix theory would work. In fact, it did work for a time, as individuals and departments put the project interests first. Still, by the early 1980s, the matrix interface conflict was widely recognized. Today we realize that the matrix theory has failed, and we are applying the current antidote of total quality management. This new approach that is sweeping the industry, when correctly implemented, does solve the matrix interface conflicts problem.

All major studies conducted during the past ten years have come to the same conclusion: our failure to make the matrix work is the single biggest problem faced by the industry today. A 1982 study by Folger & Company, *The Project Management Speaks Back*, surveyed project managers working in matrix organizations of major contractors. This was one of the study's major conclusions:

> The consensus opinion is that most problem areas are internal, rather than external in nature. Over 90% of the Project Managers stated that *conflicting interests and struggles with department managers* was their chief problem.

Many companies tried team-building programs in an effort to develop and build esprit de corps and a project team environment; they met with only limited success. At the same time, many companies implemented policies that were directly opposed to individual initiative and working-togetherness.

Total quality management can solve many of these problems and can result in a positive working environment and company culture. Still, many of the old policies will need to be eliminated. The following section refers to these old policies as matrix interface conflicts.

III. TYPICAL MATRIX INTERFACE CONFLICTS

The matrix interface conflicts discussed in this section are common in the engineering/construction industry and are to be found in even the best of today's leading companies.

A. Production Versus Engineering—The Hidden Company Charter

In large operating companies, the production company has the budget for all capital work, and the engineering department provides the technical service and support for designing and building the capital projects. The production company can delegate some of its financial responsibility for project development to the engineering department while still holding final financial approval. On the other hand, the engineering department is separately charged with technical/economic viability and must act as *technical auditor* of the production company. This gives the production company protection through a check-and-balance system, but it can and does create people problems because of the divided responsibility.

Numerous studies have shown that the biggest single project-related problem in our Western business environment is likely to be the lack of scope definition control at the front end. This results in a constant stream of technical changes and their accompanying budget overrun and schedule slippages. The conflicting responsibilities of operations and engineering directly contributes to this problem. Although common understanding, project discipline, and a commitment to excellence can allow individuals to rise above such situations, failure is the more common result. This is not to suggest, however, that the check and balance/divided responsibility structure should be changed. It only means that senior management must reinforce the project management function, raising its status so that the image and responsibility of the two groups become more balanced. The gap between the two groups can also be reduced by continuing training of project management personnel and by continuously upgrading the training program.

B. A 10% Estimate and a Fast Track—Impossible

Company policy requires a 10% quality estimate for the funding of capital projects. Yet at the same time, company management usually wants projects developed yesterday! These two objectives are totally incompatible in plant-type projects, and to accomplish the impossible, engineering lies about the quality of the estimate. *In a total quality management culture, this policy must change.*

C. Management Controls the Contingency—Unbelievable

Company policy does not give the project manager direct control or management of the project contingency, while at the same time it holds the project manager directly responsible for every dollar of the approved funds. This situation is sometimes a result of management ignorance in confusing contingency with management reserve. Even in our enlightened technical age, many individuals still do not understand or accept the need for project contingency. *In a total quality management culture, this policy must change.*

D. The Purchasing Interface—Mostly Negative

Companies do not allow project and discipline engineers full access to commercial information on equipment/material bids. This practice has led many procurement departments to assume a pre-eminent position in which the tail wags the dog. Ideally, the engineering and project management functions should include both business analysis and decision-making roles, particularly for all equipment and material purchasing. This is becoming a common practice in TQM. *In a total quality management culture, the purchasing department should be in a support position and provide a full-service function.*

Management commonly explains this situation by saying that project personnel who are aware of prices cannot be trusted to maintain confidentiality or to give an unbiased, technical assessment. Many owner project managers resolve this purported problem by delegating this function to outside contractors (usually on a reimbursable basis) over whom they have more direct control. *In a total quality management culture, this policy must change.*

If management feels that discipline or project engineers cannot properly evaluate technical and commercial considerations or that they cannot be trusted with sensitive commercial information, then management should train them to properly handle these responsibilities and provide a culture where these qualities are commonplace.

E. Project Cost Accounting—What For?

In many instances, accurate and timely cost accounting to an appropriate project code of accounts is not available. It is often late (by three to four weeks after the cutoff date), inaccurate, and without an adequate breakdown. To compensate for this situation, many project engineering groups set up their own independent program of cost accounting, even though having a duplicate system is expensive and takes precious time and resources to maintain.

Most accounting groups belong to the company's financial divisions where the financial vice-president or controller is in charge. These individuals are rarely conscious of or sympathetic to project management needs (since matrix interface conflicts occur at senior management levels as well as at intermediate and junior levels). *In a total quality management culture, this policy must change.*

Resolving this problem is relatively simple (although "relatively simple" is a complete understatement, since management of most companies does not have the understanding or the will to implement this solution). A project cost accounting group of two or three individuals is established within the accounting department and receives daily direction from project engineering.

F. Mandated Lump-Sum Contracting Regardless of Environment—Lunacy

Again, this situation is created by company policy. It exists regardless of the business environment, regardless of the project objectives, and regardless of resources, experience, and capability. Contracting flexibility is absolutely essential for effective project management and a cost-effective program. For example, a lump-sum approach in a seller's market can be expensive, yet in certain countries the lump-sum requirement is mandated by government regulation. Obviously, any company working in such a country must obey the law, often to its own detriment. *In a total quality management culture, this policy must change.*

IV. ROTATIONAL PROJECT ASSIGNMENTS—PROJECT CONTROL MANAGER

One of the essentials of TQM is that the manager must be a role model. This means being a technical expert over the discipline being managed. Yet in many companies, the manager of project control is not an expert in estimating, or cost control, or planning and scheduling. In key positions such as director of projects, manager of purchasing or director of construction, this lack of technical expertise is rarely a problem as only senior people with extensive project experience are assigned. Of course, the manager of design cannot possibly be an expert in all engineering disciplines, but the assignment of nonexpert managers to the project control function is a common practice and, in a TQM culture, must be curtailed.

V. THE PROJECT CHARTER—PART OF THE SOLUTION

Apart from, or in addition to, TQM, an ever-expanding technique for overcoming matrix interface conflicts is to develop a charter for each project. This formal, internal document establishes the project objectives and execution plan, and outlines major responsibilities of the key parties involved. The key parties are then required to sign the charter, which signifies their acceptance of and commitment to the project program.

VI. THE IMPACT OF PERSONNEL SKILLS—A COMPANY STUDY

Much has been written about the factors affecting efficient execution of engineering/construction projects, yet all major studies have come to the same conclusion that two fundamental factors are involved:

- adequate or high-level personal skills, and
- quality project management methods and techniques.

A. Impact of Personal Performance

Some companies have analyzed the impact that personal skills and performance play in executing engineering/procurement/construction projects. During James A. Bent's career in the engineering department of Mobil Oil Corporation, senior management initiated a study to determine why then-current project performance was poor, resulting in major cost overruns and schedule slippages in the capital projects program. The study was carried out by four of the company's most senior project managers, who arrived at the following conclusions:

- The design basis contributed 80% of cost.
- The complete project management program contributed 20% of cost.

Figures 2.1 and 2.2 show a breakdown of the study team's findings.

B. Major Effects on Engineering/Procurement/ Construction Project Costs

As the Mobil Oil study determined, once the process design technology has been selected and major project elements/objectives are established, the maximum amount represented by the total project program would be 20%. An effective project management program includes a combination of project execution methods/procedures and personnel skills. Effective methods, good personnel skills, and adequate resources are essential if quality project management is to exist. The Mobil Oil study team determined that high-quality methods and advanced personnel skills can result in savings totalling up to 20% of a project's cost. Conversely, a poor program and low skills can increase costs by 20% or more, and

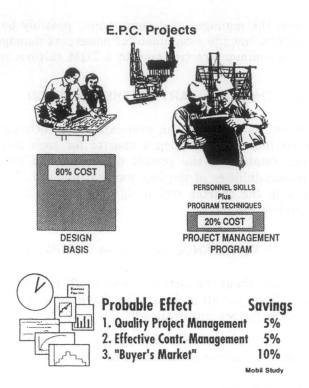

E.P.C. Projects

80% COST

DESIGN
BASIS

PERSONNEL SKILLS
Plus
PROGRAM TECHNIQUES

20% COST

PROJECT MANAGEMENT
PROGRAM

Probable Effect   **Savings**

1. Quality Project Management   5%
2. Effective Contr. Management   5%
3. "Buyer's Market"   10%

Mobil Study

Figure 2.1   Major effects on project cost.

Detailed Effect

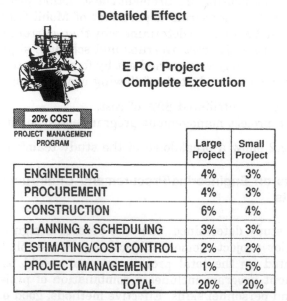

20% COST

PROJECT MANAGEMENT
PROGRAM

**E P C Project
Complete Execution**

	Large Project	Small Project
ENGINEERING	4%	3%
PROCUREMENT	4%	3%
CONSTRUCTION	6%	4%
PLANNING & SCHEDULING	3%	3%
ESTIMATING/COST CONTROL	2%	2%
PROJECT MANAGEMENT	1%	5%
TOTAL	20%	20%

NOTES:

1. On large projects, the value of Project Management is low (1%) as the effect of a single Individual is less than that of a cohesive Project Team.

2. On small projects the reverse is true, and the effect of the Project Manager can be the biggest single factor.

Mobil Study

Figure 2.2   Project management program.

it is probable that the downside risk of cost increases is much greater, sometimes in excess of 100%.

Figure 2.1, showing the overall impact, tempers the potential 20% savings with the probability that half the savings (10%) might be influenced by a buyer's market. It should be noted that a proper assessment of market conditions must be reflected in the project estimate and contracting strategy so as to prevent contracting pricing surprises at the later stages of the project.

Figure 2.2, showing a detailed breakdown of the 20% savings, allocates individual percentages to the six major phases of an engineering/procurement/construction project and shows the probable variations for large and small projects. The most significant variation is shown for project management, where the impact of an individual on a small project can be the largest single factor. This is especially true when the project manager is the designer, estimator, and scheduler, as is often the case in the multiple small projects environment.

VII. THE IMPACT OF PERSONNEL SKILLS—THE CONSTRUCTION INDUSTRY INSTITUTE STUDY

In mid-1987, the Construction Industry Institute (CII) undertook a *Project Management Practice Assessment Survey* to evaluate industry skill levels and practices. That study, and its results, are discussed below. The general purposes of the study were to:

- select, categorize, and measure key project management practices;
- relate these practices to other key project characteristics;
- determine current levels of project performance in project cost, schedule, technical quality, safety, and profit; and
- relate the levels of skills utilization to project performance.

A. Key Project Management Categories

To meet the general purposes of the study, eight project management categories were determined to be essential and were examined in detail; they were:

1. strategic project organizing (front-end planning),
2. contracting practices,
3. design effectiveness,
4. project controls,
5. management of quality,
6. materials management,
7. human resource management, and
8. safety.

The detailed composition of each category is shown in Table 2.1.

The Construction Industry Institute then surveyed owners, architect/engineers, and contractors to determine their assessment of utilization and importance of the identified categories. Table 2.2 contains information about survey participation.

Table 2.1 Project Management Categories of CII/CICE Principles and Recommended Practices

Management category	Scope statement
Strategic project organizing	This category focuses on principles/recommendations related to project organization, establishing objectives, scope definition, control, establishing communications/information processes, and constructability planning.
Contracting practices	This category focuses on those principles/recommendations related to contracting strategy (planning, packaging, etc.) and the utilization of specific contract provisions and/or clauses for contracts controlled by the initiating party.
Design effectiveness	This category covers principles/recommendations relevant to the evaluation of the design effort, incorporating constructability concepts into design, and control of design activities.
Project controls	This category focuses on principles/recommendations related to control integration, decision making, scope control, control techniques, and estimating practices.
Management of quality	This category is concerned with principles/recommendations related to the implementation of quality assurance/quality control and the documentation of quality effectiveness.
Materials management	This category focuses on principles/recommendations related to planning and utilization of materials management on projects.
Human resource management	This category is concerned with principles/recommendations related to the quality of site supervision, field work force motivation, training, and site labor practices (substance abuse, overtime, etc.)
Safety	This category covers principles/recommendations related to safety communications, specific practices, and management attitude toward safety.

Table 2.2 Survey Distribution and Receipt

Participants	Sent out		Returned		% Companies responding	% Surveys returned
	Companies	Surveys	Companies	Surveys		
Owners	220	2204	105	773	47.7	35.1
Architects/engineers	630[*]	3230	113	465	17.9	14.4
Contractors	1450	7306	210	664	13.7	9.1
Total	2300	12740	428	1902	18.6	14.9

[*] Some duplication exists because several of these companies provide engineering and construction services. They completed both the A/E and contractor questionnaires.

B.   Cost Benefits of Enhanced Skill Levels

Three areas were determined to be crucial if a project is to be cost effective; they were:

1.   making technology enhancements,
2.   properly managing the design-construct interface, and
3.   using good project management practices.

This section will concentrate on the third area.

It is difficult to quantify potential savings achieved by using good project management techniques and practices. This is because real-life cost-benefit analyses of different alternatives are not always based on hard facts but on subjective assessments made of historical data. Further, many assumptions that are critical to the assessment are subjective in and of themselves. Nevertheless, assessment of dollar savings is considered to be a worthwhile undertaking.

The Construction Industry Institute asked 35 industry experts to assess potential gross savings and the associated benefit-to-cost ratios. Table 2.3 shows the result of these assessments. The numbers are rounded and do not imply precision. The number one category of choice was "Strategic Project Organizing," closely followed by "Design Effectiveness," "Human Resource Management," and "Project Controls." It should be noted that project control skills are also at the core of the first three categories, thus emphasizing the need for these skills. This subjective evaluation led to the conclusion that the potential return (integrated effect) of 25%, with a cost-to-benefit ratio of 15:1, is most attractive.

C.   Techniques Utilization—Owner, Architect/Engineer, Contractor

Another part of the CII study analyzed the respondents' utilization of the key techniques listed in Table 2.3. The respondents were asked to assess those practices that were performed directly by the company by which the respondent was employed. Owner/respondents were also asked to assess those practices required of their engineering design and construction contractors. Table 2.4 shows that overall utilization ranged from 66–70%, with safety highest at 76–82%, and project control being second highest, at 72–77%.

Based on this data and taking a conservative approach, a savings of only 10% would result in $24 billion in savings per year. (The construction industry is estimated to be $240 billion per year). At the indicated 70% utilization level of high-quality project management skills by the owner (Table 2.4), the resulting savings would be an annual $15 billion.

D.   Current Project Performance

In addition to conducting the skills impact study, CII also asked owner/respondents to indicate performance of their current projects. These results are shown in Table 2.5. Current cost and schedule performance (at 61% and 66%, respectively) is the result of an industry out of control. This is the other end of the scale from the subjective cost savings discussed previously.

Table 2.3 Cost Benefit of Maximum Utilization of CII/CICE Principles and Recommended Practices

Management category	Potential gross savings for maximum utilization[a] (%)	Range of gross savings for maximum utilization (%)	Benefit to cost ratio
Strategic project organizing	15	10–20	20:1
Design effectiveness	10	5–20	15:1
Human resource management	10	5–20	15:1
Project controls	10	5–15	10:1
Management of quality	8	5–10	10:1
Materials management	5	3–8	10:1
Contracting practices	5	3–10	10:1
Safety	5	2–8	10:1
Integrated effect[b]	25	15–30	15:1

[a] Gross savings of total project cost.
[b] The integrated effect does not reflect a summation of savings for all eight management categories. It represents a composite of the subjective assessments of all the respondents (adjusted for other data sources). Each respondent had its own weighting process to derive the percent savings for the combined maximum utilization of all categories.

Table 2.4 Comparison of Owner, Architect/Engineer, and Contractor Utilization

Management categories	Average percent utilization of principles and recommendations		
	Owner	A/E	Contractor
Strategic project organizing	70	69	73
Contracting practices	69	—	58
Design effectiveness	69	69	—
Project control	74	72	77
Management of quality	72	60	62
Materials management	63	58	62
Human resource management: all	58	—	64
Human resource management: union	73	—	76
Safety	82	—	76
Total	70	66	68

Table 2.5 Summary of Current Owner Performance

Performance measure	Average percent frequency of meeting or performing better
Cost	61
Schedule	66
Technical	80
Quality	Not tested
Safety	Not tested
Profit	Not tested

VIII. SUMMARY

Even though a full TQM program is time-consuming, expensive, difficult to implement, and requires tremendous company-wide changes, correct application of such a program can lead to major improvements and cost benefits. The impact of a good TQM program on the project control function is direct, immediate, and significant.

Project control personnel are generally not the project leaders or decision makers but are the generators and compilers of the project status and, especially, of change and variation. To accomplish this generation and compilation, project control personnel need to be very effective communicators. Establishing facts, forecasts, and trends, as well as developing people skills and listening capabilities are essential.

One of the key objectives of TQM is to create a culture where team-building, togetherness, and good communication channels are commonplace. *Thus, there is a direct correlation between project team-building and effective cost-schedule control.*

3

Estimating Keys—Establishing a Realistic Cost Baseline

I. INTRODUCTION

A. Quality of Estimates

Estimating is roughly divided into conceptual and detailed. The general range in the quality of these two phases of estimating is about 40% to 10%, respectively. The measure of the quality of an estimate is usually categorized by the amount of contingency that is contained in the estimate. For example, a 10% estimate would have a 10% contingency. Due to the high development cost and the time necessary to produce a 10% quality estimate, many companies approve the funding and full execution of design-based projects at the ±20% estimate quality. It is possible, in the *special equipment* areas and building industry, to produce 10% quality estimates from preliminary design information. With plant projects, the construction phase can have a 10% estimate, but only when design is 80% complete or greater.

The accuracy of estimates is largely dependent on the quality of the estimating program and the experience of the estimator. Quality is also dependent on estimating labor and time. The relationship is not linear. Reasonable investments of time and resources will, usually, provide better cost estimates. Further improvements become increasingly expensive, with only modest improvements in accuracy resulting from substantial expenditures of time and resources. A point is soon reached where estimate quality is almost completely controlled by problems of forecasting future economic conditions, local project conditions, and quality of project performance. No significant improvement in estimate quality can be made thereafter, except by incorporation of actual design and cost information as the work develops. Often, senior management does not understand or prefers not to accept this reality and insists that a 10% quality be produced when the requisite information is not available. Thus, many projects are funded/approved with an inadequate cost baseline.

B. Purpose of Estimates

The primary purposes of performing an estimate are:

* to establish cost levels for economic evaluation and financial investment, and
* to provide a baseline for cost control as the project develops.

Most projects are estimated by system, designed by system, constructed by area, and started up by system. As most conceptual estimating bases are structured on a system basis, rather than on an area basis, it requires considerable effort at an early estimating stage to develop an estimate on an area basis that, in turn, maximizes the controllability aspect. This is especially important for construction, where work is executed on an area basis.

When there is a lack of time, it is probable that the early conceptual estimate would be a capacity-cost or curve-type estimate for direct costs, with indirect costs on a percentage basis. Even though lacking time, the estimator should be encouraged to put as much quality (definition) into the estimate as possible, as this estimate may become the control base for the project.

C. Typical Estimating Categories

Estimating can be broken down into the following categories:

* conceptual

    - proration, budget, rough order-of-magnitude, etc.
    - cost capacity curves
    - equipment ratio (curves)

* detailed (quantity/unit jobhours/unit costs).

In practice, most estimates are composed of all or many of the above categories; even a detailed estimate may have factored elements. See Chapter 16 for detailed discussions of estimating during the feasibility-conceptual phase.

II. PRORATION ESTIMATES

This method takes the cost of a similar, previously built facility, and prorates the cost for the new facility, based on changes for project conditions, capacity, escalation, productivity, design differences, and time. This method is based on some historical data and a lot of statistical relationships and assumptions. It is, therefore, usually accurate to about ±40%.

III. COST CAPACITY CURVES (OVERALL)

A historical database is developed for similar plants where the total cost is related to capacity. This method is usually more accurate, generally around ±30%, but does depend on the quality of the database. Economies of scale are also built in with capacity databases. This method is also used, at a lower level of detail, for individual pieces of equipment and/or process/utility systems.

These conceptual estimating systems are generally used to give a quick and early indication of required investment level. The resulting evaluations should only be used for budget purposes and investment possibilities. The information is not sufficiently accurate to make firm investment decisions. Sometimes investment decisions are made on this preliminary information, where economic viability is not the first priority. Projects to meet environmental standards, stay-in-business criteria, or research and development programs fall into this category. Another purpose of these early estimating programs is to provide technical and economic information on investment and resource requirements to advance the technical basis and estimating quality to a higher level. Thus, many projects are funded on a partial or phased approach.

IV. EQUIPMENT RATIO (CURVES)

This method calculates all other costs as a percentage of the major equipment cost. Ratio methods are effective with a good database. The accuracy of this method is generally ±20%. This quality of estimate is usually the minimum requirement for a full investment decision of a design-build project.

This type of estimate should be produced after completion of conceptual design and process selection and is an update of the conceptual estimate prepared during feasibility studies.

The following shows the general design/scope basis:

- overall process flow diagrams,
- heat and material balances,
- onsite and offsite facilities and layouts (power, steam, air, electricity, water),
- preliminary plot plant/building layouts,
- equipment list (by size and category),
- preliminary execution plant/organization/resources/schedule, and
- completed survey of appropriate estimating data.

This is an equipment and bulk ratio estimate for direct labor and material costs. Indirect costs are factored from direct costs. A further statistical breakdown is made to development engineering and construction jobhours for scheduling and resource evaluation.

V. QUANTITY/UNIT COST OR DETAILED ESTIMATES

This method is the most accurate, generally ±10%, and is costly and time consuming, as detailed takeoffs must be made of all material units and associated labor in the project. This method requires that engineering be sufficiently advanced so that accurate material quantity takeoffs can be produced. It also requires detailed historical data for applying unit jobhour rates and costs to the estimated quantities.

This type of estimate can be developed only when the process design has essentially been completed. It will also require a significant amount of detailed engineering to be completed so that material takeoffs can be developed for civil work, mechanical, piping, electrical, etc.

The following are typical for a design-build project:

- approved process descriptions—feedstock and product slate;
- licensor engineering (Schedule A package);
- approved flowsheets;
- heat and material balances;
- approved process piping and instrumentation diagrams (P&IDs) (process and utilities);
- approved plot plans;
- general specifications;
- equipment specifications and data sheets;
- completed site-soil survey and report;
- site development, grading drawings, building layouts/specifications;
- underground piping and electrical layouts;
- concrete foundation layouts;
- above-ground piping layouts;
- one-line electrical drawings;
- milestone schedule;
- detailed project-owner conditions and requirements;
- project-owner conditions and requirements;
- environmental and governmental requirements;
- equipment quotations (transportation costs);
- bulk material takeoffs;
- labor cost-productivity data;
- layouts for construction temporary facilities;
- organization charts (project, engineering, and construction);
- personnel schedules and labor histograms; and
- construction equipment schedules.

A detailed estimate is quantity based, and reflects separate unit costs for material, labor, and jobhours. Construction is based on an area breakdown rather than on the system basis of a conceptual estimate. This estimate can be an updated, trended version of the first conceptual estimate and subsequent updates. In most cases, however, it is a completely separate exercise, since the format and work breakdown structure are different from and more detailed than that of a conceptual estimate. In particular, the construction estimate is done on an area basis with takeoffs by work units and jobhour rates.

On a design-build, reimbursable-type project, this estimate can be developed between six and eight months after the contract is awarded, as that amount of time will be required to provide an adequate level of completion of detailed engineering (about 50%).

The most significant element of a high-quality estimate is the maximizing of quantities and minimizing of factors and statistical relationships.

VI. FUDGING THE DETAILED ESTIMATE

Many companies have a policy that requires a detailed 10% estimate before the project appropriation will be approved. These same organizations (typically manufacturing companies), also require that the project be started yesterday. Manufacturing and plant management often insist on these two objectives, even though they are incompatible. In most cases, the practical resolution of this management inconsistency is for the estimate to be *fudged*. In other words, the estimate shows a 10% contingency *below the line*, with a similar amount of money buried *above the line* in individual categories where the risk is deemed to be the greatest. While this process meets the company financial approval policy, it nevertheless is poor management practice and provides an inadequate basis from which to execute and manage the project.

It is also quite common for some companies to execute projects on a crisis management basis, an approach that will in most cases increase a project's capital costs. Still, since the project may be able to reach the marketplace at an earlier time, this approach may increase the economic return as well.

VII. DESIGN CONSTRAINT ON ESTIMATING QUALITY

Figure 3.1 shows the quality of estimating in relation to its completion of the engineering, procurement, and design (EPC) progress curves. This information is based on historical experience and shows that for a 10% quality estimate, the percentage completions for a design-build project should be:

Engineering	85%
Procurement	90%
Construction	20%

With an outstanding historical database and high-quality personnel, it is possible to provide a 10% estimate with lower percent completions of EPC.

VIII. PROJECT MANAGEMENT ESTIMATING RESPONSIBILITY

Since many companies have a formal estimating section, the relationship between the estimator and project manager should be clearly defined and properly understood by all parties. The project manager should direct development of the estimate before it is issued and should ensure the estimate properly reflects:

- project objectives and their priorities,
- design scope and design specifications,
- maximizing of quantities and minimizing of factors (numbers of drawings and construction work units),
- correct evaluation of design and labor productivities,
- current project and site conditions (access, congestion, etc.),
- proposed execution plan/contract strategy,
- schedule requirements (economic versus acceleration), and
- adequate contingency evaluation.

As can be seen from the above list, the project manager is actively involved in developing the estimate and is ultimately responsible for the final product.

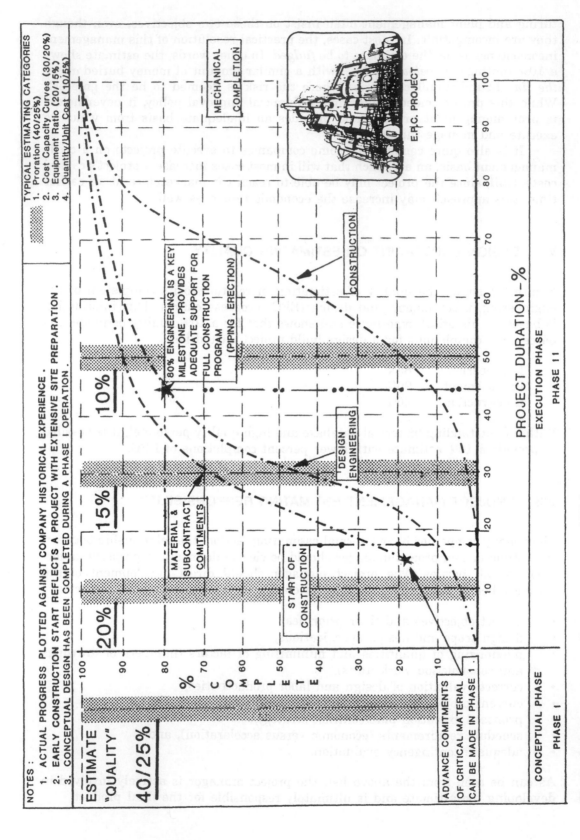

Figure 3.1 Estimating quality versus status of EPC.

IX. SCOPE REVIEW

To ensure that the scope definition is of the required quality, the estimator/project manager should make a detailed review of all basic design documents, their revision numbers, and dates of issue, as follows:

- check that all major equipment is included and is listed by equipment number;
- review all items shown on plot plans, flow sheets, P&IDs, and equipment lists to ensure their inclusion in the estimate;
- check equipment and system capacities, flow rates, temperatures, and pressures for deviation;
- check whether owner costs are to be included or shown separately;
- evaluate deviations in the scope, design, or estimating basis from those assumed in the earlier estimate, and include them on a *puts-and-takes* list; and
- have specialist engineers assigned to the project review and verify the design scope.

X. PROJECT CONDITIONS REVIEW

Before developing the line-by-line details of the estimate, an overall evaluation should consider the following:

- project location (i.e., site characteristics such as high winds, weather, soil conditions), and local affiliate/governmental practices or regulations;
- schedule (i.e., start of engineering, start of construction, mechanical completion, and milestone dates);
- labor basis (e.g., subcontract or direct hire);
- economic outlook;
- contracting mode and execution plan; and
- compatibility of the estimate with contract conditions.

XI. REVIEWING SIGNIFICANT OVERALL RELATIONSHIPS

A comparison should be made of significant relationships. This can be for both owner scope and contractor scope, such as:

- engineering jobhours per piece of equipment,
- contractor's home office and engineering costs as a percentage of total cost,
- contractor's fee as a percentage of total cost,
- indirect construction costs as a percentage of direct labor cost,
- percentage breakdown of engineering jobhours by prime account,
- percentage breakdown of construction jobhours by prime account,
- all-in engineering jobhour rate,
- all-in field jobhour rate,
- escalation allowances for material and labor,
- productivity factors for engineering and construction, and
- currency exchange rates (for overseas procurements).

XII. MAJOR OR ENGINEERED EQUIPMENT AND MATERIAL

The cost of major or engineered equipment can be established by actual quotations or from historical data. The method used depends on the type of equipment involved and its relative cost. For example, quotations should be obtained for large compressors, but small mixers may be estimated from catalogs or estimating manuals. A *cheapest source* program should be used for source guidance on worldwide purchasing.

A. Developmental (or Growth) Allowances for Fast-Track Projects

Estimates based on vendor quotes, catalog prices, or initial inquiries should include an allowance for future increases in scope. When the design is at an early stage, costs can rise as much as 15% from an original purchase price as a result of design changes. Verify that the estimate has included an appropriate design allowance (typically 5–10%) for future changes.

B. Some Major Equipment Considerations

The following are considerations for engineered or major project equipment:

- vessels (towers, reactors, drums)—adjust for shop fabrication versus field fabrication, and the need for lifting lugs (field erection);
- heaters and furnaces—evaluate the degree of prefabrication before field erection;
- boilers and superheaters—check the proposed field erection program; and
- storage tanks—ensure that tank foundations are adequate for duty and soil conditions.

C. Project-Schedule Conditions That Could Influence Prices

Care should be exercised to ensure that schedule conditions do not adversely influence price. For instance, evaluate:

- market conditions;
- purchasing preference/plant compatibility/maintenance costs;
- schedule acceleration (premium costs);
- escalation/currency exchange rates;
- freight, duties, taxes; and
- size of order/quantity discount.

XIII. BULK MATERIALS—MAJOR CONSIDERATIONS

A. Concrete Foundations (Equipment)

- Spot-check design quantities for large equipment.
- Determine average jobhours per cubic yard installed (with rebar, formwork, and embedments).

B. Roads and Paving

- Determine cost per square foot installed.

C. Fireproofing

- Ensure adequate allowance for cutouts, penetration seals, and rework.

D. Buildings and Structures

- Examine all-in square foot costs of building.

E. Site Preparation

- Check soil conditions (i.e., type, frost depth, de-watering, sheet piling, and drainage requirements).
- Consider possible underground obstructions.
- Be aware that on large, grassroots projects, earth-moving quantities are often underestimated.

F. Piling

- Check the type of piles (e.g., precast, in situ, timber) and the cutting of pile caps.
- Determine who will do the layout work (i.e., the prime contractor or a subcontractor).

G. Piping Estimating Methods

Following are four methods of preparing a piping estimate. The method chosen depends on detail and accuracy of the estimate.

Estimating by length. This method is based on historical data and assumes an average number of fittings and flanges for a standard piping configuration. Costs are on a unit length basis by pipe size and schedule. Fabrication is separated from field installation. It is necessary to add only the cost of such items as valves, pipe supports, and testing to arrive at a total direct cost for the piping system. Care should be taken to check allowances for unusual complexity of piping arrangements (especially onsite units or revamps).

Estimating by weight. In this method, piping materials are assumed to have a value approximately proportional to their weight. Pipe is assigned a cost per pound for material and a number of jobhours per ton for fabrication and erection. Adjustments should be made for unusual materials and labor productivity for the plant location.

Estimating by ratio. This method calculates piping as a percentage of the major equipment cost. Ratio methods can be used only with an appropriate database. This is not a very accurate method and is usually applied only to conceptual estimates.

Estimating by unit cost. This method is more accurate than estimating by ratio, but it is costly and time consuming, since detailed takeoffs must be made of all labor and material units in the system. This method requires that engineering be well advanced before accurate takeoffs can be produced. It also requires detailed historical data for pricing installation labor.

After an estimating method is selected and applied, a piping estimate review should be performed to examine the method and extent of the takeoff by sampling line takeoffs and comparing actual quantities and costs with estimates. Such items as the basis of fabrication and the impact of special materials should also be reviewed.

H. Electrical

In estimating electrical work, a schedule of the number and size of motor drives is a basic requirement. Motor control center and power distribution items usually constitute a major part of the electrical work. Since their prices can vary considerably, budget prices should be obtained from potential suppliers. The cost of power cable should be estimated in reasonable detail. A plot plan layout is useful in assessing quantities, while material unit prices may be estimated from historical data. Minor/miscellaneous services, such as emergency lighting, fire alarms, intercoms, power outlets, and telephone systems, can be assessed approximately or represented as an allowance. Plant lighting may be estimated on an area or unit length basis. A gross estimate of electrical work based on horsepower can be inaccurate. The estimate should take into consideration local electrical codes and area classification. Climatic conditions may require a different type of cable and hardware, and therefore could affect cost.

I. Instrumentation

The following methods are generally used to estimate instrumentation:

Factors. With an adequate database, instrumentation can be factored relative to the installed major equipment cost. Additional points for consideration are local electrical and environmental codes, and the degree of computer control/technical upgrading.

Cost per instrument loop. This can be done by using previous return data to establish costs for typical loops based on instrument type and materials of construction, and multiplying these by the estimated number of loops in the system. Loop configurations should be developed by the instrument engineer.

Total installed cost per unit. Instruments are priced from a preliminary list by means of quotes, catalog prices, or price data. Auxiliary material and installation costs (such as tubing, wiring racks, supports or testing) are assessed for each instrument based on experience and judgment.

Detailed estimating. This is the most accurate approach and requires a detailed instrument list priced from past data or quotes. Labor jobhours for each instrument are added. Instrument tubing and wiring should be established by detailed takeoff. Auxiliary material and labor cost can be taken as a percentage of the total instrument cost.

When the instrument estimate is complete, it should be reviewed and the process and instrumentation diagrams examined for numbers and complexity of instrumentation such as distributed control systems (DCS) and/or extensive alarms for safety and security. Conflicts between owner and contractor specifications should also be checked.

J. Insulation

Review requirements for heat conservation, winterizing, cold insulation, and personnel protection for equipment and piping.

K. Painting

Painting is not usually large enough to justify a detailed estimate, but any prorated method and values used should be reviewed.

XIV. DIRECT CONSTRUCTION LABOR

A. Equipment Installation (Jobhours)

A check of jobhours required for equipment installation may be made either by jobhours per weight and type of equipment, or jobhours per piece and type of equipment.

B. Bulk Materials Installation (Jobhours)

The following major items should be checked:

- jobhours per cubic yard for excavation (machine, hand, or weighted average);
- jobhours per cubic yard for foundation concrete (including forming, pouring, reinforcing steel, and embedments). Review dewatering, sheet piling, and shoring requirements for a civil program;
- jobhours per ton of structural steel (for field fabrication and erection); and
- jobhours per foot of piping by size and pipe schedule.

C. Productivity (Jobhours)

Depending on the quality of the estimating base, the preceding jobhours are normally factored for time and project location. A geographic productivity system is essential for a quality estimating program. General items (such as handling, scaffolding, testing, and rework) are on a jobhour percentage basis for a detailed estimate, and are included in jobhour rates for a conceptual estimate.

D. Labor Costs

Current labor agreements and conditions, productivity factors, labor availability, site conditions, and project conditions should be reviewed. Total jobhours as well as the craft jobhour distribution should also be reviewed:

- subcontract versus direct hire, and what is covered in the all-in subcontract wage rate, especially field indirects;
- average wage rate;
- inclusion of appropriate fringe benefits, taxes, and insurance; and
- allowances for premium pay on overtime and shift work.

XV. CONSTRUCTION INDIRECT COSTS

Where possible, ensure that estimates have dimensional sketches showing layouts of temporary facilities that can then be quantified for estimating.

A. Temporary Facilities

Review estimates for:

- temporary utility lines and utilities consumed during construction;
- temporary roads, parking, warehousing, and laydown areas;
- fencing and security;
- temporary buildings, furnishings, and equipment;
- personnel transportation and equipment-receiving facilities; and
- erection-operation of a construction camp, if required.

Most of these items are estimated on a cost-per-foot and square foot basis.

B. Construction Tools and Equipment

Discuss and check the methods used by the construction group in establishing equipment requirements. Check for:

- list and scheduled duration of all major equipment;
- small tools (normally estimated as cost per labor jobhour or percent of direct-labor costs);
- availability of equipment, and start and finish of rental period;
- major and minor equipment maintenance;
- purchased equipment, and rented equipment and its source;
- special and/or heavy lift requirements; and
- construction equipment cost per direct-hire jobhour.

C. Construction Staff

Examine the site organization chart and assignment durations of personnel. Also review:

- relocation costs, travel and living allowances, fringe benefits and burdens, and overseas allowances;
- total staff jobhours related to total labor jobhours; and
- supervision cost related to the construction labor cost.

D. Field Office Expenses

Review the estimates of field office supplies, copying, telephone and facsimile, office equipment, and consumables. These items are usually estimated as cost per labor jobhour or as a percent of direct field costs.

XVI. CONCEPTUAL ESTIMATING—ENGINEERING

A. Key Elements

Conceptual estimates vary widely in quality, but generally in the range of 40% to 15% quality, where quality is defined as the amount of estimating contingency. For developing a 20% quality estimate, the following are key elements of engineering costs:

- engineering scope review,
- project conditions review,
- key database—hours per piece of equipment,
- engineering costs (Fig. 3.2),
- large project experience—size effect (Fig. 3.3),
- breakdown of home office hours (Fig. 3.4),
- home office expenses (Fig. 3.5),
- typical data points (Fig. 3.6), and
- project support for approved projects—owners (Fig. 3.7).

B. Engineering Scope Review

To ensure that the scope definition is of the required quality, the estimator/project manager should make a detailed review that all preliminary-basic design documents are available and up-to-date (revision numbers and dates of issue); the estimator/project manager should:

- check that all major equipment is included and is listed by equipment number;
- review all items shown on plot plans, flow sheets, P&IDs, and equipment lists to ensure their inclusion in the estimate;
- check equipment and system capacities, flow rates, temperatures, and pressures for deviation;
- identify unique, special, or high cost equipment; and
- check whether owner costs are to be included.

C. Project Conditions Review

Before developing the details of the estimate, an overall evaluation should consider the following:

- project location (site characteristics, labor productivity, and local conditions);
- schedule (anticipated project start, fast track, economic, acceleration);
- grassroots installation, revamp, and shutdown elements; and
- economic outlook, engineering and construction resources, vendor-material capacity.

The overall evaluation also should recognize innate characteristics of the project. Both equipment items and construction elements can add to design engineering hours. Special handling or foundation requirements of equipment can lead to more than normal engineering hours. Similarly, construction preplanning constructability considerations and unusual installation requirements can, again, add to engineering hours. Schedule acceleration and lack of skilled resources can impact on engineering productivity.

D. Key Database—Hours Per Piece of Equipment

The most widely used database for conceptual estimating of engineering is:

Engineering Hours per Piece of Equipment

The engineering hours represent total design hours, including supervision, starting from a completed basic design package to mechanical completion. Feasibility-conceptual design hours are not included; this category of work is covered with a follow-up paragraph. Also not included are project management and other home office services. The total pieces of equipment are represented by the equipment list, which is a standard industry document. The following is a typical database (U.S.):

Category	Engineering hours
Petrochemical, utilities	800–1200
Chemical	600–800
Pulp and paper	500–700
Pharmaceutical	400–500
Metals-mining	300–400
Food	200–300
Mechanical handling	100–200

Selecting a specific number from the above ranges will depend on engineering complexity and special considerations, as previously outlined, of the individual project. For example, considering the petrochemical range of 800–1200 engineering hours per piece of equipment, the following would be typical:

Cracking (complex)	1200
Merox (simple)	800

However, good judgment is required to properly assess the engineering complexity and associated project conditions to determine the correct rate. In exceptional cases, it is possible for a project to be outside the normal range. In addition, it is possible that overseas engineering would require an added factor of 10–20%.

The following descriptions of the figures noted above provide further guidance.

E. Engineering Costs

Figure 3.2 is a chart showing a family of curves for differing work categories/industries, where engineering is shown as a percent of the total project cost. The most widely used curve is Curve C, chemical process plants, where the midpoint of the curve shows engineering at 10% of the total cost. Curve E is used for small

Engineering costs are approximately 10% of TIC for intermediate and large projects. However, experience indicates that this number can increase as shown in curve F due to size effect, lack of skilled resources, and poor planning.

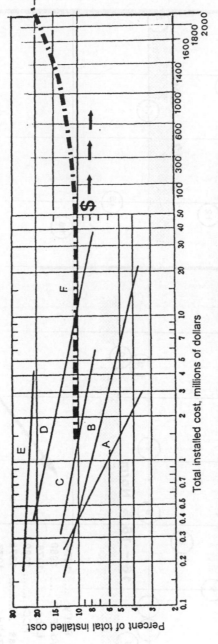

Engineering costs as average percent of total installed costs: A, office buildings and laboratories; B, power plants, cement plants, kilns, and water systems; C, battery-limits chemical process plants; D, complex chemical and grass-roots chemical plants and pilot plants; E, small revamp/retrofit projects; F, large process projects.

Figure 3.2 Engineering costs.

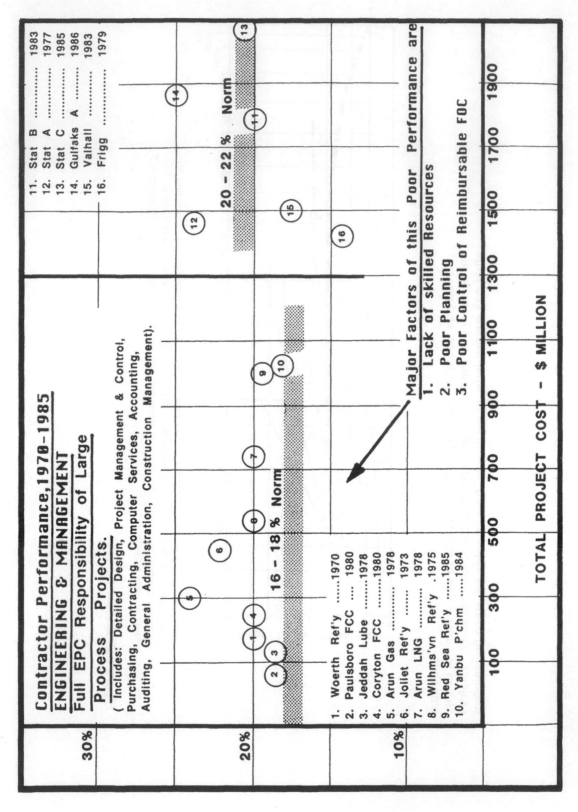

Contractor Performance, 1970–1985
ENGINEERING & MANAGEMENT
Full EPC Responsibility of Large Process Projects.

(Includes: Detailed Design, Project Management & Control, Purchasing, Contracting, Computer Services, Accounting, Auditing, General Administration, Construction Management).

1. Woerth Ref'y 1970
2. Paulsboro FCC 1980
3. Jeddah Lube 1978
4. Coryton FCC 1980
5. Arun Gas 1978
6. Joliet Ref'y 1973
7. Arun LNG 1978
8. Wilhms'vn Ref'y .1975
9. Red Sea Ref'y 1985
10. Yanbu P'chm ...1984

11. Stat B 1983
12. Stat A 1977
13. Stat C 1985
14. Gulfaks A 1986
15. Valhall 1983
16. Frigg 1979

16 – 18 % Norm

20 – 22 % Norm

Major Factors of this Poor Performance are
1. Lack of skilled Resources
2. Poor Planning
3. Poor Control of Reimbursable FOC

30%

20%

10%

100   300   500   700   900   1100   1300   1500   1700   1900

TOTAL PROJECT COST – $ MILLION

Figure 3.3 Large project experience—size effect.

revamp/retrofit projects, and it should be noted that the percentage more than doubles, to 20% plus. Curve F applies to large process projects. Key features of this curve are illustrated by Figure 3.3.

F. Large Project Experience—Size Effect

Figure 3.3 shows data of 16 large projects, over the period 1970–1985, where the size or scale effect of the previous curve (unit cost reduces as size increases, as per 6/10th rule) is reversed. This means that the curves, instead of steadily reducing as size increases, can and do increase when projects reach a very large size and/or are outside the experience of the assigned project personnel. This is known as the *size effect* and is, in fact, a further variation of the learning curve.

G. Breakdown of Home Office Hours

Figure 3.4 is a typical breakdown of total home office hours for a contractor who has total responsibility (i.e., full scope) for executing the project. Other conditions are:

- work completion is mechanical completion,
- data does not include feasibility-conceptual design,
- data applies to projects over $5 million,
- owner's support hours are not included,
- project's experience over the period 1970–1990, and
- adjustment for increased sophistication of instrumentation is included.

Application

This information can be used to check an estimate or a contractor proposal of home office hours. It can also be used for early evaluation of home office labor and schedules when only total costs or hours are available.

Example:

1. For a typical project, we can assess the percent of piping hours. This is derived by totalling the hours required for piping engineering activities (plant design, 16.4%; piping engineering, 2.1%; bill of materials, 2.1%; and model, 0.4%, for a total of 21%).
2. As a percentage of engineering only, piping becomes 21% divided by 0.67, or 32%. As overall engineering and piping design are often on the critical path, individual evaluations are frequently required. Where information is lacking, the following percentages should be used:

- engineering hours as a percentage of total home office (65%), and
- piping hours as a percentage of engineering (35%).

H. Home Office Expenses

Figure 3.5 illustrates many home office services considered part of the expense category that are often forgotten or underestimated.

- Full scope EPC responsibility for contractor
- Data does not cover conceptual design (Phase 1)
- Data is based on historical experience

- Work completion at mechanical completion
- Data applies to intermediate / large projects ($5 MM plus)
- Owner's support jobhours not included

Design & Drafting	Full-scope %	% Jobhours
Civil and structural	25.00	10.0
Vessels	7.50	3.0
Electrical	15.00	6.0
Plant design (piping)	41.00	16.4
Piping engineering	5.25	2.1
Bill of material	5.25	2.1
Model	1.00	0.4
Sub-Totals	100.00	40.0

Administration - indirect drafting		4.0
Engineering		
Instrument (engineering and drafting)		3.0
Mechanical (rotating machinery, plant utilities,metallurgy, etc.)		3.0
Mechanical (consultants)		0.2
Project management		7.5
Project engineering		6.0
Project (operating expenses, services administration)		3.0
Process design		3.0
Process technology services		0.1
	67% Engineering	
Project services		
Estimating and cost control		4.0
Proposals		--
Computer control		--
Computer systems		1.0
Initial operations -- office		0.2
Technical information		0.2
Scheduling		2.0
Procurement		
Purchasing		5.0
Inspection and expediting		5.0
General office		
Stenographic		4.5
Accounting		7.0
Office services		2.0
Labor relations		0.1
Construction (office)		2.0
Total		100.0

Figure 3.4 Typical contractor home office breakdown.

Figure 3.5 Home office expense breakdown.

I. Typical Data Points

Figure 3.6 shows engineering hours, at both conceptual and detailed levels, for various categories of drawings and documents.

J. Project Support For Approved Projects—Owners

Figure 3.7 is a set of curves used to estimate the owner cost of its project team throughout the project execution phase. The two curves-databases are for standard approach cost-reimbursable projects and for a lump-sum approach. Obviously, the size of an owner project team should be less on a lump-sum job. As illustrated, the curves are based on actual project experience.

K. Other Home Office Services Costs

As covered by Figure 3.4 and Figure 3.8, this category is generally 3% of total project cost.

XVII. CONTINGENCY—ESTIMATING ALLOWANCES

The contingency or estimating allowance is usually a function of:

• design definition (such as process, utilities, facilities, and revamp);
• estimating methods (database and level of detail);
• time frame and schedule probability;

The following are typical data points for evaluating
estimating levels or monitoring project performance.
This data applies to large U.S. process plants.
Adjustments should be made for overseas locations
and for small projects.

Engineering
• Jobhours per drawing (total drawings).........150 to 160
• Jobhours per piece of equipment.............1000 to 1200
• Jobhours per piping isometric
      without CAD..8 to 10
      with CAD...4 to 5
• Jobhours per P & I diagram.........................400 to 500
• Jobhours per plot plan...............................200 to 300
• Jobhours per material requisition......................8 to 10

Figure 3.6 Typical data points—engineering.

PROJECT SUPPORT - Approved Projects.
• Hours per $ million (Base Project Cost).

This chart reflects a "Full, independent
Project Task Force Approach", as used
by leading international oil companies.
It does not reflect a "Partnering" Approach

Covering design, project management, project control, procurement services, construction &
contract administration. With a Managing Contractor, project team is in a monitoring role. For
"Small Projects Program", includes detailed engineering and "full" management of the work.

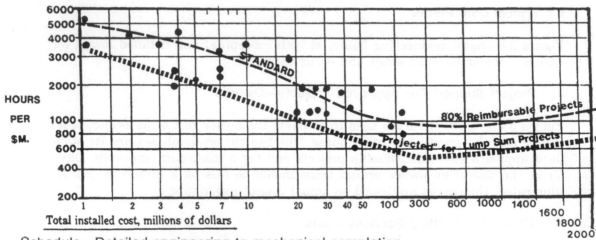

• Schedule - Detailed engineering to mechanical completion
• Confidence level at plus/minus 20% with 80% probability.

Figure 3.7 Owner engineering project department standards.

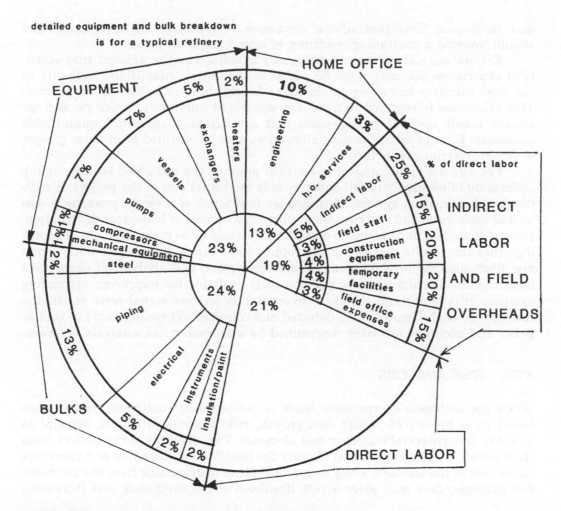

Figure 3.8 Typical prime contractor cost breakdown.

- new technology and prototype engineering;
- remoteness of job site, and infrastructure requirements;
- engineering physical progress (percentage complete); and
- material commitment.

Determining overall estimate reliability is made more difficult by the fact that some segments of a project may be completely defined at the time of estimate, and others only sketchily defined; some may be estimated by reliable methods and others by methods that produce less accurate results.

It is necessary, therefore, to separately quantify the degree of reliability of each major, independently estimated segment or unit of an estimate. This can be done by using guidelines for classifying the degree of definition and the quality of methods/data used. These in turn establish appropriate estimating allowances and accuracy ranges for each of the segments.

After a project is approved and work begins, changes begin to take place in facility definition, estimating methods, knowledge of project conditions, and forecast time-span. This then allows successive re-appraisals of contingency and should produce a continuing reduction of estimating allowances.

Estimating allowances or contingency is defined as the amount that statistical experience indicates must be added to the initial, quantifiable estimate so the total estimate has an equal chance of falling above or below the actual cost. This allowance is required to cover oversights and unknowns, which on average always result in final project costs that are higher than initial quantifiable estimates. If required, estimating allowances may be modified to produce greater or lesser overrun probabilities.

For any individual project or series of projects, the estimated cost, including estimating allowance, will fall under or over the actual cost of the project. A well-developed estimating system, when applied to a series of projects, produces a pattern of underruns and overruns that approach a normal or bell-curve distribution. Overestimate and underestimate amounts are caused by so many unrelated happenings that the results resemble those obtained by chance. Major systematic errors are eliminated in developing an estimating system, and analysis of departures from normal distribution is one of the tools available for improving estimating systems. It is important, therefore, to constantly analyze actual costs versus the estimate so such biases can be detected and corrected. These elements of contingency and accuracy are often determined by a computer risk analysis program.

XVIII.   RISK ANALYSIS

When the estimator's experience leads to conventional contingency application based on a history of project cost growth, risk analysis principles attempt to quantify this potential for direct cost elements. The members of the project team then review these elements and identify the likelihood of increases and decreases in the cost of the element, along with probabilities of deviations from the estimate. For example, they may place a 70% likelihood of the earthwork cost increasing

Table 3.1 Example of Results from a Computerized Risk Evaluation Program

Target estimate = $1,225,000

Probability or accuracy

Certainty	Contingency
90	162,600
80	108,400
70	81,100
60	49,400
50	21,500
40	900
30	−28,900
20	−54,200
10	−200,950
0.5	−268,947

because soil borings are not available. For instance, the earthwork cost might increase by as much as $100,000. The cost might also be $30,000 less than the estimate due to competitive market conditions.

After the project team establishes the highest and lowest possible values for each element in the estimate, these numbers are entered into a computerized risk evaluation program that performs a *Monte Carlo simulation*. In this process, a random sample value for each element is taken, and the resulting total estimates are arranged in order from highest to lowest. The deviations of each total value from the original estimate are presented in Table 3.1. The certainty column represents the likelihood that the actual cost will not exceed the original estimate after the indicated amount of contingency has been added.

This information is provided to management to determine the risk it is willing to take on the project. For instance, in this example, if management selects a 70% risk level, the estimate would be $1,306,100 ($1,225,000 + $81,100). Risk analysis is discussed in greater detail in Chapter 11, "Range Estimating."

XIX. ESCALATION

Escalation is usually included as a separate line item. Escalation rates for material and labor costs should be separately identified.

XX. CURRENCY EXCHANGE CONVERSION

Currency conversion rates can fluctuate widely over the life of a project. It is recommended, therefore, that the estimator use the rate established at the time of appropriation and track subsequent deviations as a one-line item. Corporate and affiliate financial groups should be consulted when establishing currency conversion rates for the estimate.

XXI. CONSTRUCTION LABOR PRODUCTIVITY

Good assessments of labor productivity are essential for a quality cost estimate. Cost control, planning, and scheduling can be ineffective without an adequate evaluation of labor jobhours.

Figure 3.9 shows major elements of a labor productivity system. The recommended *additional factors* and associated curves have been developed from historical data. However, on a specific project, any one or even several conditions can have an abnormal effect on productivity.

A. Project Condition Analysis

The productivity analysis starts from a general area factor. Additional allowances, based on the recommended ranges are then made for the listed conditions. Judgment and experience are necessary for determining these additional allowances. Even if no previous experience of the area exists, this procedure can still be very effective since the *condition productivity adjustments* are often of a greater magnitude than area productivity differences.

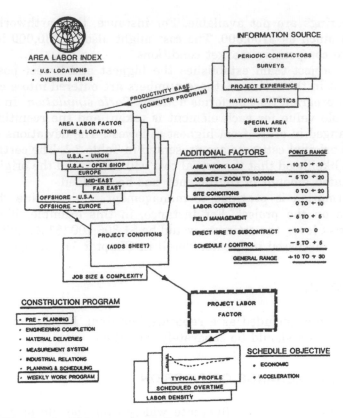

Figure 3.9   Flowchart of labor productivity system.

Figure 3.10 depicts productivity adjustments for:

- area workload and peak construction labor,
- job size, and
- extended workweek.

The data and analysis are generally based on construction of new facilities at an existing plant, with minor hot work restrictions and an average labor performance. Guidance for major revamps is given in the historical data section.

B.   Area Workload/Peak Construction Labor

As seen in Figure 3.10, in a normal economic environment, as area workload and peak labor increase, productivity will generally decrease. It is assumed that the estimating database, at a productivity of 1.0 (on the curves), is based on the normal area workload, with a peak labor level of 750.

If area workload is low, then the labor peak can rise to 2000 before a productivity loss adjustment needs to be made. Conversely, if area workload is low and project size and schedule requires a labor peak of 1000, then a productivity improvement (0.96) is determined from the curves.

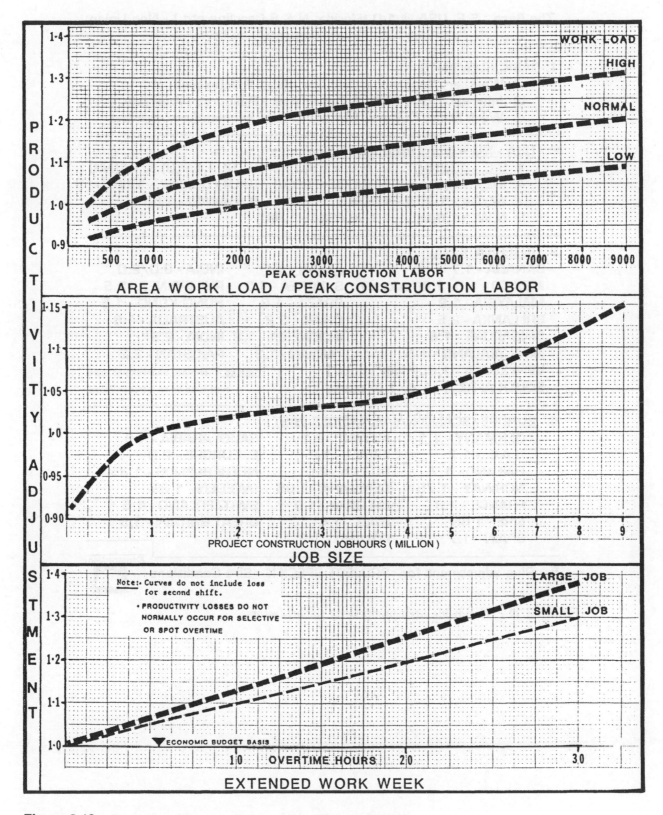

Figure 3.10   Project condition analysis—construction productivity.

The Base : S.E. USA @ 1.0 (Florida, N & S Carolina etc.) - Non Union; and has best **"Basic Elements"**, as listed ....

1. BASIC ELEMENTS
a) Weather/Geographic-Cold, Ice, Heat, Rain(excess) etc.
b) Inherent Trade Skills & Capabilities - Resources Available.
c) Work Ethic, Culture, Militant Trade Unionism(very negative).

2. USA - NU = Non Union, U = Union
1. S.E. - 1.0(NU)   2. N.E. - 1.6(U)   3. TX. - 1.2(U) & 1.1(NU)
4. ILL. - 1.6(U)   5. GA/AL - 1.3(U)   6. UT/CO/WY - 1.5(U)
7. AZ/CA - 1.2(U) - 1.1(NU)   8. HAWAII - 1.5(U)
9. N.W. - 1.6(U)   10. ALASKA - 1.9(U)

3. INTERNATIONAL
CANADA - 1.7	JAPAN - 1.1	INDIA - 3.0/2.0
MEXICO - 1.5/1.8("sticks")	S. KOREA - 1.3	PHILIPPINES - 2.5
MALAYSIA - 1.9	BRAZIL - 1.8	HONG KONG - 1.5
SRI LANKA - 2.5	CHINA - 2.2	AUSTRALIA - 1.6
SINGAPORE - 1.6	INDONESIA - 1.9	INDIA - 2.5
NEW ZEALAND - 1.5	TAIWAN - 1.3	PAKISTAN - 2.2
VENEZUELA - 1.65/1.8("sticks")		"sticks"= Remote, undeveloped area,
SOUTH AFRICA - 1.4/1.9("sticks")		with little or no Infrastructure.

4. EUROPE/SCANDINAVIA/MIDDLE EAST
GERMANY - 1.1	KUWAIT - 2.1	SAUDI ARABIA - 1.6
U.A.E. - 1.7	ISRAEL - 1.8	E. EUROPE - 2.0
FINLAND - 1.7	RUSSIA - 2.0	NETHERLANDS - 1.35
NORWAY - 1.75	DENMARK - 1.35	SWEDEN - 1.35
FRANCE - 1.3	BELGIUM - 1.3	SWITZERLAND - 1.5
EIRE - 1.65	ITALY - 1.4	GREAT BRITAIN - 1.5
SPAIN - 1.7	POLAND - 1.9	

5. Plus adjustment for PROJECT SPECIFIC Site Conditions

Figure 3.11 Worldwide area productivity (labor) factors.

C. Job Size

In addition to staffing levels, as shown in Figure 3.10, job size (in jobhours) has a significant effect on productivity. This curve shows that the estimating database reflects a project size of one million jobhours (direct work only). Thereafter, as job size increases, productivity decreases.

D. Extended Workweek—Productivity Loss

Figure 3.10 shows extended workweek curves for small and large jobs. The productivity loss adjustment applies to total jobhours and not merely to the additional overtime hours. Note, however, that during periods of major low employment, such as recessions and depressions, it is possible that labor productivity will not follow these curves, since the fear of unemployment can be an overriding consideration.

E. Area Factors

Figure 3.11 contains a series of worldwide area productivity factors. As shown, the basic elements are weather, trade skills, and work ethic. These are, however, conceptual numbers, and considerable judgment is required for their proper application. The data has been developed over many years from the author's company and individual contracts at the locations illustrated.

XXII. PRE-ESTIMATING SURVEY AND CHECKLIST

Figure 3.12 shows the major items to be developed and/or considered before developing the estimate. In conjunction with the pre-estimating survey, a comprehensive checklist can be a significant aid in ensuring that all appropriate details have been covered. The following is not a complete list, but it will significantly assist with major considerations:

- plan the estimate,
- cover all items,
- serve as a base for your database, and
- cover the three p's—political, procurement, process design.

A. Political Considerations

Political considerations can be broken down as follows:

- local political and social environment,
- regulatory and permitting requirements,
- business environment,
- tax structure and expense versus capital costs allocation, and
- overseas (nationalistic/logistics/infrastructure).

Project Management and Control Manual	Section	Estimating
	Subject	

Estimating Checklist

☐ Project Execution Plan _____
☐ Process Flow Diagram _____
☐ Plot Plan _____
☐ P & ID's _____
☐ Construction Drawings _____
☐ Schedule Constraints _____
☐ Equipment List _____
☐ Transportation and Freight _____
☐ Spare Parts _____
☐ Vendor Representatives _____
☐ Purchasing Plan _____
☐ Contracting Plan _____
☐ Permitting Requirements _____
☐ Code Requirements _____
☐ Safety Requirements _____
☐ Gravel Requirements _____
☐ Demolition _____
☐ Revamp _____
☐ Foundation Requirements _____
☐ Painting and Coating _____
☐ Insulation _____
☐ Fireproofing _____
☐ Fire Protection _____
☐ Instrumentation _____
☐ Communications Equipment _____
☐ Furniture and Fixtures _____
☐ Geotechnical Requirements _____
☐ Non Destructive Testing _____
☐ Welder Certification and Testing _____
☐ Functional Check Out _____
☐ Start Up _____
☐ Operating Fluids _____
☐ Utility Tie-Ins _____
☐ Hot-Taps _____
☐ Construction Mobilization
   and Demobilization _____
☐ Labor Productivity _____
☐ Payroll Benefits and Burdens _____
☐ Cleanup _____
☐ Scaffolding _____
☐ Material Handling _____
☐ Temporary Facilities _____

☐ Jobsite Utilities _____
☐ Construction Supplies _____
☐ Equipment Rental and Leases _____
☐ Security _____
☐ Fabsite Leases and Fees _____
☐ Taxes _____
☐ Fees _____
☐ Documentation and As-Builts _____
☐ Home Office Field Assistance _____
☐ Design Status and Pending Changes _____
☐ Vendor Shop Visits _____
☐ Expediting _____
☐ Quality Assurance _____
☐ Inflation and Escalation _____

Allocated Costs

☐ _____ _____
☐ _____ _____
☐ _____ _____
☐ _____ _____
☐ _____ _____
☐ _____ _____
☐ _____ _____
☐ _____ _____
☐ _____ _____
☐ _____ _____
☐ _____ _____
☐ _____ _____
☐ _____ _____
☐ _____ _____

Other

☐ _____ _____
☐ _____ _____
☐ _____ _____
☐ _____ _____
☐ _____ _____

☐ Risk Analysis _____
☐ Contingency _____

Date		Page

Figure 3.12 Estimating checklist.

B. Procurement Program Considerations

A careful review of the procurement program is essential, since the equipment/material costs can comprise more than 50% of a project's total cost. The following are typical considerations:

- quality vendors list (information/experience of suppliers);
- domestic versus worldwide purchasing plan;
- import duties, taxes, delivery charges;
- currency considerations and exchange rates;
- vendor service people requirements;
- plant compatibility of existing versus new;
- ease of maintenance, operating costs;
- spare parts requirements;
- inspection and expediting requirements; and
- critical purchasing plan (schedule priority).

C. Detailed Checklist for Estimating

- climate:

  - humidity,
  - temperature,
  - prevailing winds,
  - seasons,
  - storms,
  - snow accumulation and ice conditions,
  - rain (average and seasonal),
  - days lost due to weather,
  - shelters required,
  - special methods of construction necessary, and
  - indoor/outdoor equipment.

- earthquake factors
- access:

  - distance from metropolitan area;
  - roads, water, air, railroads;
  - road conditions;
  - road clearances (tunnels);
  - road and bridge capacity; and
  - ice conditions.

- offshore facilities:

  - water depth,
  - wind forces,
  - wave forces,
  - sea floor conditioning, and
  - soil conditions.

- environment:

  - attitude of the community,
  - present and future zoning,
  - other industry in the area,
  - environmental restrictions,
  - environmental impact study,
  - required permits—local/state/federal/other,
  - legal counseling,
  - delays in obtaining permits and associated costs in terms of escalation,
  - requirements for and costs of pollution control (noise/air/water/waste disposal), and
  - considerations of alternate site.

- political aspects:

  - political climate of the proposed site and its prospects for future stability;
  - governing authority's encouragement/discouragement of investment, attitude toward business, tax structure; and
  - on overseas projects, the degree of government involvement and the terms of payment and the probability of delayed payment.

- procurement:

  - source of information about vendors;
  - location of vendors;
  - methods of transporting equipment and material;
  - availability of a minimum of three bidders;
  - vendor reliability and experience;
  - origin of material and equipment;
  - import restrictions;
  - import duty;
  - availability of equipment on reasonable delivery schedules;
  - terms and conditions;
  - discounts for large purchases;
  - consideration of whether purchase orders will be firm, cost plus, or with specified escalation;
  - warranties;
  - service that suppliers can provide during construction and operation, and the cost of those services;
  - provisions for inspection and expediting;
  - export packing requirements;
  - spare parts and their cost;
  - currency in which purchases will be made;
  - exchange rates;
  - payment schedules;
  - marshalling yard requirements;
  - loading and unloading requirements;
  - lightering;
  - demurrage costs;
  - higher costs due to congested harbors; and

- – use of trading companies.
- process design:
  - – plant capacity;
  - – plant product;
  - – by-products;
  - – flowsheets available;
  - – utility flowsheets;
  - – plant layout;
  - – material specifications (standard or exotic); and
  - – mechanical specifications (pressures, temperatures, flows, corrosion).
- process specifications:
  - – local code requirements,
  - – state code requirements,
  - – federal code requirements,
  - – client/engineer's specifications,
  - – architectural requirements,
  - – metric/English measurements, and
  - – pollution control.

XXIII.   STATISTICAL/HISTORICAL DATA

This section includes references, which will be helpful in developing a cost estimate.

A.   Typical Prime Contractor Cost Breakdown

Figure 3.8 illustrates the overall breakdown of project cost based on historical data for projects built in the United States on a prime contract basis during the past 20 years. Since most costs have retained the same parity over time, the only significant change to this experience is the increase in instrumentation costs.

Application

When only an overall cost is known, this breakdown can be useful in providing overall data for a quick evaluation of engineering and construction costs and jobhours. This can enable cost claim and schedule evaluations to be made.

Example:

Assume that a project has an estimated overall cost of $100 million.

1.   Figure 3.8 shows that home office costs are roughly 13%, or $13 million. By a further assumption that the contractor's home office all-in cost is $50 per hour, we can derive a total number of home office jobhours:

$$\text{Number of jobhours} = \frac{13,000,000}{50} = 260,000$$

Thus, a gross schedule and jobpower evaluation can now be made.

TYPICAL PROJECT COST BREAKDOWN
(CONTRACTOR - TOTAL SCOPE PROJECT)

GRASS ROOTS - LARGE	ITEM	SMALL REVAMP
%		%
10	ENGINEERING (D & D)	20
3	HOME OFFICE (SUPPORT)	5
47	MATERIAL (DIRECT)	40
21	CONSTRUCTION DIRECT	17
19	CONSTRUCTION INDIRECT	18
100 %		100 %

DIRECT CONSTRUCTION LABOR HOURS

10	SITE PREPARATION	1
12	FOUNDATIONS/UNDRGRDs	8
7	STRTL. STL. BLDGS.	5
10	EQUIPMENT	12
35	PIPING	48 (INCL. FAB.)
11	ELECTRICAL	10
6	INSTRUMENTS	8
4	PAINTING	3
4	INSULATION	3
1	HVAC/FIREPROOFING	2
100 %		100 %

Figure 3.13   Typical project cost breakdown.

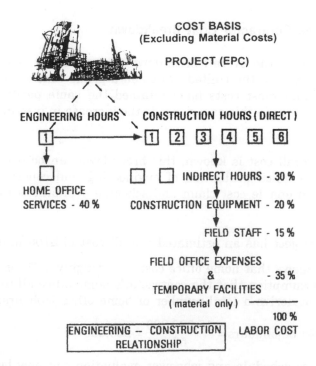

Figure 3.14   Cost basis—engineering/construction relationship.

2.   Figure 3.8 shows that direct field labor costs are roughly 21% or $21 million. By a further assumption that the direct field labor payroll cost is $25 per hour, we can derive a total number of direct field jobhours:

$$\text{Number of jobhours} = \frac{21,000,000}{25} = 840,000$$

Applying known and historical relationships allows gross evaluations for engineering and construction durations to be made. These, in turn, can be used to prepare labor histograms and progress curves. Note that for larger projects, the percentage of home office and field overheads may increase.

B.   Typical Project Cost Breakdown

Figure 3.13 shows the same overall cost breakdown as Figure 3.8 and compares it with small revamps. The probability for "Grassroots—Large" is 80%, whereas the probability for "Small Revamp" is 70%, as the variation in small projects costs is much wider. The data also includes a breakdown of construction direct labor hours, by discipline. As with the previous figure, these breakdowns/relationships can be helpful in evaluating labor hours, schedules, and labor requirements.

C.   Cost Basis—Engineering/Construction Relationship

Figure 3.14 shows construction indirect and direct costs. A typical breakdown of the major indirect costs is also shown. Individual companies might allocate their indirect costs somewhat differently, but a high degree of conformity still exists in the contracting side of the industry. Again, it is emphasized that the stated information only applies to a full EPC project.

A further emphasis of this figure is on the engineering/construction labor hour relationship. This is shown as the ratio of 1:6 and is for large projects. Small projects have a 1:3 ratio. In other words, one engineering labor hour automatically generates 6 direct construction labor hours. This is a very useful rule of thumb, although the ratio does vary as the design/construction complexity and size vary.

This relationship highlights the need for design engineers to realize that as they are designing, they are also generating the construction labor hours. Full realization of this fact should lead designers to more carefully consider the question of constructability of their designs. This is the design process of working to construction installation considerations, as well as working to standard design specifications. Constructability considerations can result in significant savings in construction labor. Such considerations are essential in the following types of construction/conditions:

*   heavy lifts,
*   prefabrication and pre-assembly,
*   modularization,
*   offshore hookup work,
*   site problems of limited access, and
*   lack of resources at the jobsite.

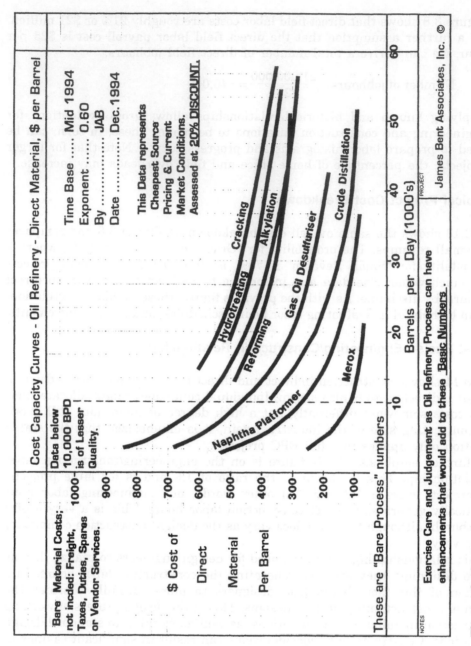

Figure 3.15 Cost capacity curves—material costs per barrel.

The relationship between engineering and home office support services (such as project management, project control, procurement, computer, and clerical) is shown also at 40%. This relationship is in labor hours.

D. Cost Capacity Curves

Figure 3.15 is based on 30 years of experience and shows costs for various oil refinery process units. These are *bare bones* rates, and the following adjustments need to be carefully considered:

- bare process numbers (there are usually enhancements/additions);
- assumes cheapest source;
- assumes 20% discount from current pricing levels; and
- does not include inspection, expediting, spares, vendor services, etc.

This is a project approach to estimating, where current market experience is built into the budget/estimate as targets for the project manager. Overall project experience has been that, in most cases, the lower costs of cheapest source and 20% discount are not achieved due to low project-purchasing skills. With project-specific information and assuming fair-to-average project skills, the above adjustments can be more than +50% to +60%.

Figure 3.16 shows labor hours for various oil refinery process units. These are bare bones rates, and the following adjustments need to be carefully considered:

- area factor, (see Figure 3.10; base is Southeast USA)
- area workload, (see Figure 3.9; –10% to +10%)
- job size (large), (see Figure 3.9; –5% to +20%)
- site conditions (extreme), (0 to +20%)
- labor conditions (militant unions), (0 to +10%)
- field management (quality), (–5% to +5%)
- direct hire to subcontract (US only), (–10% to 0)
- schedule/control skills, (–5% to +5%)

If the above specifics cannot be developed, use a general number from the range of +10% to +30%.

E. Subcontract Estimating

As unit price estimating databases are difficult to obtain, an effective technique is to develop an all-in dollar rate to apply to hourly estimating databases, which are readily available. Figure 3.17 shows the major components for developing an all-in dollar rate.

F. Productivity Loss for Extended Workweek

Projects are occasionally placed on extended overtime so the schedule may be shortened. In many of these cases, under normal economic conditions, productivity will be reduced and costs will increase. If this condition was not part of the original estimate, an assessment of the increased cost, as well as the schedule advantage, should be made. The schedule evaluation should recognize increased jobhours in

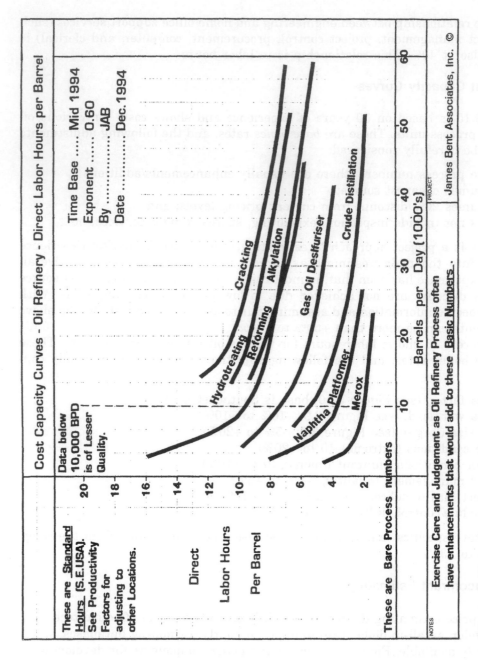

Figure 3.16 Cost capacity curves—jobhours per barrel.

• Convert $ units to hours by building an all-in labor rate as follows:	
1. Bare Labor Rate *CAREFULLY consider type & size of job and add the following items to labor rate. Then convert into $ rate.*	Base $
2. Construction Staff	10 to 20%
3. Construction Equipment	10 to 40%
4. Field Expenses	30 to 40%
5. Home Office Overhead & Profit	10 to 20%
ALL-IN LABOR COST	60 TO 120%

Figure 3.17 Subcontract administration.

the duration calculation. It is also possible that absenteeism will increase, sometimes to such an extent that there is no schedule advantage for the increased workweek.

Figure 3.18 presents data compiled from the sources indicated and plots labor efficiency against overtime hours worked, based on 5–, 6–, and 7–day workweeks. This data applies only to long-term, extended workweeks. Occasional overtime can be very productive with no loss of efficiency. The figure shows a recommended range of productivity loss by project size (small to large). It should be noted that studies of the Construction Users Anti-inflation Roundtable (now the Business Roundtable) have concluded that prolonged periods of scheduled overtime can actually result in a net loss of productivity (i.e., no gain from the expenditure of overtime hours). Depending upon the schedule, this can begin to occur after 6–8 weeks and, if the extended overtime continues, the schedule can actually be lengthened rather than being improved. Scheduled overtime should therefore be used with great caution.

Application

Figure 3.18 can be useful in an overall evaluation of the impact overtime hours will have on schedule and cost. It can establish an increase in total labor hours required for a loss in efficiency due to an extended workweek. However, judgment should be used on an individual location basis. Some areas, particularly less developed countries, work 60-hour weeks that are only as productive as 40-hour weeks.

Example:

Assume that a project has a total construction scope of 1 million jobhours and is based on a 5–day, 40–hour workweek. If the same workweek is increased by eight hours, look to the chart for eight hours of overtime, and using the NECA 5–day, large project curve, read across to an efficiency of 90%. This indicates that, due to a loss in efficiency, 10% more hours will be required to accomplish the same amount of work. Thus the jobhours will be estimated at 1 million × 1.10 = 1,100,000 jobhours. Schedule and cost evaluations can now be made for an additional 100,000 jobhours, but at an increased level of work. Obviously, a schedule advantage accrues.

Note, however, that the curves contained in Figure 3.18 do not include efficiency losses for a second shift, which can be about 20%. However, shift work

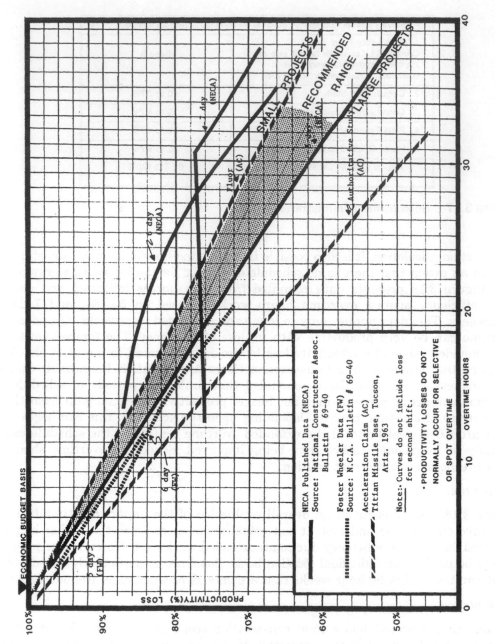

Figure 3.18 Productivity loss for extended workweek.

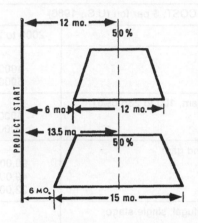

Escalation of "rate costs" for a new time period

Originally, the costs for direct labor and field indirects were developed by estimating the monetary rate that would be appropriate at the point in time when 50% of the costs were expended. Assume all the costs of construction are proportionally expended, as per the sketch, for this example.

With a three-month schedule extension and the start of construction occurring at the previously planned date of six months, then the center of construction costs is now 13½ months.

The general approach to escalation, then, is to take 50% of the schedule slippage as the "escalation time period" and apply the appropriate escalation rate to the construction costs for that time. An adjustment may be appropriate as escalation may not apply to some minor fixed costs (such as mobilization) which could occur in the originally planned time period.

Escalation = Total Construction Costs x Escalation Rate x 3 months ÷ 2

Use same technique for engineering costs.

Figure 3.19 Escalation calculation technique—construction labor.

losses depend on the type of work, company organization, and experience. In the offshore industry, where shipyards traditionally work on a shift basis, losses can be minimal or even zero.

G. Escalation Calculation Technique

A widely used method of calculating escalation is to develop a scheduling assessment of the projected midpoints of engineering, procurement, and construction and then escalate these costs with individual escalation rates from the point of estimating to the projected midpoints. Figure 3.19 illustrates this technique with a construction example and also provides a simple but effective method for calculating escalation costs for a schedule extension.

H. Conceptual Database Modules

Figures 3.20A and B illustrate major equipment costs, labor unit rates (direct labor jobhours per ton), and some dollar rates for fabrication and hookup of major types of modules); these include:

- module fabrication (can be used if design drawings are available and detailed quantities taken off),

CONCEPTUAL ESTIMATING FACTORS		
MAJOR EQUIPMENT, MATERIAL COST, $ per ton (U.S., 1986)		
1. Vessels		2000 to 2300
2. Heat Exchangers a) Single Pass b) Two Pass		4000 6000
3. Packaged Boilers (saturated steam, 100 to 235 psig) a) 30,000 lbs/hr b) 225,000 lbs/hr		5000 9000
4. Fired Heaters - (dual fired, oil and gas) a) 750,000 Btu/hr b) 8,000,000 Btu/hr c) 12,000,000 Btu/hr		90,000 49,000 22,000
5. Pumps (including driver) - centrifugal, single stage, API-610, 3550 rpm a) 10 hp, 100 gpm b) 100 hp, 500 gpm c) 200 hp, 2000 gpm		 14,200 11,500 9,500
6. Compressors - Centrifugal a) 300 hp b) 3500 hp c) 2000 hp d) 4500 hp	Low Pressure 13,000 8,000 --- ---	High Pressure --- --- 13,000 9,000
7. Compressors - Reciprocal a) 1500 hp b) 6000 hp c) 12,000 hp	1,000 psi 16,000 14,000 10,000	6,000 psi 22,000 19,000 13,000
8. Bulk Material a) Structural b) Piping c) Electrical d) Instruments e) HVAC f) Surface Protection (per wt. of structural/mechanical) g) Quarters Fittings		700 to 1,000 1,000 4,000 10,000 to 20,000 1,000 20 9,000

Figure 3.20A   Conceptual estimating factors—major equipment, USA/modules.

- overall module fabrication (broad data when only overall module weight and/or category of module is known), and
- site installation/hookup (broad data when only overall module weight is known).

As indicated in Figure 3.20A, the labor unit rates are for a 1986 European location. The rates may have to be adjusted for a United States location due to higher productivity levels at U.S. yards. For well-established U.S. yards (such as those in Oklahoma and the Gulf states), the fabrication productivity (jobhours) can be better by 30–40%, and installation/hookup by 20–30%. However, for many

MODULES, direct labor jobhours per ton (Europe, 1986)	
9. FABRICATION	
a) Structural - Primary Steel	75
- Secondary Steel	100
- Outfitting Steel	200
b) Major Equipment	15 to 20
c) Piping Prefabrication	250
d) Piping Installation	200
e) Electrical	600 to 700
f) Instruments	700 to 800
g) HVAC	300 to 400
h) Surface Protection (per wt. of structural/mechanical)	17
• Painting	520
• Fireproofing	110
• Insulation	850
10. OVERALL MODULE FABRICATION	
• Size Category	
a) Larger Modules (over 2000 tons)	150 to 200
b) Small Modules (50 to 500 tons)	50 to 130
• Facility Category	
c) Process Module (6000 tons)	200
d) Wellhead Module	140
e) Mud Module (6000 tons)	180
f) Utility Module (6000 tons)	150
g) Services/Quarters Module (4000 tons)	150
h) Drillers Office (1000 tons)	130
i) Helideck (1300 tons)	120
j) Air Control Module (200 tons)	200
k) Derrick (1300 tons)	45
l) Main Lifeboat Station (300 tons)	100
m) Flare Boom	200.
n) Main Pipe Rack Module	175 to 195
o) Shipping	$200 to $1200 per ton
11. SITE INSTALLATION / HOOKUP	
a) Large Modules (over 2000 tons)	25 to 45
b) Small Modules (50 to 500 tons)	75 to 100
c) Load Out and Sea Fasten - Labor (per wt struct/mech)	2 to 3
- Equip. Rental (per wt struct/mech)	$20

Figure 3.20B Conceptual estimating factors—major equipment, Europe.

small, inexperienced U.S. yards, these improvements should not be expected. This data applies to land-based modules as well as offshore modules.

I.   General Data

The following figures illustrate historical relationships and data that can be used in estimating: Figure 3.21, time-escalation data of the the Nelson-Farrar Cost Index; Figure 3.22, economies of scale; Figure 3.23, an alkylation process cost capacity curve; Figure 3.24, typical factors in shutdowns and retrofits; Figure 3.25, breakdown/factors of a good construction estimate; Figure 3.26, factors of

Refinery construction (1946 Basis)

	1962	1976	1991	1992	1993	July 1993	June 1994	July 1994
Pumps, compressors, etc.	222.5	538.6	1,177.8	1,216.4	1,254.6	1,254.0	1,278.6	1,278.6
Electrical machinery	189.5	287.2	548.1	550.4	555.5	555.9	561.8	562.2
Internal-comb. engines	183.4	348.3	794.4	809.2	820.6	822.6	834.0	836.5
Instruments	214.8	466.4	844.7	865.5	879.3	877.3	886.6	887.4
Heat exchangers	183.6	478.5	772.6	746.6	704.1	704.1	677.9	682.1
Misc. equip. average	198.8	423.8	827.5	837.6	842.8	842.8	847.8	849.4
Materials component	205.9	445.2	832.3	824.6	846.7	850.1	865 2	869.0
Labor component	258.8	729.4	1,533.3	1,579.2	1,620.2	1,624.8	1,660.4	1,664.9
Refinery (Inflation) Index	237.6	615.7	1,252.9	1,277.3	1,310.8	1,314.9	1,342.3	1,346.5

Refinery operating (1956 Basis)

	1962	1976	1991	1992	1993	July 1993	June 1994	July 1994
Fuel cost	100.9	384.5	443.8	425.9	421.5	423.6	449.0	454.7
Labor cost	93.9	145.5	280.8	281.1	286.2	275.0	260.3	261.7
Wages	123.9	314.3	787.4	824.9	868.0	855.2	871.3	871.4
Productivity	131.8	216.1	280.6	293.8	303.4	311.0	334.8	333.0
Invest., maint., etc.	121.7	252.6	511.4	519.2	524.3	526.0	536.9	538.6
Chemical costs	96.7	195.2	228.5	218.8	210.0	209.4	207.3	210.4
Operating indexes Refinery	103.7	209.3	392.2	393.3	396.3	392.9	394.2	396.3
Process units*	103.6	267.1	418.6	415.1	416.9	414.9	423.1	426.2

Reproduced by permission from <u>Oil & Gas Journal</u>, 3050 Post Oak Blvd., Houston, Texas 77056. The Nelson-Farrar refinery construction and refinery operating cost indexes are published regularly in <u>Oil & Gas Journal</u>.

Figure 3.21   Nelson-Farrar cost indexes.

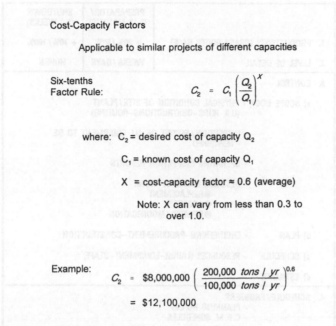

Cost-Capacity Factors

Applicable to similar projects of different capacities

Six-tenths
Factor Rule:

$$C_2 = C_1 \left(\frac{Q_2}{Q_1} \right)^X$$

where: C_2 = desired cost of capacity Q_2

C_1 = known cost of capacity Q_1

X = cost-capacity factor ≈ 0.6 (average)

Note: X can vary from less than 0.3 to over 1.0.

Example:

$$C_2 = \$8,000,000 \left(\frac{200,000 \ tons / yr}{100,000 \ tons / yr} \right)^{0.6}$$

$$= \$12,100,000$$

Figure 3.22 Cost-size scaling (economy of scale).

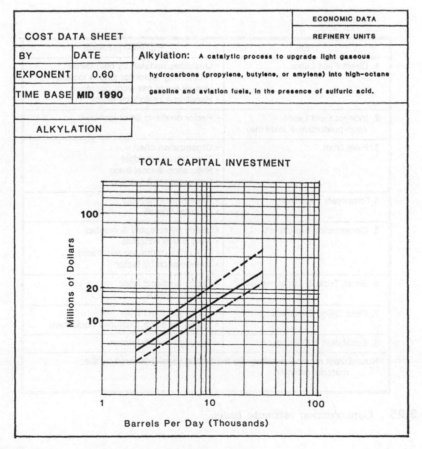

Figure 3.23 Cost capacity curve—alkylation unit.

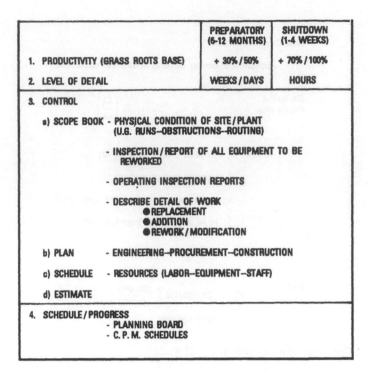

	PREPARATORY (6-12 MONTHS)	SHUTDOWN (1-4 WEEKS)
1. PRODUCTIVITY (GRASS ROOTS BASE)	+ 30% / 50%	+ 70% / 100%
2. LEVEL OF DETAIL	WEEKS / DAYS	HOURS

3. CONTROL

    a) SCOPE BOOK - PHYSICAL CONDITION OF SITE / PLANT
                  (U.G. RUNS--OBSTRUCTIONS--ROUTING)

                - INSPECTION / REPORT OF ALL EQUIPMENT TO BE
                  REWORKED

                - OPERATING INSPECTION REPORTS

                - DESCRIBE DETAIL OF WORK
                    ● REPLACEMENT
                    ● ADDITION
                    ● REWORK / MODIFICATION

    b) PLAN   - ENGINEERING--PROCUREMENT--CONSTRUCTION

    c) SCHEDULE - RESOURCES (LABOR--EQUIPMENT--STAFF)

    d) ESTIMATE

4. SCHEDULE / PROGRESS
                    - PLANNING BOARD
                    - C. P. M. SCHEDULES

Figure 3.24   Estimating shutdowns/turnarounds, and retrofits.

ITEM	ESTIMATE BASIS
1. Direct Field Labor	• Quantities updated by field takeoff • Productivity factor for time & location • Unit jobhours per work operation • Handling & rework by factor
2. Indirect Field Labor (non-productive & lost time)	• Factor on dierct labor jobhours
3. Field Staff	• Organization chart • Time frame schedule • Relocation & local living • Replacement & training
4. Temporary Facilities	• Dimensioned layouts • Quantity takeoff
5. Construction Equipment	• Listing by category & number • Time frame schedule • Unit rates (rental vs. purchase) • Maintenance by factor
6. Small Tools & Consumables	• Factor on direct labor • Loss allowance
7. Field Office Expenses	• Factor on direct labor • Listing for office furniture / equipment
8. Escalation & Contingency	• By judgement & formula
Note: Direct material purchase by the field is usually covered by the material estimate.	

Figure 3.25   Construction estimate basis.

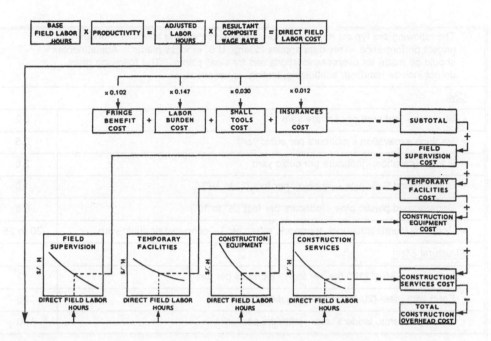

Figure 3.26   Construction overhead costs—cost estimating logic.

typical construction overhead costs; and Figure 3.27, historical data (direct hours per work unit) for construction work.

J.   Buildings Estimating Data

Figures 3.28 through 3.31 present a range of conceptual estimating data.

K.   Productivity Guidelines (Construction)—Curves

Figure 3.32 provides general guidelines for establishing a direct-labor productivity profile for the construction phase of a project. These guidelines cover incremental and cumulative profiles. Since bad weather can have a significant impact on productivity, separate guidelines are provided for incremental productivity planning.

This profile should be developed as soon as the physical site conditions are known and a detailed construction schedule is available. The horizontal axis should be translated from percent complete to a calendar time frame.

Direct construction labor can represent 20% of a project's total costs, so it is important that labor productivity be tracked as early as possible. Productivity can be measured properly only if construction progress is evaluated with physical quantities and associated work measurement units (i.e., in an earned value system).

Application

These guidelines show incremental productivity for the major phases of construction. The mobilization phase (the first 15%) is shown with a reduced productivity of 10% from the construction estimate. It improves during the next two phases, by 5% dur-

The following are typical data points for evaluating estimating levels or monitoring project performance. This data applies to large U.S. process plants. Adjustments should be made for overseas locations and for small plants. The following rates do not include handling, scaffolding, testing, or rework.

Civil

• Site strip - jobhours per square yard	0.2
• Machine excavation - jobhours per cubic yard	0.5
• Hand excavation - jobhours per cubic yard	2.0
• Underground C.S. pipe - jobhours per foot (2" to 10")	1.0
• Underground plastic pipe - jobhours per foot (½" to 10")	0.5
• Concrete foundations (incl. formwork, rebar, etc.) - jobhours per cubic yard	20 to 25

Structural Steel

• Erect heavy steel (100 lbs per foot) - jobhours per ton	12
• Erect light steel (20 lbs per foot) - jobhours per ton	36
• Install platforms, ladders, etc. - jobhours per ton	40

Equipment

• Install pumps (0 to 10 hp) - jobhours per each	20
• Install pumps (10 to 100 hp) - jobhours per each	45
• Install compressors (large) - jobhours per ton	20
• Install heat exchangers (shell-and-tube) - jobhours per each - jobhours per ton	6 0.7
• Install towers / vessels - jobhours per ton	2
• Install vessel internals - jobhours per ton	120

Piping (including pipe supports and testing)

• Prefabricate - all sizes - jobhours per ton	80 to 100
• Erect piping (0" to 2½") - jobhours per foot	0.6 to 1.0
• Erect piping (3" to 8") - jobhours per foot	1.5
• Erect piping (10" to 20") - jobhours per foot	2.0
• Erect piping (3" to 8") - jobhours per ton	200
• Erect piping (10" to 20") - jobhours per ton	250

Figure 3.27 Typical data points—construction.

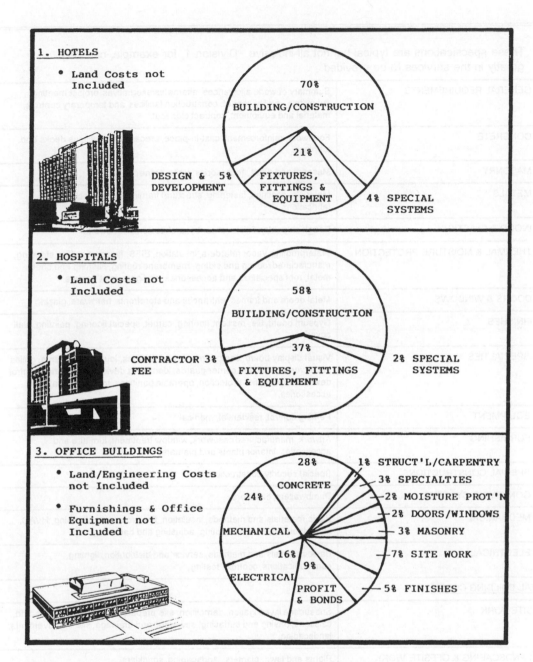

1. HOTELS

- Land Costs not Included

70%
BUILDING/CONSTRUCTION

21%
FIXTURES, FITTINGS & EQUIPMENT

DESIGN & DEVELOPMENT 5%

4% SPECIAL SYSTEMS

2. HOSPITALS

- Land Costs not Included

58%
BUILDING/CONSTRUCTION

37%
FIXTURES, FITTINGS & EQUIPMENT

CONTRACTOR FEE 3%

2% SPECIAL SYSTEMS

3. OFFICE BUILDINGS

- Land/Engineering Costs not Included
- Furnishings & Office Equipment not Included

28% CONCRETE

24% MECHANICAL

16% ELECTRICAL

9% PROFIT & BONDS

1% STRUCT'L/CARPENTRY
3% SPECIALTIES
2% MOISTURE PROT'N
2% DOORS/WINDOWS
3% MASONRY
7% SITE WORK
5% FINISHES

Figure 3.28 Typical building cost breakdowns (pie charts).

These specifications are typical but not all-inclusive. Division 1, for example, can vary greatly in the services to be provided.	
1. GENERAL REQUIREMENTS	Summary of work, allowances, alternates/alternatives, project meetings, submittals, quality control, construction facilities and temporary controls, material and equipment, contract closeout.
3. CONCRETE	Formwork, reinforcement, cast-in-place, precast, cementitious decks and toppings.
4. MASONRY	Masonry and grout, accessories, units, stone.
5. METALS	Materials, coatings, fastening, structural framing, joists, decking, fabrications.
6. WOOD & PLASTICS	Rough carpentry, finish carpentry, architectural woodwork.
7. THERMAL & MOISTURE PROTECTION	Waterproofing, vapor retarders, insulation, EIFS, fireproofing, fire stopping, manufactured roofing and siding, membrane roofing, flashing and sheet metal, roof specialties and accessories, skylights, joint sealers.
8. DOORS & WINDOWS	Metal doors and frames, entrances and storefronts, hardware, glazing.
9. FINISHES	Gypsum board, tile, resilient flooring, carpet, special flooring, painting, wall covering.
10. SPECIALTIES	Visual display board, compartments and cubicles, louvers and vents, grilles and screens, wall and corner guards, identifying devices, pedestrian control devices, lockers, fire protection, operable partitions, toilets and bath accessories.
11. EQUIPMENT	Parking control, residential, medical.
12. FURNISHING	Artwork, manufactured casework, window treatment, furniture and accessories, interior plants and planters.
13. SPECIAL CONSTRUCTIONS	Special security construction.
14. CONVEYING SYSTEMS	Dumbwaiters, elevators.
15. MECHANICAL	Basic materials and methods, insulation, fire protection, plumbing, HVAC, air distribution, controls, testing, adjusting and balancing.
16. ELECTRICAL	Basic materials and methods, service and distribution, lighting, communications, controls, testing.
TOTAL BUILDING COST	
2. SITEWORK	Subsurface investigation, demolition, site preparation, earthwork, piles and caissons, paving and surfacing, sewerage and drainage, site improvements, landscaping.
A. LANDSCAPING & OFFSITE WORK	Plants and lawn, planters, landscaping, sprinklers.
B. ARCHITECT FEES	
C. ENGINEERING - Design; Acoustical; Lighting.	
D. PROJECT MANAGEMENT	
E. CONTINGENCY	
TOTAL PROJECT COST	

Figure 3.29 Conceptual estimating—office buildings—CSI division descriptions.

CATEGORY	SQ.FT. PERSON
1. OFFICE STAFF(General)	195-220
2. SENIOR STAFF	220-250
3. EXECUTIVE STAFF	250-650
4. PRESIDENTIAL STAFF	850-1250
5. EMPLOYEE LOUNGE, DINING ROOM & GENERAL FILING	110-130
6. GROSS AREA TO NET AREA-ACTION/OFFICE WORK SPACES (If only gross area is known)	70-75%

These are recommended LEVELS, but some, temporary reduction can be accomodated for the short term. But not recommended for the long term as "over-crowding", high noise levels and inefficient air conditioning can lead to regulatory problems and low productivity.
OVER-CROWDING PRODUCTIVITY LOSS CAN BE 1-5%(Tot. Staff)

Figure 3.30 Conceptual estimating—office buildings—building space/personnel density.

DATA BASE - AS OF 1990 (California-State)
GENERAL. Multi story, single concrete building in an urban area.

$ Per sq. ft. C.S.I. Divisions	BASIC, SIMPLE SINGLE STORY "Minimum"	UTILITARIAN 2-3 STORY "Average"	LUXURY H.Q. "SPECIAL" "Luxury"
1. General Reqm'ts (Minimal Services)	1.0	3.0	5.0
3. Concrete	8.5	13.5	18.5
4. Masonry	4.0	5.0	11.5
5. Metals	4.5	4.5	12.0
6. Wood & Plastics	2.5	2.5	3.5
7. Therm. & Moist. Prot'n	1.5	1.5	2.5
8. Doors & Windows	3.0	3.5	6.5
9. Finishes	10.0	10.0	14.5
10. Specialties	-	0.5	1.5
11. Equipment	-	0.5	3.0
12. Furnishing	-	-	14.0
13. Special Constr'n	-	-	0.7
14. Conveying Systems	-	1.0	1.5
15. Mechanical	9.5	10.5	11.5
16. Electrical	10.5	10.5	13.5
TOTAL BLD'G COST	$55.0 sq.ft.	$66.5 sq.ft.	$120.0 sq.ft.
2. Site Work(%of TBC)	5-20%	10-25%	TBD
A. Lndscp'g-Offs. Wk(%TBC)	0.1-0.5%	0.1-0.5%	0.1-0.5%
SUPPORT COSTS (% TBC)			
B. Architect Fees	6-10%	6-10%	6-10%
C. Engineering- Design	0.5-0.75%	0.5-0.75%	0.5-0.75%
- Acoustical	0.2-0.5%	0.2-0.5%	0.2-0.5%
- Lighting	0.1-0.5%	0.1-0.5%	0.1-0.5%
D. Project Management	2-5%	2-5%	2-5%
TOTAL PROJECT COST			

Figure 3.31 Conceptual estimating—office buildings—database—dollars per sq ft.

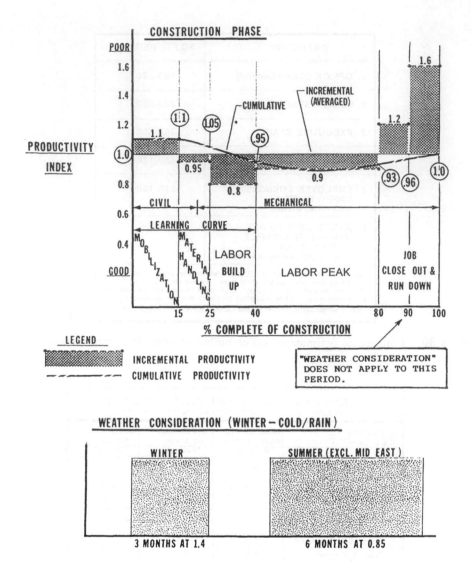

Figure 3.32 Productivity guidelines—construction.

ing the material handling phase and by 20% during the labor buildup phase. At labor peak (40% of construction), the incremental productivity is shown as .09, which is still good. For the last 10% of construction, productivity rapidly deteriorates.

The cumulative curve is calculated and is shown as tracking from poor to good and ending at 1.0. Additional factors for weather would be superimposed on the top profile. If the winter occurred at 40% of construction, the 0.9 could be multiplied by 1.4, resulting in a projection of 1.3 for the period. If the other periods were as shown on the chart, then the overall productivity of 1.0 would not be achieved. The weather consideration does not apply during the last 20% of the job.

Evaluation of labor productivity can be made early in the project, and the method/guidelines provided here can greatly assist in monitoring and forecasting productivity levels. It should be emphasized that a schedule/activity jobhour

1. INHERENT NATURE OF THE WORK
   - Punch List
   - Check Out
   Low Budget Value

2. OPERATIONAL / MAINTENANCE CHANGES
   - No Extra Budget

3. POOR LABOR ATTITUDE

4. POOR MANAGEMENT / PLANNING
   - "Crash" Program
   - Over-staffing / High Costs

Figure 3.33 Productivity guidelines—poor rundown productivity.

weighting system can also produce productivity analyses, but the results are normally of a lower quality.

L. Poor Rundown Productivity

Experience and historical data have allowed reasons for poor productivity during the last 20% of a project to be determined, including:

- Major work during this period is punchlist/checkout work that has a lower budget value. Thus the earned value system measures low productivity.
- Remedial work and changes required by the operational and maintenance staff generally take place during this period. This type of work does not usually fall into the category of official change orders, since it would then result in increased budget.

Figure 3.33 illustrates these major elements.

M. Construction Progress—Direct and Indirect

Figure 3.34 illustrates a typical relationship between direct work and indirects. The data it reflects have been compiled from experience. This figure shows that the indirect curves are essentially constant throughout the execution of a project. Early buildup for field organization and installation of temporary facilities is matched by a late buildup for final job cleanup and demobilization.

Direct work progress is a measure of physical quantities installed, and, as shown, the direct work curve is identical to the historical/standard construction curve previously illustrated. Indirect construction progress cannot be assessed by measuring physical quantities and is usually measured in jobhours. This typical curve shows the rate at which these jobhours would normally be expended.

Indirect and direct construction curves for any project can be compared with the curves in Figure 3.34. During construction, actual performance should be compared with these profiles as well, since doing so can provide an early warning that the expenditure of jobhours is deviating from the norm.

The percentage breakdowns for craft indirects, field administration, and direct supervision can be used to check estimates and performances of individual categories. On very large projects, individual control curves can be developed for specific categories.

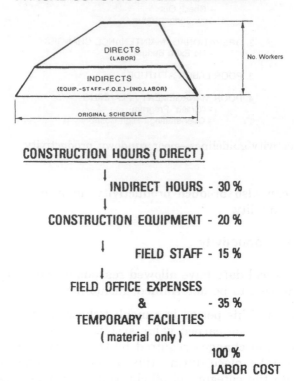

TYPICAL CONSTRUCTION BREAKDOWN

CONSTRUCTION HOURS (DIRECT)

↓ INDIRECT HOURS - 30 %

↓ CONSTRUCTION EQUIPMENT - 20 %

↓ FIELD STAFF - 15 %

↓ FIELD OFFICE EXPENSES
&   - 35 %
TEMPORARY FACILITIES
(material only) ————

100 %
LABOR COST

Figure 3.34   Construction progress—direct and indirect.

N.   Design Allowance Evaluation and Chart

A selective assessment of the major equipment carrying significant *design allowances* should be made, and each month a forecast of the purchase orders for this equipment should also be made. As engineering advances, the amount of DA should decrease. Design allowances can reflect between 5% and 15% of the estimated equipment cost and, as such, can constitute a significant percentage of total project costs. The design allowances should be shown in the estimate.

Figure 3.35 can be used as a guideline and shows a required, overall design allowance over the life of a project. The figure shows a design allowance of 10%, reducing to zero by the time engineering is 95% complete. It should be emphasized that the design allowance is a major feature of an estimate for a fast-track project. The allowance is a known condition (not contingency) and is used to cover cost increases of early purchases that are impacted by later engineering changes.

XXIV.   ESTIMATING THE CASE OF USED VERSUS NEW EQUIPMENT

This approach is for the case of complete, single, or multiple process units and does not apply to maintenance-replacement work. The author has been involved

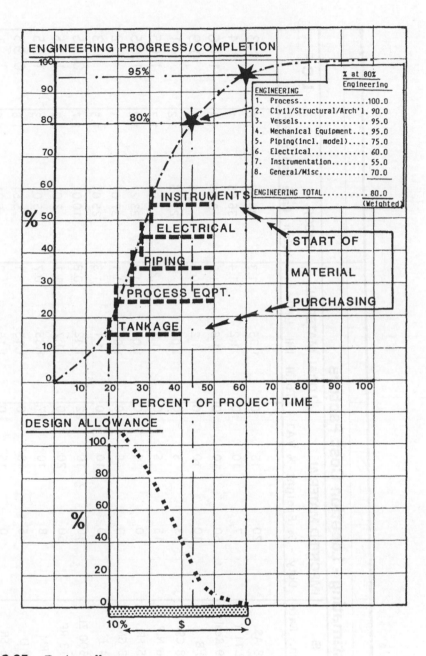

Figure 3.35 Design allowance.

in three cases of used versus new equipment. In each case, new equipment was chosen, as the cost differential was insignificant due to:

- high to exorbitant cost of the used equipment (never at scrap value);
- high cost of dismantling, refurbishing, upgrading;
- loss of equipment guarantees;
- limitations in performance efficiencies and additional production cost;

Conceptual Estimating - Location Cost Factors

Country	$ Exch. Rate	IMPORTED MATERIAL Duty - %	Freight-%	VAT	LOCAL MATERIAL Total	IMI	LMI	LABOR $ Labor Rate	Labor P F	LCF
Australia	1.38 A$	20	15	0		1.35	1.3	20.66	1.6	1.3
Japan	102.11 Y	7	10	0		1.2	1.4	36.05	1	1.4
Brazil	1.09 Real	50	10	0		1.6	1.2	3.05	1.8	1.06
Venezuela	170 B	10	10	0		1.2	0.9	2.6	1.7	0.9
Canada	1.38 C$	3	5	0		1.08	1	21.05	1.2	1
Mexico	3.36 N.P.	15	5	0		1.2	1.25	5.01	1.7	1.05
France	5.55 FR.	0	5	0		1.05	1.2	23.21	1.2	1.15
Germany	1.63 GM	0	5	0		1.05	1.1	37.01	1	1.2
United Kingdom	1.54 £	0	10	0		1.1	1.3	25.01	1.5	1.3
Poland	22,600 ZL	10	10	22		1.42	0.8	10.01	1.7	0.95
India	31.13 RP	50*	20	0		1.7	1.1	2.01	3	1.15
China(Guandong)	8.7 RM	8	10	17		1.35	0.7	5.01	3	1.05
U.A.E.	3.67 D	5	10	0		1.15	1	5.51	1.6	0.95
Saudi Arabia	3.75 SR	0	15	0		1.15	1.2	5.01	1.6	1
South Africa	0.229 R	2	10	0		1.12	0.9	12.56	1.3	0.92
Pakistan		50	20					4.01	2.2	

PF - Productivity Factor

LMI - Local Material Index
LCF - Location Cost Factor

IMI - Imported Material Index

*Maximum rate: many duties are lower. India may grant concessional duties for major projects. Duties as low as 20% have been granted for recent power projects. Also, while there is no VAT tax in India, a national sales tax of 4% plus state sales taxes of 4 to 15 % apply to local material.

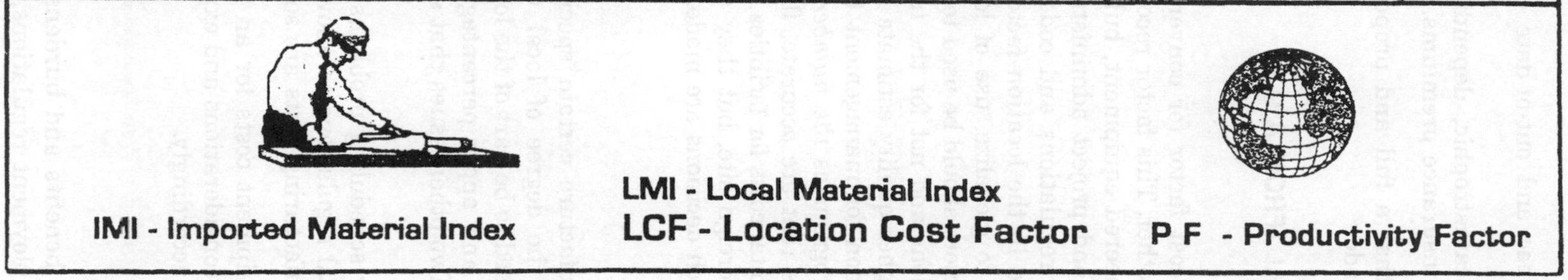

Figure 3.36 Conceptual estimating—location cost factors.

- increased maintenance risks;
- additional engineering cost because of out-of-spec and out-of-date details from existing design standards; and
- higher risk of operational failure, sometimes catastrophic, depending on process, impacting on operating economics and insurance premiums.

Costs should be developed for these conditions so that a full and proper cost comparison of used versus new equipment can be made.

XXV.   LOCATION COST FACTORS—INTERNATIONAL (FROM A U.S. COST BASE)

A location factor is an instantaneous, overall, total cost factor for converting a base project cost from one geographical location to another. This factor recognizes differences in productivity and costs for labor, engineered equipment, bulk materials, commodities, freight, duty, taxes, indirects, and project administration. (The cost of land, scope/design differences for local regulations and codes and differences in operating philosophies are not included in the location factor.)

With the current rush of industries attempting to globalize, use of location factors has become increasingly important. Location factors should be used to factor a base estimate for comparing costs at differing locations and not for the funding estimate for the selected location. After selection, a higher quality estimate should be developed for project funding. However, it is common for management to lock onto the estimate generated by the location factor program as *the number*. This drives development of location factors toward methods that are accurate, flexible, easily managed and that allow a quick turnaround. Capital costs for facilities, alone, do not dictate if a company will build a plant at a foreign site, but they are an integral part of the total business economics from which decisions are made.

A.   Key Variables and Adjustments—Material

Many countries do not have the capability to manufacture certain specialized equipment and routinely import this material. So the degree of local versus imported materials and the relative cost differences need to be part of the location factor calculation. Many companies use U.S. costs and apply percentages for freight, import duties, and customs and broker fees. Two other issues that should be considered are:

- importing of certain items because of quality or scheduling problems; and
- importing and paying the associated costs (if local regulations so allow) materials from another country because local manufacturing costs are so high.

These items can greatly affect the material and equipment costs for an actual project. The location factor needs to reflect the above considerations and expected or known strategies with the factors being adjusted accordingly.

B.   Key Variables and Adjustments—Labor

Local monetary rates, productivity differentials and benefits and burdens vary enormously by individual country. Governmental employment regulations, rules for foreign workers, travel and support costs, religious and cultural differences must be carefully evaluated. All can impact a location cost factor program. Figure 3.36 shows the detail and location cost factors for a range of countries.

XXVI.   SUMMARY—BASIC SCOPE APPRECIATION

The use of historical data, typical relationships, statistical correlations, and practical rules of thumb can greatly add to the effectiveness of estimating and project control programs. Such information can provide guidance in:

- developing/evaluating schedules,
- assessing labor requirements,
- determining appropriate productivity levels,
- improving cost/schedule assumptions,
- carrying out trend analyses,
- establishing project cost,
- evaluating status and performance of the work, and
- recognizing the scope of the work at all times.

The last item highlights the key to effective project control: *scope recognition*. This equates to the ability to properly establish the scope at the outset through use of a good estimate, and thereafter to constantly recognize the true scope of the work as the project develops and is executed. The testing and measuring of actual performance against experience can be a valuable source of verifying status, determining trends, and making predictions. Naturally the application of historical data to a specific project must always be carefully assessed.

The previous data and figures represent historical and typical cost/schedule rules of thumb that can greatly assist in establishing and developing scope during all phases of a project. This information is particularly useful at the beginning of a project when engineering is at a low percentage of completion, since it results in the generation of a broad and preliminary cost estimate and overall schedule.

4

Scheduling Keys—Establishing a Realistic Schedule Baseline (Typical and Standard Schedules)

I. INTRODUCTION

This chapter concentrates on the practical and realistic development of critical path method (CPM) schedules through the use of standard schedules. It includes only brief comments on CPM fundamentals. Specifically covered are three of the most advanced and important scheduling programs:

- fast-track—economic program,
- fast-track—trapezoidal technique, and
- construction complexity—labor density program.

The success and probability of a project schedule can be greatly enhanced with correct application of these three programs.

II. MAJOR CPM SCHEDULING OBJECTIVES

Figure 4.1 lists the major objectives of CPM scheduling. Many planning engineers believe that the end product of their work is the schedule. This figure focuses on the need for the schedule to be readily understood by the executing project staff. All too often, a finished CPM schedule is complicated, badly laid out, and incomprehensible to the user. The planner should be required to produce communicating schedules. Even though CPM schedules can be produced manually, the great strength of this technique is its ease of use with computer programs.

A specific application of this principle is the use of a summary schedule to represent the complete project schedule. Following development of CPM scheduling in 1958, the size of complete project schedules mushroomed to networks in

- PLAN
- SCHEDULE
- CONTROL
- COMMUNICATE TO THE EXECUTORS
- ORGANIZE
- IMPLEMENT

Figure 4.1 Objectives of CPM.

excess of 10,000 activities, as it was presumed that greater detail gave greater quality of critical path analysis. It then took more than 10 years for the industry to realize that such detail created a paper nightmare, high technique cost, plus a resulting loss of quality and user confidence.

With skill and a scheduling database, a summary or milestone project master schedule can first be developed and then followed by detailed networks for separate project phases. With limited skill and no database, the project master schedule is generally developed after the detailed networks, with a program summary technique.

Figure 4.2 shows the most widely used CPM formats.

A. Arrow Diagram

The arrow diagram tends to be preferred by management, as it was the first method to be developed. This method is a little cumbersome, as it requires a greater use of dummies, but it has a more visual/graphical appeal than the precedence method.

B. Precedence Diagram

The precedence diagram is a more efficient method as it requires fewer dummies. But its graphical content is not as good and is difficult to place against a timeframe.

C. Time-Scaled Diagram

The time-scaled diagram is a visual version of the arrow diagram and gives a clear picture of the project schedule, as it relates to progress and completion of activities. A vertical cursor line is drawn at each reporting period, and the activities status of on, ahead, or behind is clearly visible. It is a favorite tool of management.

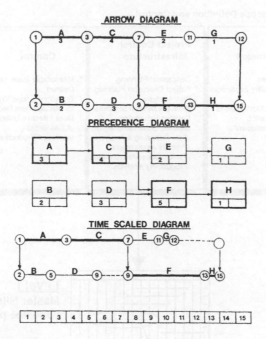

Figure 4.2 CPM formats.

III. TYPICAL SCHEDULING LEVELS

Throughout the industry, there is uniformity in scheduling levels of detail. Five levels of detail are common to many full scheduling systems. Generally, owners work up to level 3 and contractors at the full 5 levels. Figure 4.3 shows a typical owner program, consisting of the first three levels.

Level 1. This is usually a bar chart and shows all or key company projects from start to finish. Each project is illustrated by a simple bar, and important milestones are represented by symbols at the planned time of occurrence. As work is accomplished, the bar and symbols are filled in to a reporting dateline to show achieved progress. This technique can be used for all company projects but only for the construction phase.

Level 2—Total Project Schedule. This is usually a bar chart, with individual bars for major phases such as permitting, funding, engineering, procurement, construction, commissioning/start-up. Important milestones are shown as symbols along each bar. Separate bar charts showing more detail can then be developed for the separate phases.

Level 3—Project Master Schedule. This is usually a CPM schedule, but at a summary/milestone level, of an individual project from start to finish. The development of large (1000 or more), totally integrated networks for the entire project are now uncommon, as this technique has proven to be inefficient and very costly.

The standard/typical schedules discussed below illustrate the above levels. Many of them are based on experience. It is emphasized that summary networks

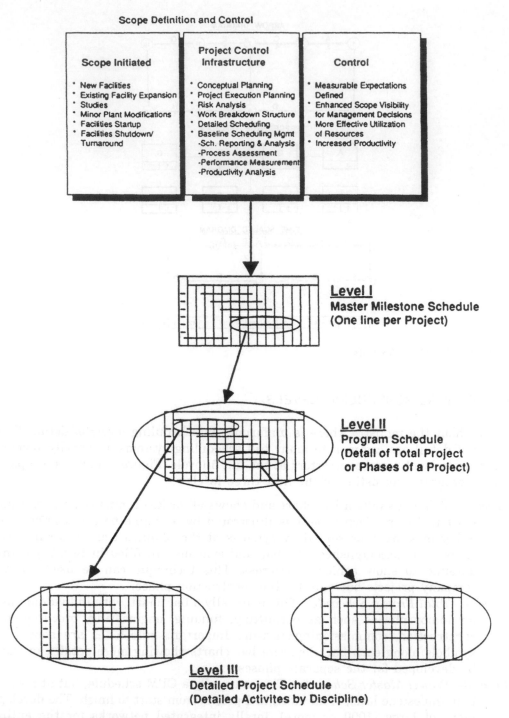

Figure 4.3 Typical owner scheduling program.

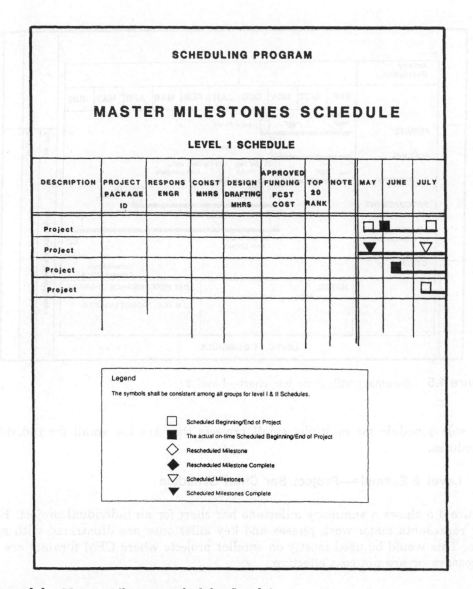

SCHEDULING PROGRAM

MASTER MILESTONES SCHEDULE

LEVEL 1 SCHEDULE

DESCRIPTION	PROJECT PACKAGE ID	RESPONS ENGR	CONST MHRS	DESIGN DRAFTING MHRS	APPROVED FUNDING FCST COST	TOP 20 RANK	NOTE	MAY	JUNE	JULY
Project										
Project										
Project										
Project										

Legend

The symbols shall be consistent among all groups for level I & II Schedules.

▢ Scheduled Beginning/End of Project
▮ The actual on-time Scheduled Beginning/End of Project
◇ Rescheduled Milestone
◆ Rescheduled Milestone Complete
▽ Scheduled Milestones
▼ Scheduled Milestones Complete

Figure 4.4 Master milestones schedule—Level 1.

showing the overall program are much preferred to bar charts, since they show the work interfaces and critical relationships. However, bar charts are still widely used, as CPM type networks require very high skill levels that may not be available to many companies.

A. Level 1 Example—Master Milestone Schedule

Figure 4.4 illustrates a multiple projects bar chart schedule, with each bar representing one project. Appropriate project information is listed, and symbols along each bar show key milestones. This schedule, a summary of more detailed projects schedules, is used as a senior management reporting tool. It can also be

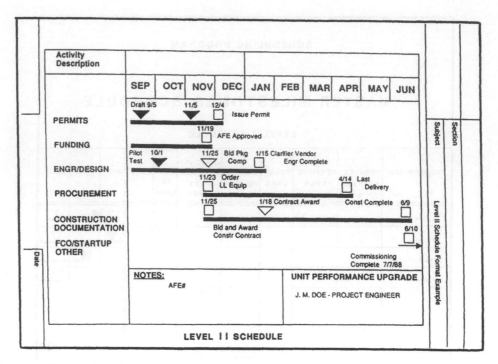

Figure 4.5   Summary milestone bar chart—Level 2.

the sole schedule for multiple, small projects that are too small for individual schedules.

B.   Level 2 Example—Project Bar Chart Schedule

Figure 4.5 shows a summary milestone bar chart for an individual project. Each bar represents major work phases and key milestones are illustrated with symbols. This would be used mostly on smaller projects where CPM formats are not necessary or are not cost effective.

IV.   STANDARD—TYPICAL SCHEDULES

The following standards are mostly Level 3 schedules and provide guidance for the overall project schedule for the execution phase (Phase 2), contract award to mechanical completion.

A.   Typical Project Master Schedule for an Intermediate Project

Figure 4.6 is a CPM schedule for an individual project and represents critical and subcritical activities. It has a unique feature in that it sequences parts of work, which enables the overlaps to be eliminated. This results in an efficient, totally linked network and becomes the working project schedule from which more detailed networks (frag- or subnets) are then developed for each work phase.

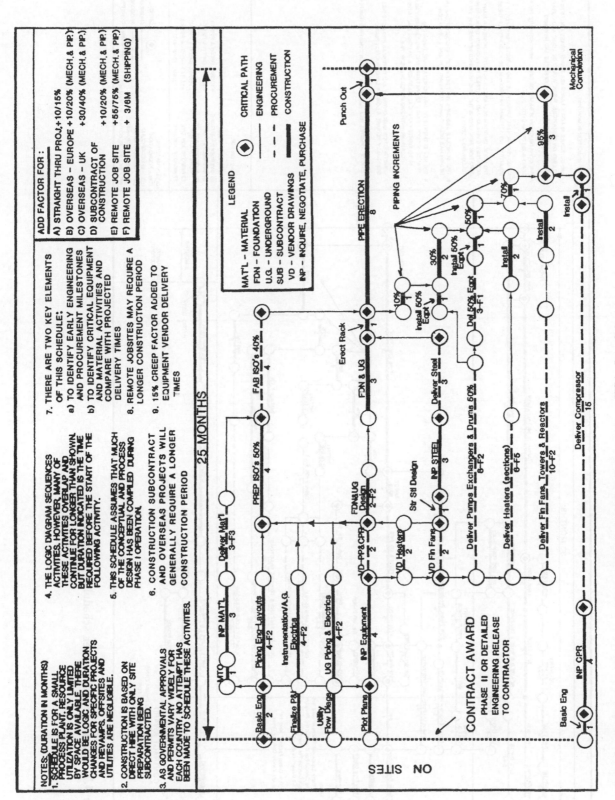

Figure 4.6 Typical project master schedule—intermediate project, USA, Level 3.

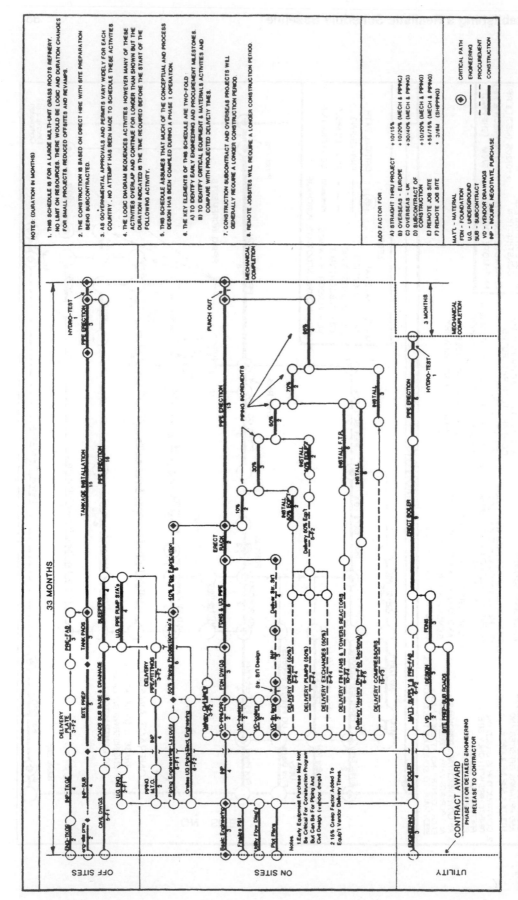

Figure 4.7 Typical project master schedule—large project, USA.

The *part activity* sequencing and durations have been determined from considerable experience/research and assume an economic (40 hours), fast-track approach. This *standard* network is for the execution phase of process plants and should be updated as economic conditions change. The notes require full consideration, since corrections must be made for subcontract construction, overseas work, and remote job sites. Another element of this standard is the sequencing of equipment delivery and installation, based on historical data, to provide sufficient nozzles for an efficient field piping program. If the plot area is known, the construction duration can be checked with the labor density-trapezoidal technique (LD/TT). If there are differences, the LD/TT duration is used and the CPM durations adjusted. Nevertheless, with skill and experience, there should only be small differences when using both techniques.

Project schedules should be drawn up as time-scaled networks. A *creep* factor is applied to equipment lead times: usually +5% for Japan, +15% for the United States, and +20 to +30% for Europe. This factor recognizes that the original vendor-quoted delivery is rarely achieved.

The purchasing/engineering cycle, usually four months for the United States and Europe, is determined with reference to the proposed project purchasing plan (cost or schedule driven). The four months represent an economic program of full, competitive bidding. The network is based on a USA (execution phase) direct-hire operation, and the *add factors* are to compensate for an overseas or subcontract operation. Environmental aspects and governmental regulations are not standard and must be treated on a case-by-case basis.

B. Typical Project Master Schedule for a Large Project

Figure 4.7 shows the same technique as Figure 4.6 but for a large, process plant project. It is also based on actual project experience/research. The same considerations as outlined in Figure 4.6 apply:

- adjust for actual project differences,
- check the standard notes for correct application,
- use the labor density-trapezoidal technique to confirm the overall construction duration,
- assess the material creep factors,
- determine if execution is on a fast-track economic basis,
- find out if construction will be subcontracted,
- identify any environmental or regulatory restraints, and
- determine if all resources are available.

C. Engineering Schedule for a Large Project—Standard

Figure 4.8 shows the key relationships and durations for Phase 2 (execution) engineering of a large, grassroots project. The database is the same as that for the project master schedule for large projects (Fig. 4.7). The same logic applies to small projects, but activity durations change significantly. The durations shown indicate only the amount of work critical to the project schedule. Activities will extend beyond the periods indicated. Engineering activity will continue through completion of construction.

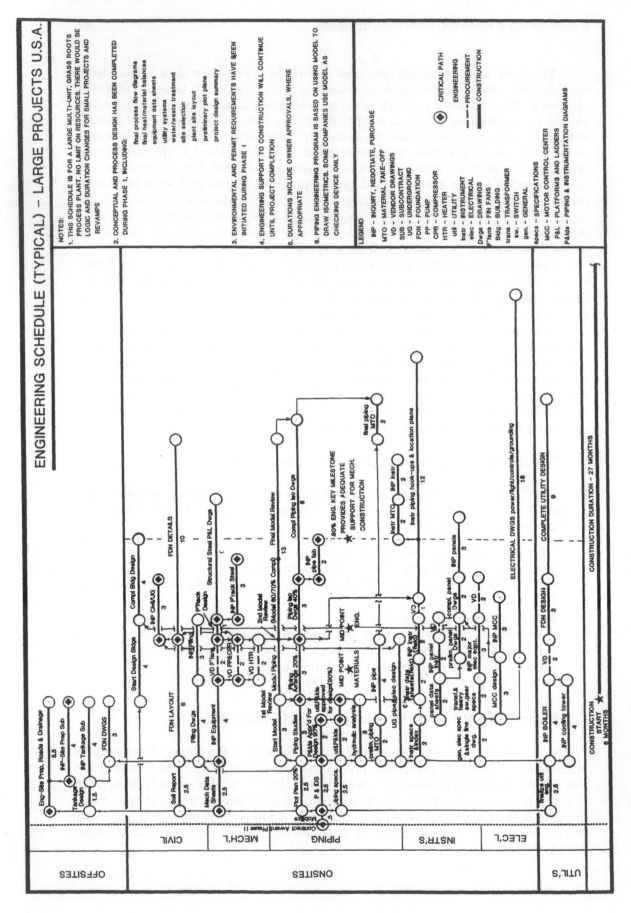

Figure 4.8 Engineering schedule—large project, standard.

Critical activities generally occurring during the engineering phase of a large project are:

- site preparation, roads, and drainage;
- receipt of critical vendor drawings;
- civil and underground pipe design;
- pipe rack design and procurement; and
- piping design.

The following factors should be considered:

- The schedule represents a Phase 2 (execution) project with conceptual and process design having been accomplished beforehand, during Phase 1.
- A large, multiunit process plant may be engineered in more than one office, and there may be multipartner joint ventures. In such instances, approval cycles and communications may be extended and may have major impact on engineering durations.
- Environmental/regulatory permit procedures may be a major factor. Normally, appropriate applications and studies for governmental approvals will have been initiated during Phase 1.

D. Construction Schedule for a Large Project—Standard

This standard (Fig. 4.9) should be used in conjunction with the overall breakdown (Fig. 4.10) and the project master schedule (Fig. 4.7). These standards provide for high-quality program development, with progress curves for engineering, material commitments, and construction. As their underlying database has been derived from actual experience, planned or actual project performance deviating from this pattern is an alarm signal for trend analysis. If the deviation is favorable, evaluation is just as important, since it can be a guide to improvement for future projects.

A key element of this standard is the sequencing of equipment delivery and installation to provide an adequate flow of nozzles for an efficient field piping program. Experience has indicated equipment-piping dependencies to be:

Equipment	Release of piping work
Pipe rack	10%
First 50% of pumps, exchanges, drums	20%
Second 50% of pumps, exchanges, drums	20%
Reactors, fin fans, towers	20%
Field-erected vessels, heaters, compressors	25%
Final punch-out	5%
	100%

This listing also shows the general sequence of delivery and installation. As can be seen, the pipe rack, pumps, exchanges, and drums will enable 50% of the

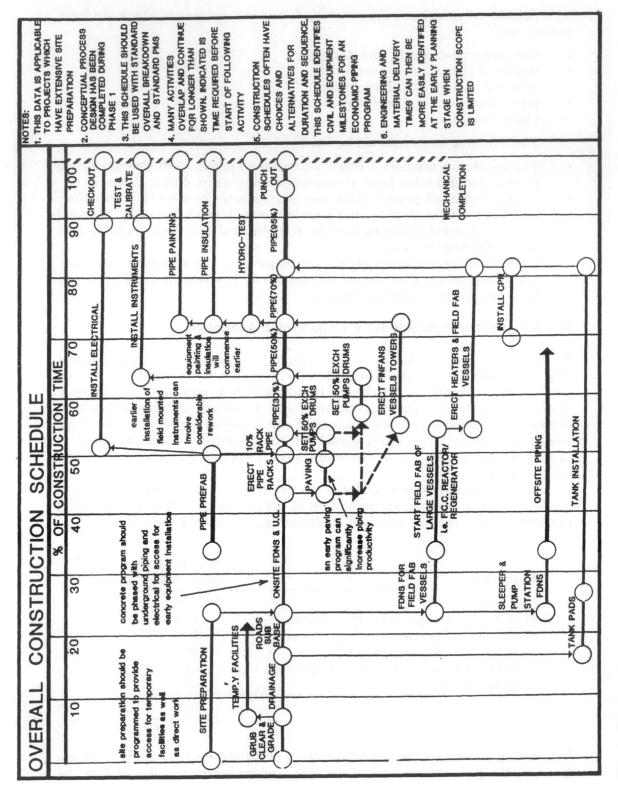

Figure 4.9 Overall construction schedule.

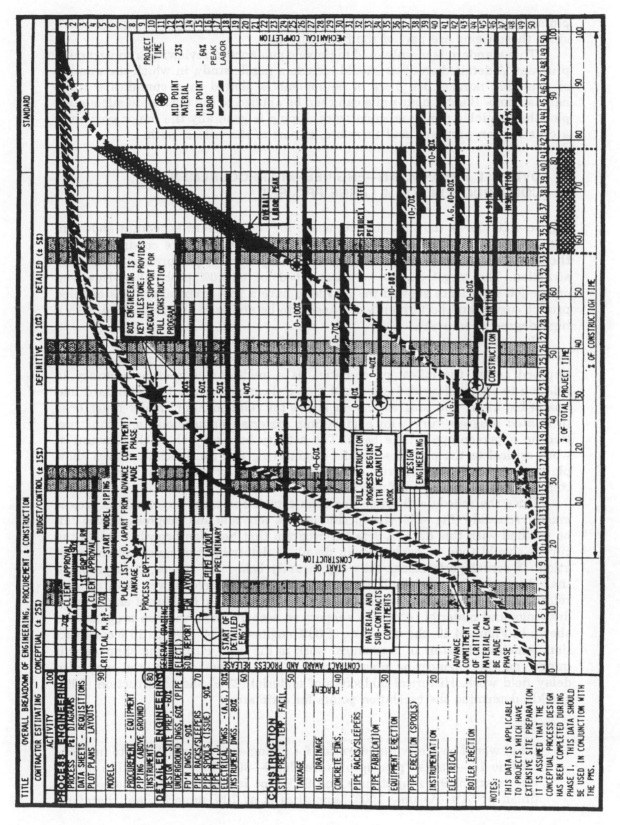

Figure 4.10 Historical database—EPC, overall curves plus key milestones.

piping to be completed and are therefore critical for a maximum piping effort, since these items are part of the first deliveries.

This standard is mainly used to provide guidance in reviewing a construction program. A detailed analysis should include a nozzle study, in which the number of available nozzles per piece of equipment is plotted in a cumulative curve against time. Budget hour units can then be applied to the nozzles to adequately plot the available piping erection work. Piping is generally the major portion of a construction program, and therefore, is part of the critical path.

V. OVERALL BREAKDOWN—ENGINEERING/PROCUREMENT/ CONSTRUCTION FOR A LARGE PROJECT

Figure 4.10 shows the historical database for the standard schedules shown in Figures 4.6 and 4.7. The actual project data and curves were conditioned and then summarized into three target curves: engineering, procurement, and construction. The target curves were developed for an 80% probability. However, on-the-job experience showed the curves to have a 95% probability. This enabled project curves and performance of projects to be analyzed and forecasted with a high degree of accuracy.

Two of these *target* curves, engineering and construction, are represented in tabular form in Figures 4.11 and 4.12. Typical estimating phases are superimposed across the milestone activities as a guide for estimating accuracy in relation to engineering definition and material commitments. This standard matrix of bar chart and progress curves shows the relationship of major milestones to related progress curves for engineering, material commitment, and construction on a percentage-of-time basis. The activity bars only show the amount of work critical to the project schedule. Engineering and construction curves show planned physical progress with construction on a direct-hire basis. The material commitment curve is on a financial basis.

Once the schedule duration has been established from the standard project master schedule (Fig. 4.7), the specific project schedule and progress curves can be verified with this data. A detailed program that lies significantly outside the curves should then be evaluated, in detail, to determine the causes of the variation. If actual performance falls significantly below the curves, it is probable that the project schedule will not be achieved. These three curves have a finite relationship to each other, where the need for an adequate backlog of engineering drawings and material is vital to support an economic construction program. As shown in Figure 4.10, the milestone of 80% engineering is *keyed* to 15% construction, which is the start of mechanical work in the field. Purchasing slippage will impact on the engineering program due to lack of vendor data and may also delay field work. Historical midpoints of material (23%) and construction (65%) quickly determine the escalation points for estimating.

Examples:

1. What would be the overall duration for piping engineering? Refer to horizontal line 16—marked pipe spools (drawings)—90%. Pipe layouts (thin dotted line) start at 9% project time and 90% pipe isometrics (heavy dotted line) finish at 61%. Assume 100% isometrics is at 70% project time; therefore,

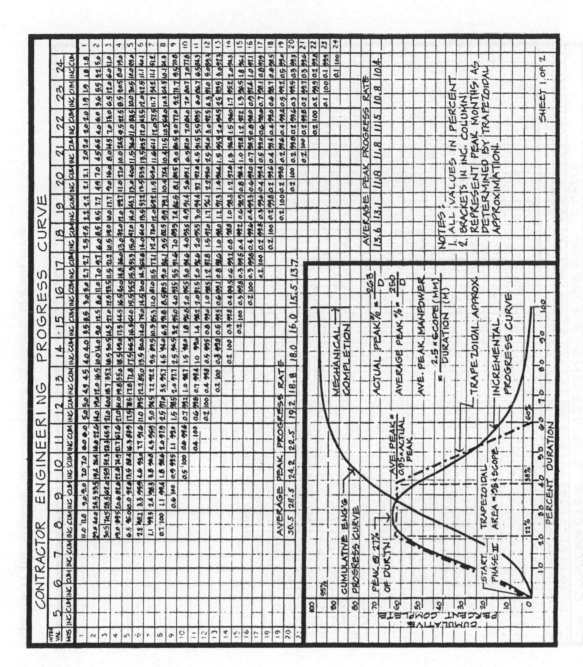

Figure 4.11 (a) Standard table—contractor engineering progress curve.

TITLE: CONTRACTOR ENGINEERING PROGRESS CURVE

SHEET 2 OF 2

AVERAGE PEAK PROGRESS RATE

	25	26	27	28	29	30	31	32	34	35	36	37	38	39	40	41	42	43	44
45	9.4	9.5	9.1	8.7	8.5	8.0	7.9	7.8	7.6	7.4	7.2	6.9	6.6	6.5	6.4	6.3	6.0	5.9	5.8	5.5

Figure 4.11 (b) Standard table—contractor engineering progress curve.

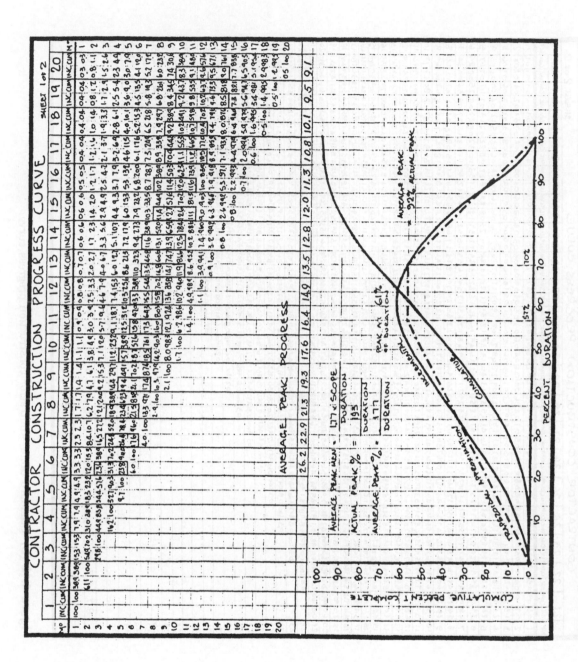

Figure 4.12 (a) Standard table—contractor construction progress curve.

CONTRACTOR CONSTRUCTION PROGRESS CURVE

SHEET 2 of 2

AVERAGE PEAK PROGRESS

Figure 4.12 (b) Standard table—contractor construction progress curve.

overall piping design is 70% less 9%, which equals 61% project duration. For a large U.S. project (Fig. 4.7 shows 33 months), piping design will, therefore, be 20.

2. What would be the probable start of detailed engineering? The key elements for the start of detailed design are approval of the process design (P&IDs) and equipment layout (plot plans). This approval would normally occur at about 70% of P&IDs and plot plans. Refer to lines 2 and 4 (P&IDs and plot plans) and approval at 9% of project time. At the same time, foundation layouts (line 14) and piping layouts (line 16) commence. This is the start of detailed design.

The main drafting effort will begin with foundation drawings, piping isometrics, and general arrangements.

Figures 4.11 and 4.12 illustrate the previously shown *target* curves of Figure 4.10 in tabular format, for engineering/project and construction durations. Note that the engineering charts show the overall project schedule, whereas the construction charts only cover the construction phase. As previously stated, these curves have a very high degree of probability and are highly recommended as a *benchmark* to monitor a project's engineering and construction progress. Each table is laid out in the same format—with incremental and cumulative progress percents against monthly periods. The engineering progress curve table starts at month 8 and the construction progress curve table starts at month 1. The boxed incremental numbers for each month show the peak period. The average percent for this period is given at the bottom of each table.

Substantial use of these curves has shown that slippage from these standards of more than 4 weeks will probably require acceleration action/cost to recover. However, if the slippage occurs above the 50% point, then recovery is virtually impossible for engineering, as there is so little design work remaining; for construction, recovery is very difficult and high cost. If recovery is not considered a viable option, then a projection of future completion, from the present position, referring to the standard curve can predict a schedule extension with a high degree of probability, sometimes years in advance of the original completion.

VI. OVERALL BREAKDOWN FOR A SMALL PROJECT—STRAIGHT THROUGH

The standard network and progress chart contained in Figure 4.13 (overall breakdown, small projects) is similar to Figure 4.10 and depicts the historical relationships of major activities. The data is based on 50 projects completed between 1955 and 1970, when this type of project program was popular. Today, this program is rarely used, and it has been replaced by the phased approach; the costs of these projects ranged from $0.5 million to $10 million. This project mode results in later midpoints of material (29% project duration) and labor (70%). The start of construction is also later, commencing at 32% of project duration.

When schedule durations have been established, progress curves can be drawn using the data from this figure. Schedules that differ significantly from the standard bar chart and progress curves should be evaluated in detail to verify the variance. When actual progress falls behind the leading curve, it is probable

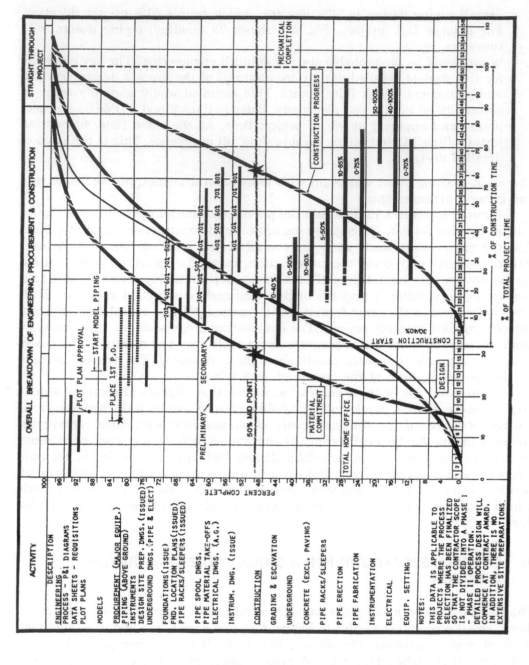

Figure 4.13 Overall breakdown—small project, straight through.

that succeeding curves will not be achieved, which could lead to a delay in mechanical completion.

Figure 4.13 shows a very aggressive construction program. The monthly peak rates of progress would generally be too high for large projects unless the "size affect": is already reflected in the project duration.

VII. ENGINEERING/PROCUREMENT CYCLE FOR A LARGE PROJECT—STANDARD

Figure 4.14 provides major milestones, logic, and durations for the engineering/ procurement cycle for equipment and bulks. Durations are based on project data/ research. Large, complex projects with worldwide purchasing, multiple engineering offices, and joint venture partners having consultation and approval requirements can increase the 12–18–week procurement cycle. Generally, projects with a minimum of worldwide purchasing have a 12–14-week purchasing cycle.

To apply this standard, assess project size, complexity, joint venture requirements, and proposed purchasing strategy and determine the duration for the specific project. Check against current project experience that the evaluation is realistic.

Only the total duration of the engineering/procurement cycle is used for overall schedules (see the standard project master schedule, Fig. 4.7) and is normally the activity designated *inquire, negotiate, purchase* (INP). Evaluation of the individual activities will be required when analyzing and monitoring detailed schedules.

VIII. TYPICAL PHASE 1 SCHEDULE

Figure 4.15 shows the general functions, logical relationships, and average durations of a Phase 1 (feasibility) operation. This Phase 1 schedule is used to provide guidance regarding typical functions carried out in a Phase 1 operation. Specific activities will depend on the individual project.

A. Typical Project Life Cycle—North Sea Platform

Figure 4.16 shows the major phases, logical relationships, and average durations for a major platform project:

- feasibility study:

    - field development,
    - technology,
    - execution plan, and
    - economics;

- process/conceptual design:

    - optimization,
    - case selection (one case), and
    - bid package;

- funding and contractor selection; and

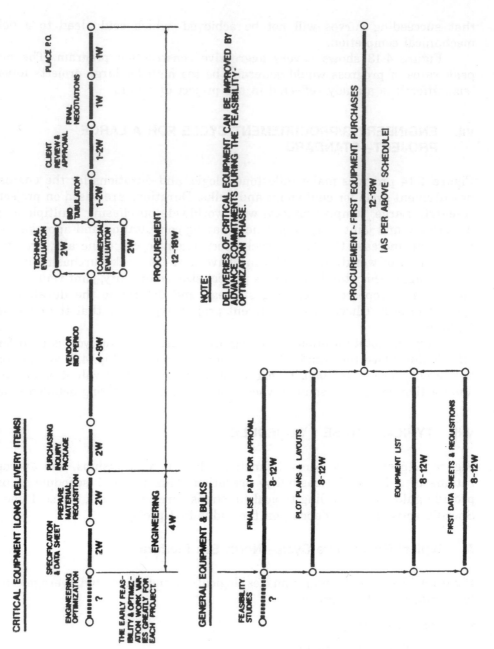

Figure 4.14 Engineering/procurement cycle—large project, standard.

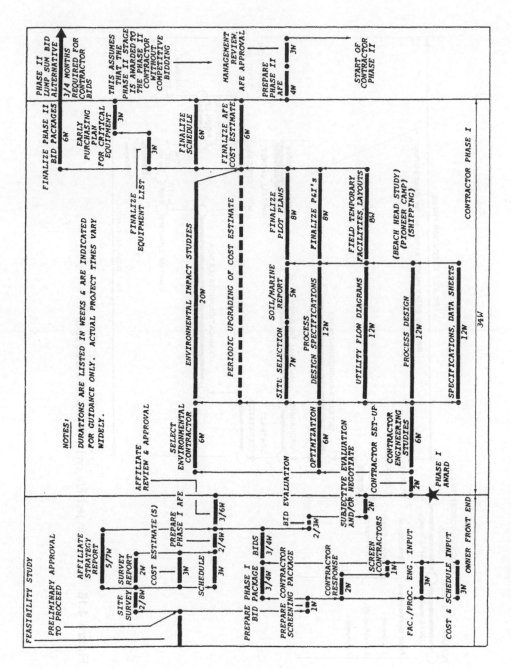

Figure 4.15 Typical Phase 1 schedule.

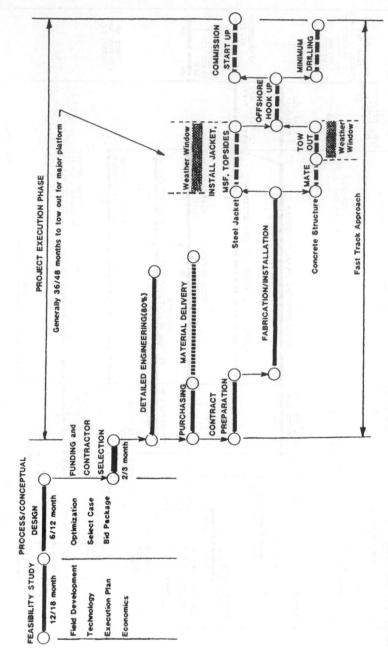

Figure 4.16 Typical project life cycle—North Sea platform.

- project execution phase:
  - steel jacket, and
  - concrete structure.

B. Project Master Schedule for Steel Jackets—Standard

Figure 4.17 depicts a project master schedule for steel jacket platforms and is based on data and experience from jacket installations in the North Sea. It shows a straight-through project, engineered in the United Kingdom. Two major critical paths should be evaluated:

- topsides engineering and module fabrication, and
- steel jacket design and construction.

The following are key considerations:

Basic engineering. (layouts, P&IDs, data sheets, specs, plot plans)

  - new technology,
  - project strategy, execution plan, and size of engineering office(s),
  - engineering capability and capacity (resources),
  - project organization and engineering consultants,
  - engineering completed before contract award;

Equipment deliveries. Add a 20% creep factor to the latest equipment delivery lead time;

Detailed engineering. Engineering needed for module quotations depends heavily on protect strategy for module contracting. Engineering completions range from 40% (a practical minimum for reimbursable bids) to 80% (essential for good lump-sum bids). Unit-price bidding requires an engineering definition of 40–60%. Projects may have a mix of contracts (e.g., lump-sum for steel, unit price for mechanical, and reimbursable for electrical and instrumentation).

For developing a relationship between project time and engineering completion, the following method is suggested:

  - Use the large project master schedule project duration from Figure 4.7 (33 months).
  - Take engineering percentages from straight-through project curves (Fig. 4.9). This reflects the situation that engineering takes longer for North Sea projects than for onshore projects.
  - Calculate detailed engineering durations before inquiry and mechanical outfitting for module bids as shown below.

Unit-price and reimbursable bidding. Using judgment and contracting strategy, assume 50% engineering required before inquiry.

  - 50% engineering occurs at 44% of project time.
  - 0% start of detailed engineering is at 15% of project time.
  - Thus, 50% of detailed engineering takes 29% of project time.
  - The project duration is 33 months.
  - 29% × 33 months = 9.9 months (10 months) for engineering before INP.
  - 80% engineering occurs at 59% of project time.

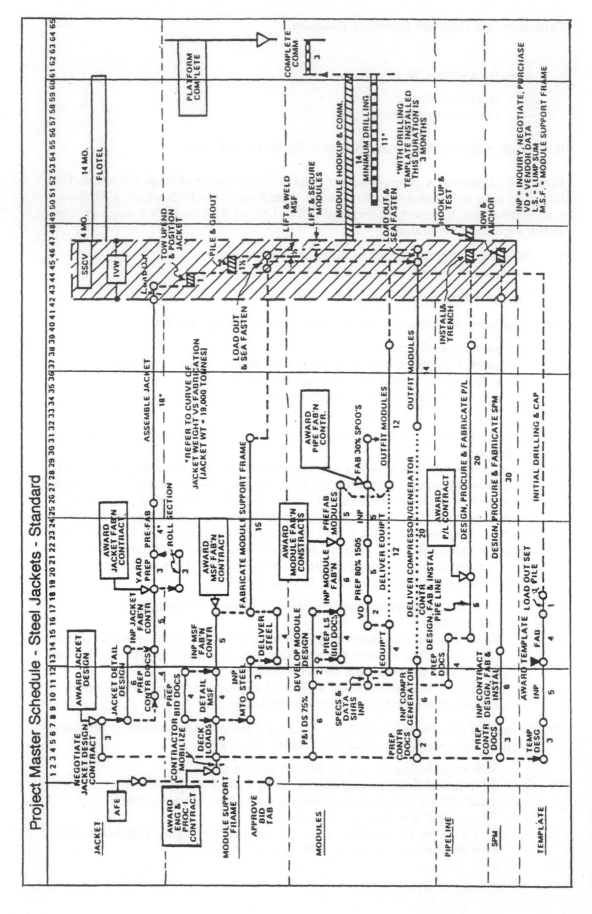

Figure 4.17 Project master schedule—steel jackets—standard.

- 0% start of detailed engineering is at *15%* of project time.
- Thus, 80% of detailed engineering takes 44% of project time.
- 44% × 33 months = 15 months for detailed engineering before module outfitting.

Module outfitting (two methods)

- First method (for process/utility modules only):

  - Evaluate individual module area by adding a 10–foot perimeter to plan dimensions. Include all additional floor levels.
  - Apply density level (180 sq ft per worker) to determine peak labor.
  - Use trapezoidal technique to determine duration.

- Second method:

  - Consider outfitting hours of total modules.
  - Using judgment, determine the peak personnel for a subcontractor.
  - Use trapezoidal technique to determine duration.

- Either method:

  - Reduce total hours by any prefabrication and support work completed away from module work/assembly area.
  - Check durations against recent module completions.
  - Consider experience and productivity of the fabricator.
  - Consider fabrication and delivery of first modules required for start of module hookup activity (fabrication durations are 14–16 months for small modules and 16–18 months for large, complex modules).

Module support frame fabrication. The project schedule is based on using a module support frame and assumes some equipment installed in module support frame

- Use trapezoidal evaluation technique with density level of 180 sq ft per worker.
- Consider shift work.
- Adjust total hours to take out prefabrication and support work completed away from the deck.
- Determine appropriate labor buildup and rundown.
- Allow for lost time and adjust peak labor if labor availability is restrained.

If equipment outfitting and deck fabrication share the same working area, outfitting labor will build up as deck labor runs down. If integrated deck sections are used instead, increase module steel fabrication by 1–3 months. Include complete outfitting activity if the module support frame requires any equipment and outfitting.

C. Project Master Schedule for Concrete Structures—Standard

Figure 4.18 shows a schedule for North Sea offshore platforms based on a concrete support structure. The latter part of the schedule shows the weather window

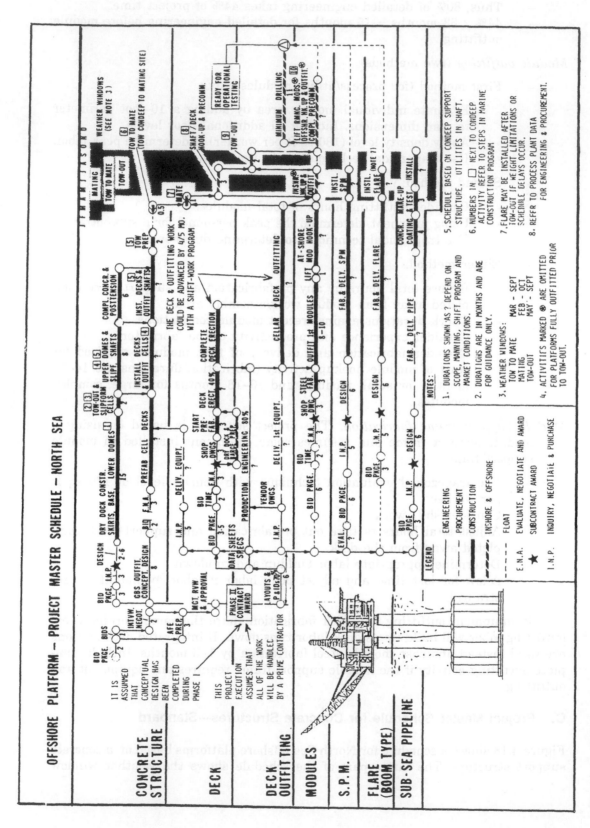

Figure 4.18 Project master schedule—concrete structures, standard.

sensitivity of some activities: tow to mate, mate, tow-out, install the single point mooring, and install the subsea pipeline. Durations marked "?" depend on the scope, labor allocation, shift work program, contracting strategy, work location, and delivery of equipment.

One major relationship relates as well to offshore platforms as it does to onshore process plants, that is, the relationship between engineering completion and the start of mechanical outfitting. For a process plant, historical data indicates that 75–80% completion of engineering is required before the start of field mechanical work begins in order to adequately support a construction program on a fast-track approach. Anything less than this can result in construction delays through lack of drawings and material. This relationship can also be applied to a platform. The equivalent basis would be 75–80% engineering completion for the start of the mechanical outfitting of the deck and modules.

A unique element in platform scheduling is the degree of onshore and offshore installation. Some tasks, such as towing and mating, can take place only in good weather (the weather window). Each activity has its own historical weather window. This is dependent on degree of difficulty, location (onshore or offshore), and financial risk (i.e., mating, February through October; tow-out, May through September; etc.).

D. Use of the Standard Schedule

After definition of the labor hours, labor availability, shift work program, contracting strategy, and equipment delivery, a project master schedule can be developed by evaluating values for the unknown durations. Construction durations can be calculated by using the labor density/trapezoidal calculation method.

Several factors can necessitate an additional allowance to calculated durations, including new technology, prototype engineering, multiple engineering offices and construction sites, heavy government involvement, and restrictive labor practices. If these factors are significant, an experience factor of 25% should be added to all calculated durations. This can be reduced to 15% for offshore construction activities. The following comments relate to durations for the major work elements:

Design—concrete structure. The duration depends on the amount of design work done before contract award. Information from the designer can help determine the schedule.

Construction—concrete structure, general. Use information from the concrete contractor when available. This is specialized construction, and historically schedules have been maintained. The durations indicated are for a very large concrete structure (100,000–150,000 m^3).

Prefabricated cell decks. Consider the number and size of the decks, the degree of prefabrication, and the total labor hours.

Installing decks, outfit cells, and shafts. Assume a density level and calculate shaft area and peak labor. The degree of complexity and effectiveness of labor classify North Sea platforms in the 200–250 sq ft per worker density category. However, due to schedule criticality, a density of 150–200 sq ft per worker is used in practice. The resulting reduction in productivity is traded off against an improved schedule.

Also consider regulatory guidelines, which may restrict the number of workers within the shafts. Assume a shiftwork program and calculate the work month (effective labor hours per month). Determine the duration using the trapezoidal method. Shaft outfitting may not be completed before tow to mate in order to meet weather window limitations. In such a case, break the outfitting activity for tow to mate, and allow demobilization and remobilization time in the mating duration.

Equipment deliveries. Add a creep factor to equipment delivery lead times.

Basic engineering (layouts, P&IDs, data sheets, and specifications). Consider the degree of experience with previous platforms, the number and location of engineering offices, project organizations, and the magnitude of Phase 1 engineering. The duration can vary from 4 months for proven technology to 8 months for new design concepts.

Detailed engineering, 80%. Use the onshore project master schedule (large project) and adjust the project duration for the overseas-UK site by reference to add factors at the bottom. This will require an add factor of 30–40% for mechanical-piping work. For example, for a standard project duration of 33 months and onsite piping of 16 months, the adjusted durations are:

– 33 months + 40% of 16 months = 40 months

From "Overall Breakdown," Figure 4.4:

– 80% engineering occurs at 43% of project time;
– 0% (start) of detailed engineering occurs at 9% of project time;
– thus, 80% of engineering takes 34% of 40 months = 14 months.

If required, add 25% experience factor to obtain the engineering duration:

– 14 months × 1.325 = 17.5 months (use 18 months)

Deck fabrication and erection. Use the trapezoidal technique with a density level of 180 sq ft per worker; consider shift work. Reduce the total hours for prefabricated work and support work completed away from the deck; evaluate the appropriate labor buildup and rundown; make allowance for lost time, and adjust peak labor if labor is not available. Add a 25% experience factor if appropriate. Concurrent equipment outfitting will impact on deck fabrication if both activities share the same working area. Outfitting labor will build up as deck labor runs down.

Deck outfitting. Use the same method as for deck fabrication. Where outfitting cannot be completed before a mating date, cut off labor with a sharp rundown (1–2 months) and carry over the remaining labor hours to onshore-offshore.

Module outfitting. There are two methods. The first method is to evaluate the module area by adding a 10–foot perimeter to the plan dimensions. Application of the density level (180 sq ft per worker) will determine the peak labor. Use the trapezoidal technique to determine the duration. The second method is to consider the outfitting labor hours of total modules, divide the work among several or many subcontractors, and, by judgment, determine the peak labor for a subcontractor. Then use the trapezoidal technique to determine the duration. The total hours should be reduced for prefabrication and support work completed away from the module saturation area. If appropriate, add a 25% experience factor to give the activity duration. Consider the fabrication and delivery of the first modules required for the start of the module hookup activity. For guidance, module fabrication durations are 14–16 months for small modules, and 16–18 months for large and/or complex modules, assuming a 40–hour workweek. Several modules in the same shop may require additional time (15–20%).

Module hookup. This starts after the first module lift. It may occur at the deck site, onshore after mating, and/or offshore. Use the trapezoidal technique to calculate the durations. The saturation labor is obtained by assuming that 40% of the module area is available (this varies with the degree of module completion). Apply 180 sq ft per worker. Buildup is short, limited only by the ability to staff and plan the work. If work is carried onshore, break the trapezoid abruptly during mating when work ceases—no rundown and buildup for simplicity. Labor hours carried onshore should be adjusted for lower productivity. This also applies to deck outfitting labor hours, which should be combined with hookup labor hours for one calculation. By judgment, assess the labor level. Labor hours carried offshore require further productivity adjustment. Further productivity reduction may occur due to drilling interference (concurrent drilling and construction). Labor buildup and peak are limited by available accommodations (temporary living quarters, hotels, permanent quarters). The use of high-cost hotels can greatly increase the availability of craft labor and reduce the schedule. The available beds must be discounted for the number of nonconstruction personnel (operational support, drilling, management, catering). Of the total platform beds, about 50% are available for construction craft labor. The workweek is seven 12–hour days less 5% absenteeism, etc. Module hookup (and outfitting) should be evaluated separately from commissioning, since peak labor cannot be used on commissioning work. Use the trapezoidal technique to determine the duration. If appropriate, add a 25% experience factor to onshore durations and 15% to onshore and offshore durations.

Mating and tow-out. Towing the gravity structure to the mating site, mating, and tow-out are all subject to weather limitations. Weather windows for these activities are:

Tow to mating site	March through September
Mating	February through October
Tow-out	May through September (the latest start of tow-out is the first week in August)

To minimize risk, it is desirable not to have all mating and tow-out operations in the same year. It is preferable to schedule tow to mate in the year before mating and tow-out. Mating durations of 1 month and tow-out of 2 months include weather contingency.

Minimum drilling. This is the time required to complete the drilling required for start-up. Eleven months are required to drill 3 wells, including the conductor driving. Further time will be required to drill all wells for full production.

IX. THE FAST-TRACK PROGRAM AND TRAPEZOIDAL TECHNIQUE

As development of the *fast-track* approach in the early 1960s became a major operational program by 1970, it led to massive schedule failures, and project schedule slippage became routine. The major reason for this failure was lack of understanding on fast-track relationships between and within engineering, procurement, and construction. When properly applied, the fast-track project schedule produces the shortest economic program.

A. Fast-Track Economic Program

Figure 4.19 shows the progress curves for engineering, procurement, and construction in the fast-track relationship for the execution phase of a project. This is the same data, in summary form, that was established by Figure 4.10. The economic basis is:

- all purchasing and contracting is on a competitive basis;
- construction is executed on a regular 40–hour workweek;
- apart from spot overtime, there are no acceleration procedures; and
- cost is the number 1 project objective.

This project program is the most widely used of all programs.

B. Fast-Track Trapezoidal Technique (TT)

Figure 4.20 illustrates this technique for the construction program. The TT is another form of a progress curve, shown as a form of triangle. As illustrated, there is a *build up* of labor to a *peak* of labor, followed by a *run down* of labor to work completion (mechanical completion). The area of the trapezoid is the scope of work in labor hours.

C. Fast-Track Schedule and the Trapezoidal Technique

Figure 4.21 shows the scheduling relationship between engineering and construction for a project on a fast-track program. The schedule relationship/duration only covers the execution phase, which is sometimes referred to as Phase 2. Phase 1 covers the conceptual design studies and case selection.

As previously stated, the schedules are illustrated as trapezoids. This data base is extremely important, since it shows that all complex work, at a total level, is executed within a buildup, a peak period, and a rundown. The specific relationships (ratios or parts) of these 3 periods has been developed from historical data for both engineering and construction programs. This base then forms one of the best methods for determining the overall construction duration, as shown in the following paragraphs. Work classified as simple (i.e., the task is performed by a single crew or squad) has no buildup or rundown; Figure 4.22 illustrates/compares the simple and complex arrangements.

D. Schedule Basic—Simple Versus Complex Task

Figure 4.22 depicts engineering and shows the calculation of peak personnel rather than calculation of the duration. The same process can be followed for construction if the duration is known. The buildup and peak are shown as 22% and 38%, and in the database of Figure 4.19 these percentages have been rounded to 20% and 40%.

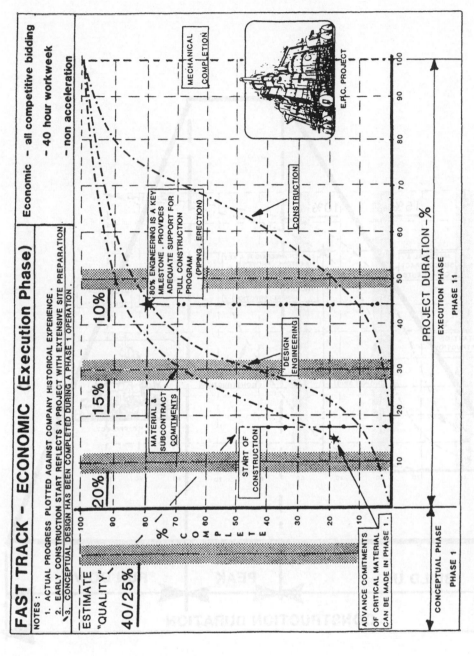

Figure 4.19 Fast-track economic program.

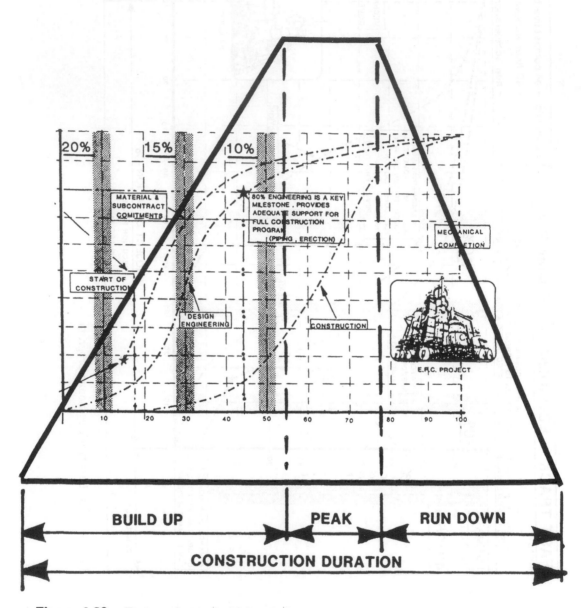

Figure 4.20   Fast-track trapezoidal technique.

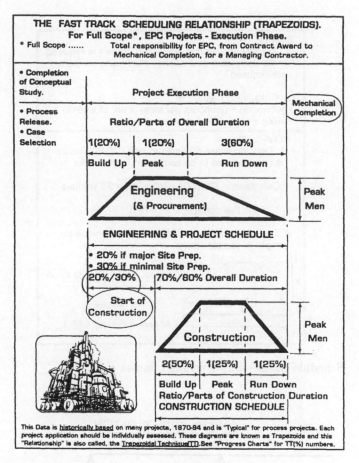

Figure 4.21 Fast-track schedule and the trapezoidal technique.

Figure 4.22 shows the calculation formula for a simple engineering task of 400 hours and the error of then applying that formula to a complex task of 20,000 hours. As shown, the calculation error of 25 engineers instead of 41 engineers is significant and results in schedule slippage.

For complex work, an alternative method to the trapezoidal calculation is to use the simple formula and multiply the result by a peak factor (usually 1.6–1.7). This is a common practice of senior professionals and usually produces adequate results. However, it will not show the required buildup of personnel over the project schedule.

E.   Construction Duration for the Trapezoidal Technique

Figure 4.23 shows the trapezoidal technique for the construction phase. The calculation procedure shows two formulas:

- Formula 1 is used to determine the peak duration, X, on the basis that the following information is known or assumed:

  - scope in labor hours,
  - effective monthly hours per person,

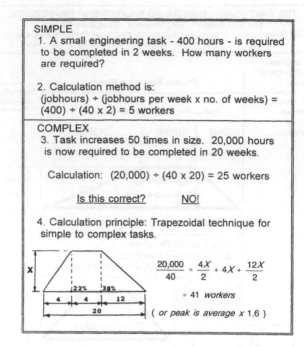

Figure 4.22   Schedule basic—simple versus complex task.

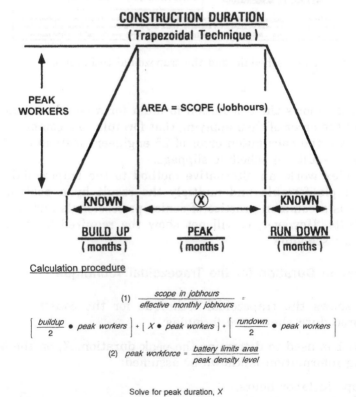

Figure 4.23   Construction duration—trapezoidal technique.

- buildup (usually developed from standard schedules or judgment),
- rundown (half of the buildup), and
- peak workers as determined by Formula 2.

- Formula 2 covers the calculation of the peak workers if the battery limits area (plot plan) is known. By evaluating a labor density level (usually in the range of 150–300 square feet per worker), the peak number of workers can then be determined.

This schedule/labor evaluation technique is a very powerful method of developing or checking overall construction duration. The key assumption, which requires good judgment, is assessing the labor density level (discussed later). If this assessment is good, then the resulting evaluation is of a high quality. This technique, however, can only be used on single projects. For smaller areas of work, the calculation may need to be modified.

F. Trapezoidal Technique—Statistical/Standard

Figure 4.24 shows the same earlier data transposed so that the X is now the overall construction duration, instead of the peak period. The figure shows the standard trapezoid for large projects (those above 5000 sq ft), and the standard for small projects (those below 3,000 sq ft). As illustrated, on small projects the buildup and rundown shortens, and the peak period lengthens. This is due to the small number of workers required.

G. Trapezoidal Technique—Worked Example, Trapezoidal Density Method

Assessing the density level is extremely important. Figure 4.25 contains a working example of the trapezoidal density method. The first step in the calculation process is to properly assess the labor hour scope. As shown, allowances were made for indirect labor working in the same area as direct labor and for the estimating allowance. An allowance for better subcontract productivity and a 12% absenteeism allowance were also made.

The buildup duration was determined from a standard schedule that showed the peak labor period would be reached at 30% of the piping duration. This activity was preceded by foundation work (3 months) and equipment installation work (2 months). The total piping duration was 15 months. In the case of subcontract labor, the labor hours are reduced to reflect a better productivity rate than with the direct hire workers (possibly as much as a 10% improvement).

H. The Trapezoidal Technique and Claims

The impact of schedule delays can be determined with the use of the TT. For many years, the owners and contractors adopted different positions in the application of the TT to schedule-related claims. Owners, some with a vested interest, stated that the delay occurred at the rundown period, where the cost impact was the lowest. Contractors opposed this position and stated that most construction schedule delays were caused by multiple changes/additions, occurring throughout the construction period. This was referred to as the *ripple effect* of change, and

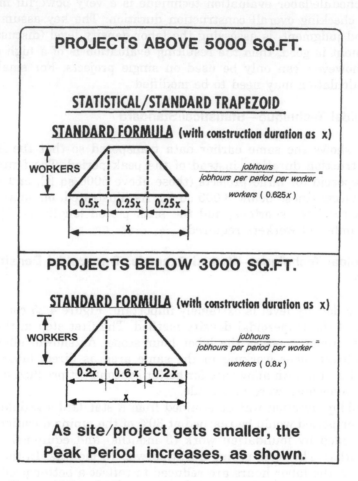

Figure 4.24   Trapezoidal technique—statistical/standard.

Refinery FCC unit:
- Plot area = 320 ft x 200 ft   =   64,000 ft^2
- Scope (direct hire)   = 445,000 jobhours
- Allowance for indirect labor in area (10%)   =   44,500 jobhours
- Estimating allowance (15%)   =   66,700 jobhours
                                               556,000 total scope for evaluation

Consider two cases: Case 1, direct-hire labor, and Case 2, subcontract labor.

Assumptions:
- Due to criticality, use density of 250 (but 300 is more probable)
- Allow 12% absenteeism to estimate effective jobhours
- Buildup duration from standard schedule (foundations + equipment + piping buildup)

Case 2 - Subcontract Labor. The project strategy, based on experience, is that local subcontractors are more productive than prime contractors (direct hire).

Scope = 556,200 jobhours for direct hire less 10% productivity adjustment for local subcontractor labor = 500,600 jobhours

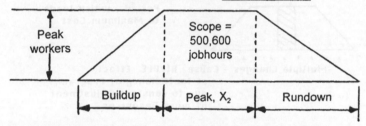

1. Labor availability: No restraint, no adjustment to staffing level.
2. Peak density level: U.S. large project, subcontract labor (from curves) = 250 ft^2 per worker.
3. Peak staffing = 64,000 ft^2 ÷ 250 ft^2 per worker = 256 workers.
4. Effective jobhours per worker-month = 40 hours per week x 88% x 4.35 weeks per month = 153 hours.
5. Buildup (by judgement) = 3 + 2 + 5 = 10 months.
6. Rundown (by judgement) = 6 months.

Solve for peak, X_2:

$$\frac{500,600}{153} = \left(\frac{10}{2} \times 256 \right) + \left(X_2 \times 256 \right) + \left(\frac{6}{2} \times 256 \right)$$

X_2 = 4.8 (say 5) months

Therefore total construction duration (subcontract labor) = 10 + 6 + 5 = 21 months

Figure 4.25 Trapezoidal technique—worked example.

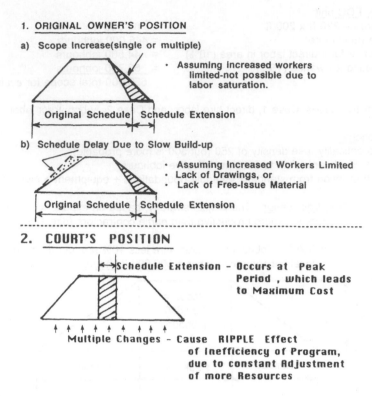

Figure 4.26   Trapezoidal technique—schedule extension for claims, direct labor.

contractors claimed the total impact of the ripple effect was a schedule delay at the peak period. Figure 4.26 illustrates these two positions. The courts finally held for the contractor's position and, consequently, for the greater cost impact.

Figure 4.27 shows a family of curves that illustrates the historical relationship of construction directs and indirects. This relationship was again examined by the courts in claims cases, and the judgment was that costs of indirects due to schedule delay would be directly proportional to the schedule extension.

Figure 4.27 shows that the indirect curves are essentially constant throughout the execution of a project. Early buildup for field organization and installation of temporary facilities is matched by a late buildup for final job cleanup and demobilization.

Direct work progress is a measure of physical quantities installed, and, as shown, the direct work curve is identical to the historical/standard construction curve previously illustrated. Indirect construction progress cannot be assessed by measuring physical quantities and is usually measured in craft labor hours. This typical curve shows the rate at which these hours would normally be expended.

Indirect and direct construction curves for any project can be compared with these curves. During construction, actual performance should be compared with these profiles as well, since doing so can provide an early warning that the expenditure of hours is deviating from the norm.

The percentage breakdowns for craft indirects, field administration, and direct supervision can be used to check forecasts and performances of individual categor-

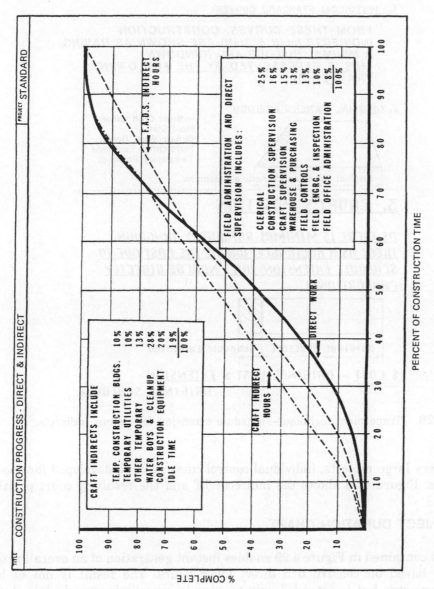

Figure 4.27 Historical curves—construction directs and indirects.

1. HISTORICAL-STANDARD CURVES

*FROM THESE CURVES, CONSTRUCTION
INDIRECTS (craft & Staff) ARE SHOWN AS HAVING
A MINIMAL BUILDUP AND RUNDOWN.
THIS IS REPRESENTED BY THE FOLLOWING
TRAPEZOID.*

2. ORIGINAL OWNER'S POSITION

- Indirect Craft Labor
- Field Staff
- Constr. Eqpt. Rental
- Field Office Expenses
- Temporary Facilities

Original Schedule | Schedule Extension

3. COURT'S POSITION

*AS THERE IS MINIMAL BUILDUP & RUNDOWN,
TREAT AS A RECTANGLE, WHERE THE COST DUE TO
SCHEDULE EXTENSION, WILL NOW BE DIRECTLY
PROPORTIONAL.*

ORIGINAL SCHEDULE | Schedule Extension

$$\text{\$ COST} = \text{ORIGINAL COST} \times \frac{\text{EXTENSION}}{\text{ORIGINAL SCHEDULE}}$$

Figure 4.28   Trapezoidal technique—schedule extension for claims, indirects.

ies. On very large projects, individual control curves can be developed for specific categories. Figure 4.28 shows the indirects TT and the resulting court position.

X.   PROJECT DURATION CHART

The chart contained in Figure 4.29 enables instant generation of an overall project duration, based on construction direct labor hours. The result is not of high quality, however, but is intended only to provide a preliminary schedule during the early development or feasibility stage of a project.

The chart provides add factors for straight-through and overseas projects, as well as curves for projects that were built in the 1950s and 1960s and for Norwegian-based projects. Additional curves can be developed for other overseas locations if necessary.

Of interest is the fact that project durations of the past (the 1950s and 1960s) cannot be repeated today. Following are some of the reasons for this:

- reduced engineering and construction productivity, which cannot always be compensated for by additional personnel;
- greater complexity for the same capacity (such as higher temperatures and pressures);
- increased engineering due to environmental/regulatory requirements;

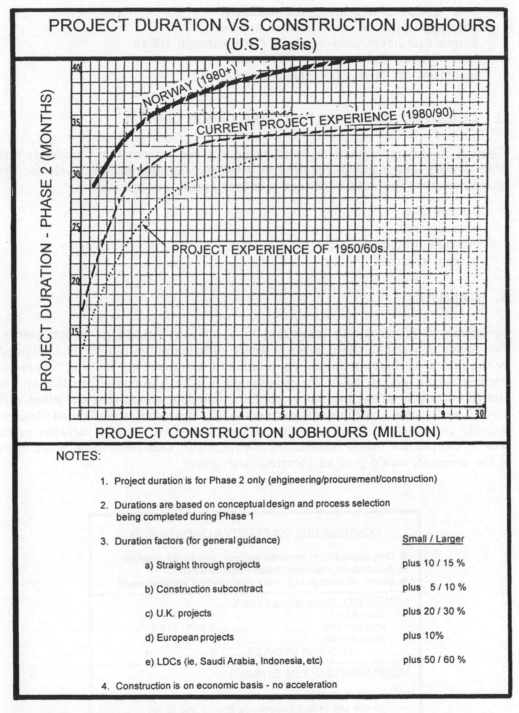

PROJECT DURATION VS. CONSTRUCTION JOBHOURS
(U.S. Basis)

NORWAY (1980+)

CURRENT PROJECT EXPERIENCE (1980/90)

PROJECT EXPERIENCE OF 1950/60s

PROJECT DURATION - PHASE 2 (MONTHS)

PROJECT CONSTRUCTION JOBHOURS (MILLION)

NOTES:

1. Project duration is for Phase 2 only (ehgineering/procurement/construction)

2. Durations are based on conceptual design and process selection
   being completed during Phase 1

3. Duration factors (for general guidance)   Small / Larger

   a) Straight through projects   plus 10 / 15 %

   b) Construction subcontract   plus 5 / 10 %

   c) U.K. projects   plus 20 / 30 %

   d) European projects   plus 10%

   e) LDCs (ie, Saudi Arabia, Indonesia, etc)   plus 50 / 60 %

4. Construction is on economic basis - no acceleration

Figure 4.29   Project duration chart—1960s to 1990s.

* poor project management/poor planning;
* poor application of the fast-track approach;
* longer equipment delivery times (1970 through 1983);
* lack of labor (1970 through 1983); and
* increased use of reimbursable-type contracts.

XI. CONSTRUCTION COMPLEXITY AND LABOR DENSITY

Figure 4.30 illustrates a typical relationship between construction complexity and labor density. Judgment is required in assessing the appropriate levels for the specific project. The data is divided into four general categories:

* simple unit,
* average unit,
* complex unit, and
* process modules.

A. Complexity

Complexity is automatically generated by a company's standard design guides. This factor is based on the number of direct labor hours (within the plot) divided by the plot area (battery limits). As noted, this assumes that there is no *pre-investment* in the design basis. Pre-investment is a fairly common practice and is carried out when forward company planning has determined that the plant will need to be expanded within a few years although, at present, the planned (design) capacity is sufficient. Design pre-investment, therefore, usually includes extra area in the plot for future installation of equipment. This extra area, while open at the moment, would give an incorrect assessment.

CONSTRUCTION COMPLEXITY & LABOR DENSITY

● Only applicable to complete process units (small or large)
● Assumes an economic design - preinvestment
● Based on average U.S. labor productivity (California/union)

COMPLEXITY (direct jobhours / sq ft)
    Simple unit...4 to 5 ⎫
    Average unit.....................................6 to 7 ⎬ x 2
    Complex unit....................................8 to 10 ⎭
        PROCESS MODULES..................................↑

LABOR COMPLEXITY (sq ft / person)

Tied to above complexity data
    Simple unit (4 to 5 jobhour / sq ft)..........150 to 180
    Average unit (6 to 7 jobhour / sq ft)........180 to 250
    Complex unit (8 to 10 jobhour / sq ft).....250 to 300
        PROCESS MODULES........................180 to 200

Figure 4.30 Chart of complexity—density factors.

This data can be used to quickly provide a reasonable labor hour estimate when only the plot/building area is known. However, good judgment is required in selecting the appropriate labor hour/square foot rate. Current studies indicate that module complexity is twice that of land-based plant projects. This factor (2) is shown on Figure 4.30 and applies only to process or utility modules. The module area (sq ft) would include all levels and allowances for mezzanines. Also, to allow for workers occupying space immediately adjacent to the module frame, a perimeter of 10–15 feet is added to the *design* module area.

B. Guide for Selection of Complexity Factor

Figure 4.31 is based on historical data and provides guidance in selecting a complexity factor for a specific project.

C. Density

From Figure 4.30, the density (sq ft per worker) is based on the maximum/economic total number of craft laborers working at peak (supervision is not included). The data is based on historical experience, and the factor is then tied to the

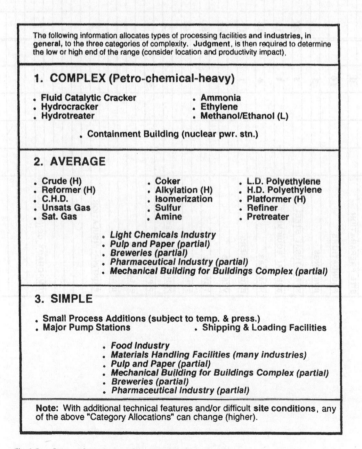

The following information allocates types of processing facilities and industries, in general, to the three categories of complexity. Judgment, is then required to determine the low or high end of the range (consider location and productivity impact).

1. COMPLEX (Petro-chemical-heavy)

- Fluid Catalytic Cracker
- Hydrocracker
- Hydrotreater
- Ammonia
- Ethylene
- Methanol/Ethanol (L)

- Containment Building (nuclear pwr. stn.)

2. AVERAGE

- Crude (H)
- Reformer (H)
- C.H.D.
- Unsats Gas
- Sat. Gas
- Coker
- Alkylation (H)
- Isomerization
- Sulfur
- Amine
- L.D. Polyethylene
- H.D. Polyethylene
- Platformer (H)
- Refiner
- Pretreater

- *Light Chemicals Industry*
- *Pulp and Paper (partial)*
- *Breweries (partial)*
- *Pharmaceutical Industry (partial)*
- *Mechanical Building for Buildings Complex (partial)*

3. SIMPLE

- Small Process Additions (subject to temp. & press.)
- Major Pump Stations
- Shipping & Loading Facilities

- *Food Industry*
- *Materials Handling Facilities (many industries)*
- *Pulp and Paper (partial)*
- *Mechanical Building for Buildings Complex (partial)*
- *Breweries (partial)*
- *Pharmaceutical Industry (partial)*

Note: With additional technical features and/or difficult **site conditions**, any of the above "Category Allocations" can change (higher).

Figure 4.31   Guide for selection of complexity factors.

ACCOUNT	CODE	%	CRAFT MIX BY ACCOUNT %											
			BOIL	BRICK	CARP	ELEC	LAB	INSUL	OPER	MILL	PAINT	PIPE	IRON	TEAM
SITE PREPARATION		9												
FOUNDATIONS		10			37		47		4				10	2
BUILDINGS		3		9	24	1	26		3			8	27	2
STRUCTURAL STEEL		5	27		4	1	4		12	1		4	47	1
SPECIAL EQUIPMENT		1	14		11	6	5		6	25		22	9	2
HEATERS		4	43		7		23	5	5			6	2	2
EXCHANGERS		1	17	7	6		6		14	5		44	7	1
VESSELS		2	64		3	1	14	1	7			4	4	2
TOWERS		2	80		3		4		8			1	2	2
TANKS		1	61		5		9		10	2		10	3	
PUMPS-COMPRESSORS		2	2		5	3	8		4	56		16	6	
PIPING		32			2		5		9			82	1	1
ELECTRICAL		7			4	86	8		1					
INSTRUMENTS		6			1	45	1		1			51	1	
PAINTING-INSULATION		9	2		7		10	59	1		16	2	2	1
TEMPORARY CONSTRUCTION		5	3		30	18	21		5			20	3	

Figure 4.32 Craft labor mix by account.

same categories as previously shown for complexity. As complexity of an area increases, the number of people who can work in that area decreases. In practical terms, this indicates that greater complexity has more equipment, hence more work (more labor hours) per square foot, thus taking up space in which people can work. As with the complexity data, good judgment is essential in selecting the appropriate density level from the illustrated range. The density level for process/utility modules is shown as a narrow range of 180–200 sq ft per person. When this range of density data is not known, a single and widely accepted number is 200 square feet per person.

XII.   CRAFT MIX BY ACCOUNT

Most construction estimates are prepared on an account basis. Scheduling labor resources at a detailed level therefore requires a craft breakdown of labor by account. Figure 4.32 shows a typical breakdown of craft labor by major account for the United States. Overseas labor practices often will not conform to this mix of labor.

5
Value Management*

I. INTRODUCTION

Every owner wants to get the greatest value possible for every dollar spent on capital construction. To reach this goal requires the application of total cost management (TCM) techniques throughout a project's life cycle. Economic analyses are used before any money is committed to determine the feasibility of pursuing the project. Value engineering (VE) may be used to select the apparently best design concept. Operability, maintainability and reliability reviews fine tune designs to make them user-friendly (and usually more cost-effective) for operators and maintenance personnel during a facility's operating days. Constructability reviews look at designs with the goal of making actual construction safer and more cost effective. Many other TCM tools come into play during actual design and construction—planning and scheduling, estimating and budgeting, change management, cost and schedule control and risk management to name a few. Once a facility is in operation, TCM techniques such as planning, scheduling, budgeting, and cost control continue to be used.

There are several TCM techniques which focus on the design itself, the goal being to create the most cost-effective design which meets all owner requirements. These are called *value management* techniques. This chapter describes these techniques.

---

* By Dr. James M. Neil, PE CCE, Boise, Idaho.

II. VALUE MANAGEMENT STUDY TYPES

A. Value Management

This is an umbrella term which encompasses the entire field of value studies.

B. Formal Value Engineering

In its purest sense, VE is a proactive, multidiscipline, highly-structured form of decision analysis which has as its objective the development of a design concept for a facility or item which will yield least lifecycle costs or provide greatest value while meeting all functional, safety, quality, operability, maintainability, durability, and other criteria established for it. However, the term is often used to describe what is more properly called *value analysis*.

C. Value Analysis

This is a structured review of a partially completed or completed design to determine if it could be improved to better meet cost or other objectives. It may or may not be as formalized as pure value engineering.

D. Value Engineering to Cost

This is a reanalysis of a partial or complete design whose estimate of cost or construction bids exceed available budget. The objective is to validate/revise functional and other requirements and redesign the facility to be within budget while meeting revised requirements.

E. Constructability Analysis

These analyses use construction knowledge and experience to make designs more constructable. Specific goals are:

- to reduce total construction time,
- to reduce labor requirements,
- to reduce costs of construction equipment and tools,
- to reduce materials costs, and
- to incorporate features in designs which promote construction safety.

III. TYPES OF VALUE

When one speaks of value, there is a tendency to think only in terms of cost. But there are other types of value that must be considered in any value management study. These types of value are found:

Cost value. This is the amount of money (or monetary equivalents) that must be spent to produce or procure an item.

Exchange value. This is the value of an item on the open market should one try to sell or trade it. The exchange value of a new car drops in comparison to its cost value as soon as the new owner takes the car from the showroom.

Use value. This is the value of an item to the user because of the functions or services it provides. A well-maintained, basic, old car may have little exchange value, but just as much use value to the owner as a newer, fancier car. That use value would equal the cost value of equally reliable replacement transportation.

Esteem value. This is a value "in the eyes of the beholder" or a consequential value derived from some investment. A gold bracelet has a definite cost value, its exchange value is usually less than its cost, it has no use value, but it has an esteem value to an individual who loves jewelry. The esteem value of a company's reputation can be converted into monetary terms only if someone sought to buy the company.

Worth. This term is sometimes used as a synonym for value. What is that painting worth (exchange value)? What is your reputation worth (esteem value)? However, value is the preferred term.

IV. LIFECYCLE COSTS

Value engineering is concerned with the lifecycle costs or value of the facility or item. Following are potential elements of a facility's lifecycle cost:

- feasibility studies,
- engineering,
- real estate,
- permits and licenses,
- special consultant fees,
- construction,
- legal fees,
- furnishings,
- operating costs,
- maintenance costs,
- insurance,
- rebuild/upgrade costs,
- interest payments,
- time-value-of-money on committed capital,
- demolition costs,
- salvage costs, and
- remedial action costs.

There may be value concerns that are not directly translatable into cost but which will be incorporated into a VE study and given a weighting factor. Examples are:

- aesthetics (exterior and interior),
- ease of future upgrading or expansion,
- convertibility to other function,
- maintenance and operator "friendliness," and
- working atmosphere for occupants.

V.   FORMAL VALUE ENGINEERING

Formal VE is a very structured process. However, it is really only an application of a well-recognized decision analysis process used in many business, military, and personal endeavors. Here are the steps:

- state goal or describe problem,
- assemble all facts pertinent to the situation,
- list assumptions necessary to proceed,
- identify all options for accomplishing goal or solving problem,
- develop a short list of most reasonable options,
- establish criteria for comparing options,
- analyze and compare available methods, and
- select best option.

A formal VE program within an engineering organization calls for the creation of a team to evolve and critically evaluate alternative design concepts. The team is led by a facilitator who:

- understands and is skillful at facilitating team brainstorming;
- understands the methodology of VE;
- is a neutral party, not a stakeholder, with respect to the subject of the study; and
- has an educational and/or experience background which relates to the subject.

Other members of the team will be the sources of ideas. They should be individuals from different design engineering disciplines, cost engineering staffs, and operations and maintenance staffs. Having a construction expert also will be helpful.

Formal VE studies require dedication of team members to the process for considerable periods. This becomes an added cost line item for design engineering and should be recognized in the design contract as a reimbursable expense. Unfortunately many owners balk at additional upfront costs and do not authorize formal VE. If not, the principles of VE can still be practiced to some degree. Basically, VE is built on the principle that two or more heads are better than one, and there may be a better approach. In other words, do not leave design decisions totally within the hands of a single person, since that person's experience may be limited and not lead to greatest value. Hence, the idea is to brainstorm ideas before putting pencil to paper. Many companies have gained considerable experience in team brainstorming through their total quality management programs and can take on VE as long as the team leader understands VE methodology.

The facilitator of the group needs to establish a time and place that the group can meet without interruption. He/she establishes the rules for brainstorming sessions and provides general direction but must not be dictatorial, impose his or her ideas on the group, or otherwise constrain the open exchange of ideas. The facilitator normally acts as recorder.

The keys to good brainstorming are:

- individuals who are willing to participate,

- an atmosphere of open-mindedness, and
- a firm rule of *no criticism* of others' ideas.

A group member should feel comfortable saying whatever comes into his or her mind even though the idea may appear ridiculous to some. The reason is that a far-out idea may stimulate someone else to think of an idea that would not have come forth without the stimulation of the first idea. This is why 4 people working in a group can produce more ideas together than 4 people working independently. The size of the brainstorming group can vary depending on the magnitude of the problem, but groups of 3–5 are normal.

VI. FORMAL VALUE ENGINEERING PHASES

Formal VE follows a structured sequence of actions. The various authorities on VE group these into phases, and, depending on the authority, from 4–8 phases may be listed. Whatever their categorization, the total included actions are essentially the same. Following, under seven headings, are these phases.

A. Goal Definition Phase

Clear definition of the goal, boundaries, and authorities of the VE study is an essential first step so that effort will not be wasted.

B. Information Gathering Phase

In this phase all information providing the basis for the project and relating to the specific location is assembled. Examples are:

- operational requirements (output, space, etc.);
- project budget;
- client-need dates;
- major equipment commitments already made or intended;
- proprietary processes or systems involved;
- client's design standards;
- design criteria;
- design calculations to date;
- alternatives already identified;
- technical memoranda pertaining to the project;
- permit requirements;
- environmental, zoning, and other regulations;
- applicable building/construction codes;
- maintenance requirements;
- other facilities potentially affected by this project;
- equipment data sheets, process flow sheets, etc.;
- cost estimates to date;
- subsurface investigation data; and
- site constraints (size, shape, existing structures and facilities, access, etc.).

C. Functional Analysis Phase

Next, the basic and secondary functions to be provided by the facility are identified. Functions are described in a generic fashion using an action verb and a noun to invite visualization of various options. Other typical verb-noun combinations are:

*   transmit load,
*   illuminate room,
*   convey liquid, and
*   heat building.

A basic function is a primary function under study. A secondary function is a consequential or potential additional function. The basic function of a lamp is to provide illumination; a secondary function may be to decorate the room.

Questions to be asked by the team in this phase include:

*   What basic functions are involved in meeting facility/product requirements?
*   Which secondary functions are associated? Which are essential?
*   How is the function accomplished?
*   What is the value of a function?

D. Creative Phase

Different ideas for performing the various basic or secondary functions are identified and listed, based on the functional analysis. For example, heating requirements for a structure can be satisfied by any number of options—heat pump, gas furnace, oil burner, space heaters, fireplaces, electric baseboard heaters, or solar panels. As another example, transport coal leads to the options of conveyor belts, trucks, rail, slurry pipeline, etc. It is not uncommon to have 5–25 ideas proposed.

It is worthwhile in this phase to itemize obvious advantages and disadvantages of each idea as a basis for initial screening in the *analytical phase*.

E. Analytical Phase

Some authorities recommend that only the group leader from the creative phase continue in the analytical phase. The reason for this is that, if the members of the team brainstorming an item know they may be required to evaluate its feasibility and costs associated with the idea, they may be reluctant to add too many ideas to the list or propose an idea that is difficult to evaluate. This phase begins with a summary review of the ideas from the creative phase to reduce the initial list to 2–5 ideas that are worthy of detailed evaluation.

With each remaining idea, the need for each function is validated and each option for performing a function is examined individually as to feasibility, cost, and value received. Then, various combinations of these ideas are tested as a system against the governing criteria for the facility. The most feasible combinations are isolated and given more detailed study. The combination with most favorable balance of cost and value is further refined through more detailed study. In this process, the value engineers will employ various combinations of economic analysis, lifecycle costing, trade-off analysis, sensitivity analysis, estimating, and other cost engineering tools.

While total lifecycle cost may be the primary criterion used to rank options, this is not always true. Another approach, when everything cannot be reduced to monetary terms, is to list various criteria, weight those criteria, and then grade each option against them.

F. Proposal Phase

In this phase, the team's recommendations are summarized for presentation to the decision makers. This presentation must have a written version that details the basis for the recommendation. There also may be an oral presentation.

This phase may be weakest phase if the preparers assume that the reader knows as much as the preparers and fail to provide the information needed to validate a recommendation, or the proposal is so voluminous that the executives with decision-making authority would have difficulty finding time to read and digest it. When preparing a proposal, assume that the reader has no background about this idea and logically build the proposal one step at a time. If there is a lot of information involved in the proposal, prepare an executive summary and put backup detail in appendices so the reader can get the flavor of the proposal very quickly and still have detail available if needed. Ensure that the proposal answers these questions:

- What criteria were being satisfied?
- What solution or solutions are proposed? If more than one are presented, provided detailed advantages and disadvantages of each.
- What will each option cost?
- What value is anticipated, cost and otherwise?
- What other options were seriously considered but not selected?

G. Implementation phase

The client's selected option is implemented. Considerable follow-up may be required with architect/engineering personnel to ensure their understanding of concepts developed.

VII. VALUE MANAGEMENT AFTER DESIGN BEGINS AND DURING CONSTRUCTION

Even though a design concept has been established through formal VE in the conceptual design phase, the principles of VE should continue during detailed design. A typical area for VE effort during design is with specifications. All too often, it is found convenient to utilize specifications directly from a computer database without questioning the need for them as written. Each should be questioned as to its need in its present form and the most cost-effective specification derived. During detailed design, various design simplifications may also result from continued value analysis.

A. Value Engineering Clauses In Construction Contracts

Some clients include *value engineering* clauses in their contracts for construction. These clauses typically set up a procedure whereby, after review of the construction documents, the contractor may submit proposals for more cost-effective designs than those of the contract drawings. If the VE proposal is accepted, the savings are shared with the contractor. As mentioned previously, such studies are more properly called *value analysis* studies since they come after completion of design.

B. Constructability Analyses

Constructability analysis is a value management technique that is most effective if initiated during the conceptual design stage and continued through detailed design and construction phases. Constructability analyses are focused on the construction phase. Basically, the intent is to save construction time and cost without compromising any other project objectives. Thus, constructability decisions are oriented toward:

- reducing total construction time by creating conditions which maximize potential for more concurrent (rather than sequential) construction, and minimize rework and wasted time;
- reducing jobhour requirements by creating conditions which promote better productivity or creating designs which demand less labor;
- reducing costs of construction equipment (and tools) by reducing requirements for such equipment, creating conditions which promote more efficient use of that equipment, and minimizing need for high-cost, special purpose equipment.
- reducing materials costs through more efficient design, use of less costly materials, and creation of conditions which minimize waste;
- creating the safest work place possible since safety and work efficiency go hand in hand.

A common view of constructability is that it only involves:

- determining more efficient methods of construction after mobilization of field forces,
- allowing construction personnel to review engineering documents periodically during the design phase,
- assigning construction personnel to the engineering office during design, and
- developing a modularization or preassembly program.

Although each of the above is part of constructability, each is only a part. Only through the effective and timely integration of construction input into planning and design as well as field operations will the potential benefits of constructability be achieved.

Who performs constructability analysis? Obviously, it must be someone familiar with the construction process. Ideally, the contractor will be allowed to co-locate construction personnel with the client's engineer. But, this has a cost and will not always be possible because of the client's handling of design and construction contracts. In such cases, reliance must be placed on the client's architect/engineer. Of course, continuing constructability analysis in the field should be a standard feature of effective project management in furtherance of its goal to provide the most cost-effective management of the project.

VIII. VALUE ENGINEERING TO COST

There are occasions when a client's available budget is less than the most recent cost estimate or less than the selected contractor's cost proposal. In these situations, the contractor may be asked to join the client in re-examining the designs for the purpose of value engineering them to fit within budget. In studying these designs, attention should be initially focused on components or features of the project which have greatest cost reduction potential:

- If an item or function is nonrevenue producing, could it be eliminated without compromising basic functions?
- If an item or system is complex, is there a simpler way?
- If an item or system has a significant cost, what are the components of this cost and can any be reduced through redesign or substitution?
- If a design for providing some function is new, is it still the best approach?
- If an item has negative constructability (including safety) implications, what can be done to make it safer or more constructable?
- If a design will force hiring of high-wage craft trades in the field, could it be redesigned to utilize workers from trades with lower wage scales?
- If specifications were copied from other projects, could they be tailored to fit this project's situation?
- If specially engineered components are involved, could off-the-shelf equipment be substituted?

    Additionally, consider the following:

- combining multiple structures into fewer structures,
- using an existing structure or system instead of constructing a new one, and/or
- using modularization or shop fabrication to reduce field labor costs.

IX. SUMMARY

Value management provides a family of useful tools to help assure lowest cost and/or greatest value for a project. It is not a discipline like structural or mechanical engineering; rather, it is a series of techniques available to all disciplines. Applying them takes additional time and money at the time, but advocates are convinced that the returns are many times greater than the initial investment. All cost engineers should include these techniques in their total cost management tool bags.

X. VALUE ENGINEERING BIBLIOGRAPHY

A. J. Dell'Isola, *Value Engineering in the Construction Industry*, 3rd Ed., Van Nostrand Reinhold, New York, 1982.

C. Fallon, *Value Analysis*, 2nd Rev. Ed., Triangle Press, Southport, NC, 1980.

Federal Construction Council, Consulting Committee on Value Engineering, *Elements of an Effective Value Engineering Program*, Technical Report No. 92, National Academy Press, Washington, D.C., 1990.

General Services Administration, *Value Management Handbook*, PBS P 8000.1A, Washington, D.C., October 31, 1978.

S. J. Kirk, and K. F. Spreckelmeyer, *Creative Design Decisions: A Systematic Approach to Problem-Solving in Architecture*, Van Nostrand Reinhold, New York, 1988.

H. E. Marshall (ed.), Roundtable on a Standard Method for Value Engineering: AACE International Annual Meeting June 21, 1994, *Cost Engineering* 37(3):9–21 (1995).

L. D. Miles, *Techniques of Value Analysis and Engineering*, 2nd ed., McGraw-Hill Book Company, New York, 1972.

D. Mitten, *Value Management for Quality and Cost Effectiveness*, Project Management Services, Inc., Rockville, MD, 1991.

U.S. Environmental Protection Agency, Office of Water Program Operations, *Value Engineering for Wastewater Treatment Works*, 430/9–84–009, Washington, D.C.

L. W. Zimmerman, and G. D. Hart, *Value Engineering: A Practical Approach for Owners, Designers, and Contractors*, Van Nostrand Reinhold, New York, 1981.

XI.   CONSTRUCTABILITY BIBLIOGRAPHY

Constructability: A Primer, Construction Industry Institute Publication 3–1, Austin, TX, July 1986.

Constructability Concepts File, Construction Industry Institute Publication 3–3, Austin, TX, August 1987.

Constructability Implementation Guide, Construction Industry Institute Publication 34–1, Austin, TX, May 1993.

Guidelines for Implementing A Constructability Program, Construction Industry Institute Publication 3–2, Austin, TX, July 1987.

Preview of Constructability Implementation, Construction Industry Institute Publication 34–2, Austin, TX, February 1993.

Implementing Constructability, Construction Industry Institute Training Package EM-11, Austin, TX, 1993.

6

Economic Evaluation in the Process Industries[*]

I. PURPOSE OF AN ECONOMIC EVALUATION

The purpose of an economic evaluation is to select the investment of wealth that will do one of the following:

- maximize the financial return on the investment, generally measured in profit;
- cost the least to correct (environmental); or
- generate the most goodwill for the least cost.

Successful management of money requires that the amount of money at the end of the year be greater than the amount of money at the beginning of the year. The greater the increase, the better the perceived management. The rewards for increasing the wealth are usually in proportion to the amount of the increase.

Many opportunities exist to make money; yet many more opportunities exist to lose money. The successful manager finds that investment that will return the most money at the end of the year. One method used by successful managers to select good investments is performing economic evaluations of the investment.

Unfortunately, not all funds expended by a business or a person result in a return or a profit. An example of this is the environmental expenditures many companies face. With the large increase in regulatory requirements, few of which return any profit, an economic evaluation is meaningful in finding the best way to meet the requirements of the law with the least expense. These evaluations

---

[*] By Dr. Klane F. Forsgren, PE, Vice President-Engineering, Sinclair Oil Corporation, Salt Lake City, Utah.

137

involve the same principles of comparison used for selecting investments, and also can suggest ways to evaluate options that minimize loss.

Another important decision faced by management is the building of good will and a good name for the company. In most instances, developing good will requires a monetary investment. The techniques used to select the best investments can also be used to evaluate the best options for developing goodwill.

II. METHODS OF EVALUATION

The method described in this chapter relates to the time value of money. The concept is that the use of money over time has a value and that a person should be compensated for the use of the money for that period of time. The method described shows ways to compare the value received for the use of the money. This value is important, since the return on an investment may be postponed for some years, is not the same every year, and may be bunched at the front or the end of the time spectrum under consideration. The method described in this chapter will help you consider many variables and put them all on an equal footing for comparable evaluation.

In order to understand the methodology of the time value of money, we must define some terms typically used in economic discussions. The textbook, *Economic Evaluation and Investment Decision Methods*, by Franklin and John Stermole, has been referenced extensively in developing the definitions. The author's experience in the oil and power industry serves as the basis of the examples.

The important definitions needed to discuss economic evaluations are the following:

*   cash flow;
*   rate of return, internal rate of return, or discount rate;
*   present value;
*   future value;
*   annual value; and
*   breakeven.

A. Cash Flow

Cash flow is the difference between the cash received (income) and the cash paid out (expenses). Cash received can come from several sources, including sales revenue, interest on money invested, and services performed. Cash out goes to expenses such as materials, labor, utilities, taxes, interest, buildings, etc. The successful business will have more cash in than cash out, or a positive net cash flow.

In most instances, the cash flow is negative for the first few years of the operation of a new business or a new process. To adequately perform the economic evaluation, the business must take this negative cash flow into consideration. The methods discussed show how to consider the total cash history of the venture in the evaluation.

A second consideration in evaluating cash flow is that the value of money changes from year to year due to inflation and deflation—like the difference between a 1995 dollar and a 1985 dollar.

B. Rate of Return

Rate of return is the interest rate that you would receive if you invested the amount of money under consideration in a bank and received the same positive net cash flow as was developed by the investment. The net cash flow occurs over a period of time, and the time value of the cash flow over that time must also be considered. *Discounting* implies reducing the value of something. Future values are discounted by a factor—the interest rate or discount rate—to their present worth or present value.

C. Present Value

Present value is the value of the investment over time in today's dollars. It includes all of the cost of the project, including management costs to get the project completed, and the hard costs for purchase of equipment, buildings, material, etc.

   Net present value (NPV) is the cumulative present worth of the positive and negative cash flows where the specified discount rate introduces the time value of money. A positive NPV means that the future cash flows of the project are higher than the minimum return established for the project. A negative NPV means that the future cash flows are not enough to maintain the return established for the project. A person or company would be better off investing in an alternate project.

D. Future Value

Future value is the value of the cash flow in the future in future dollars. This value is used in comparisons and discount rate formulas, which are available in tables of interest rates.

E. Annual Value

Annual value is the amount of annual payments, similar to a mortgage payment. The annual value is also used in discount rate formulas.

F. Breakeven

Breakeven is the time that the net positive cash flow pays off the full cost of the project. The faster a project reaches breakeven, the better the investment, or the more money that was returned in a shorter time. The time to reach breakeven can be used as a good rule of thumb in evaluating investment options, as will be seen later. Table 6.1 shows how cash flows are developed.

III. DISCOUNTED CASH FLOW METHODOLOGY

The method presented in this chapter to compare the return on an investment is the *discounted cash flow* (DCF) method. Discounted cash flow puts all investment options on a common basis of handling the time value of money with

Table 6.1 Example of How Cash Flow is Developed. The Cash in is Shown by Positive Values, the Cash out by the Negative Values. The Net Cash Flow is the Difference Between Cash in and Cash out.

Year	0	1	2	3	4	5	6
Invest	−200	−100					
Income			170	200	230	260	290
Costs							
Operating			−40	−50	−60	−70	−80
Tax			−30	−40	−50	−60	−70
Net CF	−200	−100	100	110	120	130	140

compounded interest rates. It involves taking the annual cash flow from the business, either real or projected, and applying an interest rate to each year's flow, while accumulating the previous year's cash flow with associated accumulated interest. This is the value of using compounded interest rates in the calculations.

The discount rate becomes the minimum return or interest rate acceptable for investment. The owner of the project usually establishes the minimum return or interest rate or the discount rate. What determines the selected minimum return or discount rate? There are several factors:

- the interest rate available in the money market at no risk;
- alternate investments and their return potential;
- risk associated with the project;
- accuracy of the economic analysis, generally based on the quality of the data available for analysis;
- corporate policy; and
- personal preference.

An investor can normally put money into a very safe investment, such as a certificate of deposit with a strong government, or into a very stable bank, and receive a guaranteed return on that investment for the investment period. Since the return on the money is guaranteed, the interest rates are usually low. This value would be the least return on investment that one would accept for any business proposal, since it is guaranteed. The return on another form of investment may not be guaranteed, and therefore carries risk that the investment may lose money.

In most instances, several projects compete for the same investment capital. Within a corporation, for example, several different corporate sites may be trying to add new process equipment. These projects are in competition with those who want to buy out complete companies or expand into a new market by building a completely new facility.

In this situation, a corporation will prepare a financial package, a proforma, with the projected profit for several years. Then the company will determine the DCF for each competing project. The project with the highest *internal rate of return* (IRR) will normally be selected for funding.

In preparing the proforma or financial package, companies must make several assumptions in order to predict the future cash flow. These assumptions include:

- the initial cost of the project;
- the sales projections for several years which will lead to revenue;
- the sales price and inflation over time;
- the cost of materials over time;
- the cost of operating the facility, including salaries, fringes, and management costs;
- taxes; and
- the cost of borrowing the money to complete the project.

The accuracy of the financial analysis depends on the quality of the data in the assumptions used. The poorer the quality of the input data, the greater the risk that the analysis will be incorrect. If the risk is high, then the return required in the financial analysis must be higher. This allows for error in the analysis or for the poor quality of the input data.

In many instances, the IRR is established by the corporation or the owner based on the above factors and on experience. For instance, some major corporations will not consider a project unless the *return on investment* (ROI) is at least 20% averaged over the life of the project (i.e., the IRR is 20%). When the interest rate for certificates of deposit is 6%, a 20% ROI seems large. But corporate experience has shown that many of the projects that promised to return good profits did not, in many instances due to conditions beyond the control of the corporation.

For example, when the oil embargo occurred in the early 1970s, everyone thought that oil prices would remain high. Several companies spent millions of dollars to develop alternate sources of oil or fuel, such as the oil shale of Colorado or the tar sands of Canada. The financial projections showed great profit so long as the price of oil was above $30 per barrel. History now shows that the price of oil did not stay above $20 a barrel, and the value of the investment in oil shale and tar sands was lost.

The nuclear industry is a prime example of cost of installations getting out of hand. It became so costly to build nuclear power plants due to increased governmental regulations that many were shut down when only 30–40% completed. The investment was lost.

A. How Does DCF Work?

Most business computers have DCF programs built in. You only need to obtain the net cash flow for the project for a period of several years, put that data into the computer program, and let the computer develop the DCF. The more years of cash flow available, generally the better the DCF analysis.

The computer program invests each year's cash flow at the established interest rate for the established period of time. The sum of the net profit and accumulated interest from the net cash flow is compared to the income produced by putting that same amount of money into a bank at the prescribed discount rate. The two sums are compared to determine whether the project meets the standard set by the corporation or the owner.

Table 6.2 Impact of Net Cash Flow on Internal Rate of Return (IRR) and Net Present Value (NPV)

Investment life	Case 1: 3 year	Case 2: 6 year	Case 3: 9 year
Project DCF IRR	37%	21%	14%
Project NPV @15%	$135.20	$54.70	−$11.70
+ Cash flow	$600	$600	$600
Investment	$300	$300	$300

B. What Impacts DCF Calculations?

Several factors affect DCF analysis, but four major factors are:

* the amount of the investment,
* the amount of net cash flow,
* the timing of the investment, and
* the timing of the return.

Control of these items will greatly improve the return on investment sought by the investor. The greater the cost of the project, the higher the net cash flow must be to repay the debt and to return an increase to the investor. It is also more difficult to get a good cost estimate. This is primarily because few examples of complex projects exist, and the company is more susceptible to missing an important piece of the puzzle. To compensate for these difficulties, the net cash flow and rate of return would normally have to be higher to be acceptable.

The higher the net cash flow, the faster the debt is retired and the greater the rate of return. Table 6.2 shows three different cash flows with the corresponding net present value at 15% interest and the corresponding IRR.

In Table 6.2, all three projects have the same initial investment of $300,000 spent in 2 years. Case 1 returned $600,000 over just 3 years. Case 2 returned the same $600,000, but over 6 years. Case 3 returns the same $600,000 over 9 years. Cases 1 and 2 have IRR values above the 15% standard for this investment and show a positive net present value. Case 3, with 9 years required to receive the $600,000, did not meet the 15% standard and thus had a negative net present value.

IV. A CASE STUDY: CONSTRUCT OR CONTRACT?

A good way to understand how this method of economic evaluation can be used effectively is to do a case study. The United States Environmental Protection Agency (EPA) recently required that all vehicles that traveled mainly on paved roads would have to use low-sulfur diesel (0.05 wt%). This required a decision by the refiner between these two choices:

* Should the company spend the money to build a new hydrotreater in its own facility to reduce the sulfur in the diesel fuel?
* Should the company make a contract with another refinery to have the diesel fuel produced on a contract basis?

The numbers used in this example are numbers representative of the real world, but are not intended to represent actual practice.

A. Construction Option

The following assumptions were made about the construction option:

- cost to construct is $40,000,000; $15,000,000 in the first year;
- time to construct is 2 years;
- operating time is 24 hours a day, 350 days a year;
- plant capacity is 14,000 barrels a day;
- selling price for the low sulfur diesel starts at 10 cents above regular sulfur diesel, but declines as shown over 7 years; and
- cost for operating the hydrotreater starts at 3 cents a gallon and reduces as the operators become more acquainted with the equipment and are able to optimize the operation.

The annual cost and income in U.S. dollars are shown in Table 6.3 for a 7–year period. Subtracting the cost line from the income line gives the net cash flow for this option (Table 6.4). If we set the minimum return as 15%, the net present value (NPV) of the cash flow stream is $6.93 million. The internal rate of return is 20.8%. Thus if the minimum return that is acceptable is 15%, this project should be accepted for construction. But the purpose of this evaluation is to see whether it is better to build or to contract the desulfurization. We need to run the other option and see how it compares financially.

B. Contract Option

The contract option uses the following assumptions:

- cost to construct transfer system and transportation to get diesel to the contractor is estimated to be $5 million;

Table 6.3 Construction Option for Case Study. Over 7 years, the income is substantial enough to offset the yearly cost.

Cost to construct	$40,000,000						
Time to construct	2 years						
	$15MM first year; $25 MM second year						
Operation time	350 days/year						

Annual cost and income ($US)							
Year	1	2	3	4	5	6	7
Price Δ ($/gal.)	0.10	0.09	0.08	0.075	0.07	0.07	0.065
Income ($MM)	20.6	18.5	16.5	15.5	14.4	14.4	13.4
Cost ($/gal.)	0.03	0.025	0.02	0.02	0.02	0.02	0.02
Cost ($MM)	6.2	5.1	4.1	4.1	4.1	4.1	4.1

Table 6.4 DCF Analysis for the Construction Option. Over the 7–year option, the income is substantial enough to support construction, but how does this option compare to contracting the desulfurization?

		Year							
	0	1	2	3	4	5	6	7	8
Cash flow ($MM US)	−15	−25	14.2	13.4	12.4	11.4	10.3	10.3	9.3

NPV @15% = $6.93 MM; IRR = 20.8%

- time to construct is one year;
- operating time is 24 hours a day for 350 days a year;
- selling price is the same as in the construction option; and
- cost for hydrotreating and transporting the product is set by the contract and known transportation costs. For this example, the cost of hydrotreating starts at 9 cents per gallon and decreases slightly because competition will drive the price of diesel down.

The annual cost and income from the contract option is shown in Table 6.5. Since the construction time is only one year, we have taken the annual cost and income statement for 8 years. Thus the comparison takes into consideration the timing of the projects and the timing of the expenditures, as noted before.

Comparing the two options shows which option will provide a better return on investment and potentially make more money for the owner. Comparing the analysis shown in Tables 6.4 and 6.5 gives the following:

Table 6.5 Data for Contract Option for Hydrotreating Diesel Fuel

Cost to construct	$5,000,000
Time to construct	1 year
Operation time	350 days/year

	Annual cost and income ($US)							
Year	1	2	3	4	5	6	7	8
Price Δ ($/gal.)	0.10	0.09	0.08	0.075	0.07	0.07	0.065	0.065
Income ($MM)	20.6	18.5	16.5	15.5	14.4	14.4	13.4	13.4
Cost ($/gal.)	0.09	0.085	0.075	0.07	0.065	0.065	0.06	0.06
Cost ($MM)	18.5	17.5	15.4	14.4	13.4	13.4	12.3	12.3
NCF ($MM)	0.1	1	1.1	1.1	1	1	1.1	1.1

NPV @15% = −$1.10 MM; IRR = 8.82%

Option	NPV@ 15%	IRR
Construct option	$6.93 MM	20.8%
Contract option	–$1.10 MM	8.8%

When the difference is great, the decision can be quite simple and easy. When the difference is small, you consider other factors in making the decision. The contract option only returns 8.8% on the invested dollars, and the construction option provides a 20.8% return. This suggests that the construction option should be selected.

Before a company will accept the results of the analysis, especially when the comparisons are close, it will check the quality of the input data. In order to make this as easy as possible, it helps to know which data has the greatest impact on the analysis. This is referred to as a sensitivity analysis.

C. Sensitivity Analysis of the Comparison

With the projects set up in the DCF format, you are in a position to modify the initial assumptions and determine what will be the impact. Some variables have a major impact on the financial analysis, while others have little effect.

The impact of changes to five items will be evaluated:

- impact of selling price,
- impact of initial cost of the project,
- impact of production cost,
- impact of inflation on both selling price and costs, and
- impact of taxes to show value of tax breaks to the viability of the project.

1. Impact of Selling Price

Selling price is often something you cannot control with the project. If your product is a mature product in the marketplace, the selling price has been established by history, and the likelihood of changing that price much is limited (though the oil companies have been successful at increasing prices by creating perceived shortages). Competition is likely to cause pressure on the price and make it less than you would like. Therefore, your economic analysis should consider what would happen to the project if the selling price were to be less than the projected price. Table 6.6 shows the construction option with a 2–cents-a-gallon lower selling price for the low sulfur diesel than initially projected. The costs, however, remained the same as initially projected.

This analysis shows that the IRR is just under 11%, and therefore the system will not meet the 15% standard established. The analysis would recommend not making this investment based only on the rate of return. The same analysis can be done for the contract option to find out the sensitivity to selling price.

2. Impact of Initial Cost of the Project

Table 6.7 shows the construction option with a $50 million initial cost rather than $40. The first year cost was $25 million instead of $15 million. The project then drops from meeting the 15% return criteria to only 12.9%. The investor

Table 6.6 Impact of a 20% Reduction in Selling Price on the Construction Option

		Annual costs and income ($US)						
	−1	0	1	2	3	4	5	6	7
Selling price ($/gal.)			0.08	0.07	0.065	0.06	0.06	0.055	0.055
Income ($MM)			16.5	14.4	13.4	12.4	12.4	11.3	11.3
Costs ($MM)			6.2	5.1	4.1	4.1	4.1	4.1	4.1
NCF ($MM)	−15	−25	10.3	9.3	9.3	8.3	8.3	7.2	7.2

NPV @15% = −$4.74 MM; IRR = 10.78%

would receive $3 million dollars less with this investment than with an option that will meet the 15% criteria.

3. Impact of Production Cost

An increase in the operating cost of the project—it costs more to produce the product than was projected—has the same general impact as does having a lower selling price. In both cases the net cash flow will be decreased, thus decreasing the net present value of the cash flow over time. You can run this option to see how much the decrease is, though the proof is not shown here.

4. Impact of Inflation on Both Selling Price and Costs

Table 6.8 shows the construction option with the lower initial selling price but now adds a 5% per year inflation rate for the selling price and the costs. Because the selling price is generally more than the cost, an inflation increase will help the project because the difference between selling price and cost will continue to increase as inflation increases. Under this scenario, the project that was not acceptable now becomes acceptable. The IRR is just above 15%, barely meeting the criteria standard set for the project.

5. Impact of Taxes to Show Value of Tax Breaks to the Viability of the Project

The impact of taxes is the same as modifying the cost of production. Taxes must be taken from the cash stream, and serve to reduce the value of the net cash flow. If you can, obtain some type of tax break from the local, state, or national

Table 6.7 Impact of a 25% Increase in the Initial Cost of the Construction Option

	Year								
	0	1	2	3	4	5	6	7	8
Annual cash flow ($MM US)	−25	−25	14.2	13.4	12.4	11.4	10.3	10.3	9.3

NPV @15% = −$3.07 MM; IRR = 12.9%

Table 6.8 Impact of a 5% Inflation Factor Applied to Both Selling Price and Costs

			Annual cost and income ($US)					
	-1	0	1	2	3	4	5	6	7
Selling price ($/gal.)			0.08	0.07	0.073	0.077	0.08	0.085	0.09
Income ($MM)			16.5	14.4	15.0	15.8	16.5	17.5	18.5
Costs ($MM)			6.2	5.1	5.4	5.6	5.9	6.2	6.5
NCF ($MM)	-15	-25	10.3	9.3	9.6	10.2	10.6	11.3	12.0

NPV @15% = $0.48 MM; IRR = 15.38%

governments that adds net income to the cash flow, thus improving the performance. Usually the tax break has a limited time that it is in place. Thus this method of DCF could be projected to beyond the time that the tax break is allowed to determine the impact on overall project viability.

6. Comparing the Changes

Comparing the results found from Tables 6.6, 6.7, and 6.8, we learn which variables have the greatest impact on the economic viability of the project. A 20% reduction in selling price has more impact than a 25% increase in the initial cost. Therefore, selling price is more critical.

A 5% inflation rate takes a project with a lower initial selling price, which did not meet the 15% IRR criteria, and converts it to an acceptable project. Again, inflation impacts selling price as well as costs, while the initial investment stays constant. We impact selling price with inflation while initial investment remains noninflated.

V. RELIABILITY OF THE DCF METHODOLOGY FOR ECONOMIC EVALUATION

The DCF methodology has great versatility; it quickly helps to evaluate several alternatives and to determine how sensitive the alternatives are to variations in the data used in the analysis. However, the system is very dependent on the quality of the data used in the analysis. This dependence bears the impact of cost of the process, selling price of the product, and cot of manufacture. In establishing any economic evaluation, ensure that the numerical values used in the process are as close to the real world as possible. Getting an accurate cost of the project and determining the correct selling price are generally the two places that economic evaluations fail.

When determining the cost of the project, make the costs equivalent between competing options. The project manager of the economic evaluation must demand that the project costs are complete, current, and realistic. The manager must define the total extent of the process being considered. In the case study, the desired hydrotreater must be carefully defined in terms of size, location in the refinery, needed operational flexibility, type of construction material, and unique process considerations.

The project definition must include supporting processes. For instance, the hydrotreater that removes sulfur from the diesel must have a place to deposit it. You must consider the unit size and use. You must analyze the amine system to ensure that it is adequate. Do you have hydrogen to do the treating? All of the costs of the supporting process changes must be included into the analysis to make it meaningful.

Review all of the utility needs of the new process to be sure that adequate steam, water, power, etc., are available. You will also need to review environmental issues and include waste treatment upgrades or additions, if needed.

Ensure that you include all costs in the project cost, such as professional costs of consultants, financing costs if they apply, and start-up costs, including the cost of fuels and materials needed to start up the new process.

A final caution on determining project or plant costs for the analysis: the source of the cost estimate is important. Some estimates are done from handbooks and general industry rules of thumb. These estimates are appropriate in initial screening estimates, but when the analysis approaches the decision stage, the estimates must be more accurate.

Many site-specific costs will impact the project, and you must consider them. Rules of thumb are no longer adequate. In the estimate, make refinements to the estimates based on such things as union versus nonunion labor, local labor versus imported labor, and new equipment versus used equipment.

VI. SELLING PRICE OF THE PRODUCT

Earlier in this chapter, information on determining the selling price was discussed. Developing good selling price information is crucial to the success of a project analysis. To help you develop good information, you will also need to consider the following:

- commodity versus new product pricing,
- reaction of the competition,
- government impact,
- availability of product, and
- raw material costs.

The selling price of an existing product class is established by the market. The likelihood of increasing the selling price of the product to the consumer is limited. However, tougher regulations may demand more of businesses without allowing them to pass some of the costs to the consumers. Currently, oil companies are required to add oxygenates to gasoline that is sold into ozone nonattainment air quality areas. The oxygenates are added by blending ethanol or a chemical called MTBE (methyl tertiary butyl ether) into the gasoline. These additives cost 50–75% more than the gasoline. The additive can be as much as 10% of the total blend, yet the oil company cannot get enough for the oxygenated gasoline to cover the increased cost. The consumer is used to paying the established price for the product regardless of the improvements to the product.

The whole purpose of these examples is to require that the economic evaluator pay special attention to the selling price used, and basically assume lower prices. Lower selling prices are true more often than the higher prices promoted by marketing personnel.

If the product for sale is a new product, you have more price flexibility, but again the analysis should look at generally lower prices than projected.

In established markets, economic analysis should evaluate how the competition will react. The competition will fight to keep their market share. The fastest way to respond to new competition is to lower prices. In some instances, the competition will take a loss for a period of time to keep new competitors out.

The government can have a major impact on selling price. Complaints from the public on price fixing can lead to expensive, time-consuming defense. Government requirements to modify a product for environmental reasons but with no opportunity to increase the selling price can be detrimental.

VII. NONFINANCIAL FACTORS

In performing an economic evaluation, you must be aware of the nonfinancial factors that affect the final outcome of an analysis of options. In the case study in this chapter, several nonfinancial considerations exist.

If the corporation's policy does not want the debt required to own the manufacturing facilities that do their processing, the analysis becomes one of showing such an overwhelming argument for the construction option that it will change corporate policy. Thus the criteria for this evaluation of the construction option is much higher than just to provide the best IRR. Here it would just be to see that the contract option meets the minimum requirements of the company.

The availability of quality contractors is a consideration. If the only manufacturers are new to the market with a limited track record, then look around more carefully for the construction alternative.

Transportation and transfer problems arise, perhaps with governmental participation. For instance, the ability to ship nuclear waste between states is being limited and costs much more.

Does the company want to risk the investment of \$40–\$50 million on a market that may not be real for the long term? There is a viable \$5 million option. Admittedly, the IRR is greatly reduced, but the company's position in the low-sulfur diesel fuel is protected without so much money being risked on a new plant.

VIII. SUMMARY

Economic evaluations are developed to help us make the best investment choice of the options available. Discounted cash flow analysis, which uses compound interest rates to determine the new present value of a future net cash flow, is a versatile tool; it compares many alternatives and helps to determine which economic factors provide the greatest change in the analysis.

7

Keys to Controlling and Reducing Environmental Costs*

I. OBJECTIVE

Few people realize how costly environmental cleanup is and how costly it may be to prevent environmental problems in the future. The U.S. cost of complying with environmental regulations from 1972–1992 was about $1.4 trillion. The cost will be about $1.6 trillion in the remainder of the 1990s.

This chapter identifies proven methods for remediation and discusses the expected costs attached to each method. It also identifies the role and attitude of the governmental regulators and illustrates how to save expenses through initial negotiations with the regulators. It identifies cost reduction options for closure and postclosure care.

Finally, it provides some basic guidelines to estimate remediation costs. It is extremely difficult to estimate remediation costs because a project always changes as it progresses. Still, it is important to estimate costs well to develop budgets and to adopt and monitor cost control activities.

This chapter is based on remediation activity to be conducted under the *Resource Conservation and Recovery Act* (RCRA) and *Underground Storage Tank* (UST) guidelines. These guidelines are similar to other remediation strategies and illustrate the methods proposed below for cost control of remediation projects.

---

* By Dr. Klane F. Forsgren, PE, Vice President-Engineering, Sinclair Oil Corporation, Salt Lake City, Utah.

II. PLAYERS IN THE ENVIRONMENTAL ARENA

To be an effective manager, a person must understand the rules of the game or realize the elements of the arena of competition.

The environmental arena has become a major political playing field. This has led to the enactment of more environmental legislation on both levels. With each new federal regulation, lawmakers assign a federal agency to interpret the law and develop a new set of statutes to enforce. Also, each state wants to control its own destiny. Each must pass its own laws that are in accord with or more stringent than federal laws. In many instances, state laws and state administration are at odds with federal laws and control. Sometimes business interests get caught in feuds between the state and the federal agency.

Since the government has taken responsibility for regulating many activities that affect the environment, the regulation process suffers from problems of extensive governmental paperwork, heavy public involvement in reviewing proposed legislation, long approval processes, overkill, cross-purposes, lack of funding, and inefficiencies.

However, these problems also provide opportunities to reduce costs. The main goal of the regulators is to be successful. They do not like to read large documents any more than anyone else. They generally like to be team members and be part of the solution.

One final understanding will be helpful. All the legislative and statutory rules and regulations were developed by governmental agencies that interpreted congressional law. Guidance documents provided by government agencies on how to comply with the law are large and complex. The sheer size and complexity of the rules and regulations have frightened business managers, leading them to get outside consultants to help them comply with the law.

Some regulators leave government service after a number of years and become consultants. This is their career plan. The incentive for the regulator-turned-consultant is to help the client cover all the requirements and to keep the client out of difficulty. While this approach is appropriate, many consultants do not attempt to simplify the process for their clients. Instead, the consultants produce large, comprehensive studies, permit applications, and reports. They are more likely to choose sophisticated and expensive remediation options, and agree with agency requests for comprehensive chemical testing. The comprehensive approach is the safest approach, but it is also the most expensive. The key to controlling cost is to do only what needs to be done to solve the problem. The purpose of this chapter is to provide guidance to do just that.

On the federal level, the National Environmental Policy Act (NEPA) of 1970 imposes a *trustee of the environment* duty on each generation of American people and requires federal agencies to evaluate the environmental impact of major projects and actions. The National Environmental Policy Act is managed primarily by the Environmental Protection Agency (EPA), which requires EPA to promulgate and enforce most environmental regulations. Table 7.1 lists specific legislative actions that EPA enforces, with a brief description of each action.

Other related environmental laws include:

*   Federal Environmental Pesticides Control Act, 7 U.S.C. 136 (FEPCA)
*   Hazardous Materials Transportation Act, 49 U.S.C. 1801 (HMTA)

Table 7.1 Legislation the EPA Enforces

Regulation	Description
TSCA	**Toxic Substance Control Act**, 1976, 15 U.S.C. 2601 Requires manufacturers, processors, users, and distributors of chemical substances that present an unreasonable risk of harm to pretest and notify EPA of risks before commercial production starts. Specifically restricts the manufacture and use of chlorofluorocarbons (CFCs), PCB, dioxin, etc.
RCRA	**Resource Conservation and Recovery Act**, 1976, 42 U.S.C. 6901; amended by HSWA (Hazardous and Solid Waste Amendments), 1984, 42 U.S.C. 3251 Requires *cradle to grave* management of characteristic and listed hazardous wastes. Generators, transporters, and storage facilities must obtain permits for the active treatment, storage, and disposal of hazardous wastes. Also regulates underground storage tanks.
CERCLA	**Comprehensive Environmental Response, Compensation, and Liability Act (Superfund)**, 1980, 42 U.S.C. 9601; amended in part by SARA (Superfund Amendments and Reauthorization Act), 1986, 42 U.S.C. 11001 Requires the generator to clean up inactive hazardous waste sites.
EPCRA	**Emergency Planning and Community Right-to-Know Act**, 1986, Title III of SARA (Superfund Amendments and Reauthorization Act), 1986, 42 U.S.C. 11001 Requires all appropriate businesses to have response plans for any type of emergency that could occur and make the plan public. Requires communities and facilities to coordinate emergency planning for and notification of hazardous releases.
CWA	**Clean Water Act**, 1977, 33 U.S.C. 1251, amending Federal Water Pollution Control Act, 1972, as amended by Water Quality Act, 1987, and Oil Pollution Act, 1990 33 U.S.C. 2701 Prohibits the deterioration of surface water quality by requiring Spill Prevention, Control and Countermeasure plans and National Discharge Elimination System, NPDES permits.
CAAA	**Clean Air Act Amendments**, 1970, 42 U.S.C. 7401, amending the Air Quality Act, 1967; amended by the Clean Air Act amendments of 1977 and 1990 Sets national ambient air quality standards in terms of allowable concentrations levels for five criteria pollutants: carbon monoxide, sulfur dioxide, nitrous oxide, particulates, and volatile organic compounds. Sets technology-based standards for reducing emissions. Requires each state to have its own implementation plan to reach clean air goals.
SDWA	**Safe Drinking Water Act**, 1974, 42 U.S.C. 300f, enacted as Title XIV of Public Health Service Act Requires that suppliers of water to residents meet EPA quality standards and prevents degradation to the groundwater. Sets standards for culinary or drinking water and monitors the systems through permits to ensure the water meets the standards.
OPA	**Oil Pollution Act**, 1990, 33 U.S.C. 2701 Primarily resulting from the Valdez accident in Alaska, this act requires a company to certify that it has the ability to respond to a worst case oil spill in navigable waters. The plan must be updated annually and be prominently displayed at all locations where applicable.

- Migratory Bird Treaty Act, 16 U.S.C. 703 (MBTA)
- Endangered Species Act, 16 U.S.C. 1531 (ESA)

In addition to the pollution control acts, the federal government has also passed health and safety regulations that require emergency response plans. These regulations are administered by the Office of Safety and Health Administration.

Each state has a mirror organization at the state level to administer each of the above regulations. Each state government must qualify to operate the federal program on the state level. This requires the state to write implementation plans for federal approval. It requires the state legislature to specifically authorize and fund the state agency to perform a certain function.

In many cases the desires of the individual states and the desires of the federal government are not consistent. This leads to conflict between the state and federal forces (i.e., the federal government wants to push one aspect of regulation and perhaps be lenient on another, but the state wants to push another aspect of the same regulation), often with the business caught in the middle.

In many instances, the state is more stringent than the federal agency. In other cases, the states are not proceeding fast enough in getting *primacy*, or the authorization to run the program; EPA then controls the regulations, monitoring, and penalties for noncompliance. Thus businesses end up serving two masters—in some instances two masters who are at odds with each other.

III. CREATING AN ENVIRONMENTAL REMEDIATION STRATEGY

A. General Environmental Procedure

One good point about the procedure for environmental remediation is that cleanup is basically the same regardless of which act applies—RCRA or the *Comprehensive Environmental Response, Compensation, and Liability Act* (CERCLA). For ease of discussion, the remediation activity will be based on the environmental media affected.

For cleaning contamination in soil or in groundwater, the remediation procedure (Fig. 7.1) includes the following:

1. Report any release of contaminants. A release can be identified by seeing a spill or deposit of waste, finding contamination in groundwater, discovering contamination during excavation, or suspecting contamination exists because of how the site looks or from knowledge of past activities on the site.
2. Investigate the site for contamination. Determine if contamination exists; if so, explore the extent of contamination. Write a report and submit it to the controlling agency for approval.
3. If contamination is found, conduct a further investigation to determine how widespread it is. A written report is provided to the agency for approval.
4. Once the extent of contamination is identified, write a remediation plan and submit it to the controlling agency for approval. The approval process usually includes public hearings.
5. Implement the approved plan according to a negotiated time schedule.
6. After reducing the contamination to acceptable levels, formally close the remediation work.
7. Monitor the site to ensure that it remains clean.

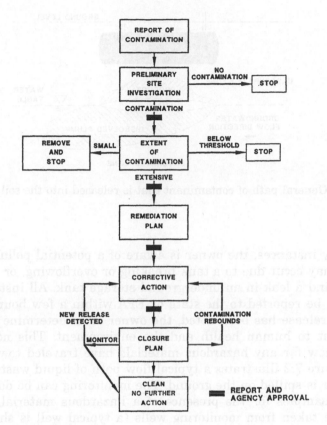

Figure 7.1   Remediation flowchart.

The Environmental Protection Agency and each state establish guidelines and regulations outlining how to complete each of these phases. The agencies also reserve approval rights on each phase. If the agency does not approve the proposed work on a particular phase, the site owner is required to rework the plan or complete agency additions to the phase before it is approved.

For the *Clean Water Act* (CWA) and *Safe Drinking Water Act* (SDWA), EPA now requires the site owner to test the water delivery system for contaminants, such as lead, copper, and chemical constituents. For air issues, there is little historical cleanup to do. The air regulation primarily controls what is being emitted today and will be emitted in the future. Operating systems require permits. The states are required to submit *State Implementation Plans* (SIPs), to show how the state will reach an air quality condition that is safe for the inhabitants. Each industry is required to limit emissions in accordance with the SIP.

B.   Elements of the Environmental Project

Reported releases generally result from readily apparent evidence. A neighbor may find oil in a water well, a city may find gasoline in a sewer system, a jogger may notice a foul smelling material seeping out along a creek bank, a property owner may find he cannot grow grass in his front yard, etc.

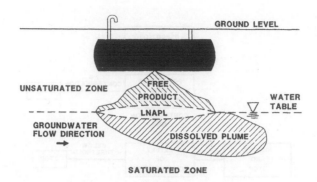

Figure 7.2 General path of contaminant that is released into the soil with groundwater present.

In many instances, the owner is aware of a potential pollution situation. A major spill may occur due to a tank breaking or overflowing, or a service station owner may find a leak in an underground storage tank. All instances of a major release must be reported to the state or EPA within a few hours.

Once a release has been noted, the owner must determine if it could cause endangerment to human health and the environment. This normally requires monitoring how far any hazardous materials have traveled towards a potential receptor. Figure 7.2 illustrates a typical flow path of liquid waste that leaks into the ground or is spilled on the ground. The monitoring can be done by extracting and testing samples for the presence of a hazardous material. These samples generally are taken from monitoring wells (a typical well is shown in Fig. 7.3) that are drilled to intersect the contaminant plume. Data on the samples must be analyzed and understood. The results of the analysis are presented in a report to the state or EPA.

Substantial effort may be required to determine how far the contamination has moved towards a receptor, either in the soil or in the groundwater. Normally the agencies will require evidence that the farthest point of contamination has been found. This means that a clean sample has been taken in each direction.

The type and size of remediation is determined by the analysis of the extent of contamination.

C. Remediation Options

Many remediation options are available, some more successful than others. In order to organize the options, this section will deal first with soil remediation options and then with water remediation options. As indicated, air problems tend to be fugitive and control of pollutants is generally done by process changes. Thus there are generally no remediation problems with air pollution.

1. Soil Remediation

Soil remediation strategies can be divided into two general categories: treatment by removal or treatment in situ (on site).

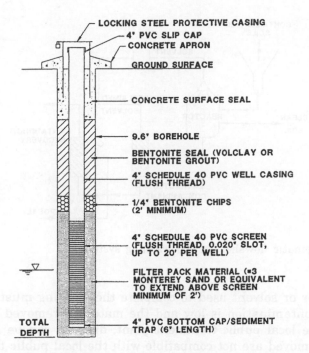

LOCKING STEEL PROTECTIVE CASING
4" PVC SLIP CAP
CONCRETE APRON
GROUND SURFACE
CONCRETE SURFACE SEAL
9.6" BOREHOLE
BENTONITE SEAL (VOLCLAY OR BENTONITE GROUT)
4" SCHEDULE 40 PVC WELL CASING (FLUSH THREAD)
1/4" BENTONITE CHIPS (2' MINIMUM)
4" SCHEDULE 40 PVC SCREEN (FLUSH THREAD, 0.020" SLOT, UP TO 20' PER WELL)
FILTER PACK MATERIAL (≈3 MONTEREY SAND OR EQUIVALENT TO EXTEND ABOVE SCREEN MINIMUM OF 2')
TOTAL DEPTH
4" PVC BOTTOM CAP/SEDIMENT TRAP (6" LENGTH)

Figure 7.3 Construction of a monitoring well. (Construction of other types of wells is similar.)

a. Treatment by Removal. Contamination at a site can be removed by physically excavating the contaminated soil and removing it to a location licensed to receive the material or to remove it to a treatment process that is licensed to treat the material. The excavation is refilled with clean material or the treated material can be replaced if it meets the cleanliness requirements of the state.

The locations that can handle removed soil are licensed landfills. The material may be used for cover. This applies to material that is not considered as hazardous or medical waste. They can also be disposed at licensed landfills where the material is deposited in a disposal area that has double impermeable liners to prevent leaching of pollutants. This is required for hazardous materials that meet EPA established land disposal restrictions.

Hazardous excavated materials may also be treated to stabilize them so the hazardous material does not escape in high enough concentrations to be of concern. These stabilized materials are generally land disposed.

The removed material may also be incinerated at temperatures above 1800°F. This procedure destroys any organic pollutant and generally oxidizes metals to insoluble oxides that cause no problem. A variation of incineration is vitrification where the contaminated material is heated with minerals at high temperature to form glass. The pollutant is thus encapsulated. The encapsulated material does not leach out and has been rendered harmless.

Soil washing is as the name suggests. The soil is excavated, put into a washing facility, and then returned to its original place after washing (see Fig. 7.4 for a schematic diagram of a soil washing system). Often detergents or solvents are needed to remove the contaminant from the soils. A major problem

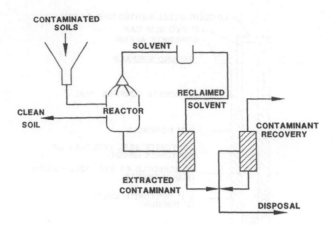

Figure 7.4 Schematic flowsheet of soil washing system.

is that the water or solvent used to complete the washing must be disposed of. If the level of contamination is low and the materials removed by washing are acceptable to the local public treatment plant, disposal can be fairly simple. If the materials removed are not compatible with the local public treatment plant, water treatment may also be needed, driving the cost up very quickly.

b. In Situ Treatment: Volatile or Biodegradable Contamination. Treating the soil in the ground or adjacent to the contamination has obvious economic advantages, but also has very strict requirements imposed by EPA and/or the state. If the contaminant is volatile or subject to biodegradation, there are four basic systems available: land treatment, soil venting, soil washing, and biodegradation in slurry. Permitting is generally required in all cases.

Land treatment involves spreading the soil out in 6–12–inch layers on top of the ground. The areas must be fenced, bermed to control runoff, and signed to keep the general public out. Access to the air will allow some of the volatile and semivolatile materials to escape to the atmosphere, but at a low enough rate as to not impact the air quality. Access to air also enhances the biodegradation that will occur naturally. Biodegradation can be enhanced by keeping the soils moist, adding fertilizer, and tilling the soils occasionally to keep the oxygen level high at the surface of the soils. Land treatment normally takes several months depending on the level of contamination and the cleanup criteria required. It is low cost, requires a lot of land, requires maintenance, and the soil is generally removed after the treatment is completed.

Vapor extraction involves leaving the soil in place. Wells are drilled at designed locations with the well screens being placed at the proper depth to be adjacent to the contaminated soil. Wells are an important part of remediation activities. Monitoring wells (as illustrated in Fig. 7.2) and pump and treat wells are similar in construction. A vacuum is drawn on the wells mechanically. The air is drawn from the wells bringing along the volatile materials from the adjacent soils that naturally vaporize into the adjacent air. Depending on the type and concentration of the volatile materials removed, the emissions may have to be further treated by condensation or by incineration. In low concentrations, the

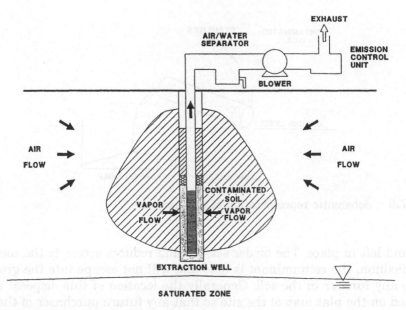

Figure 7.5 Vapor extraction system.

materials are allowed to go directly to the atmosphere. Figure 7.5 shows a complete vapor extraction system.

A second benefit to vapor extraction is that the movement of air in the zone of contamination also increases the oxygen content in the zone. This increases the biological activity in the zone and reduces the overall time of cleanup. Some claim that more remediation occurs biologically than mechanically in this type of system. The system is moderately priced and has no additional waste streams to treat, but does take months to complete the cleanup.

Bioslurry technology involves extracting the contaminated soils, putting them into a mixing container and introducing biological activity or enhancing the natural biological activity by proper chemistry. The system is generally agitated continuously to keep maximum contact with air, fertilized to keep maximum growth, and heated to keep maximum activity. This system works well for heavy contamination and restricted amounts, though some systems have been built that handled several thousands of yards of contaminated material.

c. In Situ Treatment: Nonvolatile and Nonbiodegradable Contamination.
There are two primary methods for treating this type contamination: fixation and extraction. The extraction process is similar to soil washing except that the extraction vehicle is a solvent instead of water. Metals can be removed by acid leaching, or leaching with other material such as cyanide solutions. In these instances the chemicals used in the extraction process must then be handled. Again the disposal problem has only moved from one media, soils, to another, liquids. Generally there must be an economic return from recovering the metal for this system to be justified.

Fixing the contaminant in the soil by use of cement or some other long-term chemical binder has been used very successfully (see Fig. 7.6 for a schematic representing soil fixing). The contaminated soils are treated with 10–20% of the

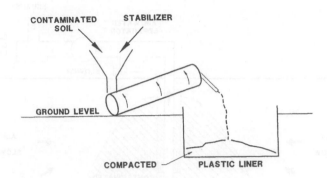

Figure 7.6 Schematic representation of soil fixation.

binder and left in place. The binder sets up and reduces access to the contaminant. In this fashion, the contaminant is held and will not escape into the groundwater or move any further in the soil. Generally the location of this disposal area must be located on the plat map of the site so that any future purchaser of the property knows there is a solidified waste deposit on the land and its exact location.

2. Groundwater Remediation

Treating groundwater is difficult, time-consuming, and costly. If the contaminant is a hydrocarbon, with limited solubility, the contaminant will float on the water as well as dissolve in the water. If the contaminant is metal, it generally has low solubility and must be removed chemically. If it is organic and heavier than water as some chlorinated solvents are, then it almost has to be removed mechanically and treated. There are many variables that can impact the cleanup and they are generally not controllable. Some general categories of remediation technologies are listed next to give the reader some idea of the methods that are used. A more thorough discussion of methods can be found in most textbooks on remediation. The methods used most are pump and treat, vapor extraction, air sparging, air sparging in conjunction with vapor extraction, and bioremediation.

Pump and treat has been used most historically, but has not been able to clean up the contamination to acceptable levels. Experience suggests that only about 30% of the contamination can be cleaned up with this method. Its main value is in hydraulic control to keep contaminated water from flowing onto other property.

Pump and treat systems involve installing wells around and in the contaminated area. Pumps are installed and the water is pumped to the surface. This creates a cone of depression in the aquifer. If there is free hydrocarbon on top of the groundwater, this cone allows free hydrocarbon to accumulate in greater depth at the cone. A pump can then remove the free water and product into a tank (see Fig. 7.7 for a schematic diagram of a pump and treat system). An alternative is to place two pumps into the well. The first pump causes the water cone and pumps only water. The second pump removes only free product.

The water removed by the pump is contaminated and must be treated before it is disposed. Treatment can be with air strippers that will remove the contaminant if it is volatile or semivolatile. It can be a rapid sand filter or settling basin if it can be chemically treated and precipitated. It can be in an aerated lagoon

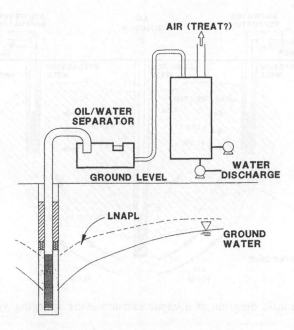

Figure 7.7 Schematic flowsheet for pump and treat system.

or basin if the material is biodegradable. Again, treatment of the extracted material becomes a major problem that generally is expensive.

Vapor extraction systems are installed in the same way as for soil treatment. Free hydrocarbon floating on the groundwater will volatilize and be removed with the air. If there is no free product some of the dissolved contamination will escape into the air and be removed. The enhanced biological activity in the soils adjacent to the water table will remove some of the dissolved contamination also.

Air sparging involves pumping air into wells that have been drilled into the groundwater table. Sparging may be used alone or it can be used in conjunction with vapor extraction. Wells are strategically located to introduce air into the water table and to remove it. As the air bubbles up through the water table, it picks up the dissolved contaminants. Also, as the bubbles come through the contamination layer on top of the groundwater, they pick up product. The air and vapors are extracted by natural convection or by the vacuum. The increased oxygen available increases the biological activity both in the water and in the soils adjacent to the water. See Figure 7.8 for a complete air sparging/vapor extraction system.

Bioremediation of the groundwater in place is accomplished by increasing the oxygen concentration in the water and adjacent soils. In rare instances the naturally occurring biological activity must be augmented by inoculation of addition bacteria that have been developed specifically for the contaminant in the water. In most instances the native bacteria is already acclimatized to the surroundings and are best able to neutralize the contaminant. In many instances the natural biological activity can be enhanced by adding nutrients that are used as food. These nutrients are generally readily available in the form of commercial fertilizers. Chemical analysis of the aquifer conditions can determine the need for oxygen and nutrients.

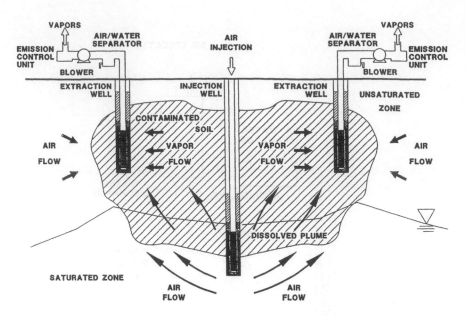

Figure 7.8 Schematic diagram of a vapor extraction/air sparging system.

D. Controlling Cost of Corrective Action Requirements

Control of remediation costs starts very early in the process. Cost savings begin with the studies that are completed to define the nature and extent of the contamination. Cost saving methods are discussed for each of the phases of the remediation process.

E. Controlling Cost Through Front-End Management

Remediation activities are almost always done with the approval of the responsible agency, EPA or state. Normally a remediation plan is submitted and approved before any activity is allowed to proceed. The plan that is approved controls what will be required and must be negotiated with great care.

   In most instances, the agency will follow the guidance manuals prepared by EPA. These guidance manuals have been developed by regulators who have not had much actual experience or by consultants who want to be sure they have covered all the possible conditions. Accordingly, the guidance documents tend to be very conservative and will overkill rather than miss something. It is wise to try to negotiate a scope of work that will accomplish the necessary objectives with the minimum amount of expense. Some specific examples are discussed below.

F. Managing Sampling and Analytical Expenses[*]

During the study phase of a remediation project, the analytical costs can be 40% of the total cost. Therefore the number of samples taken and the amount of

---

[*] Cost figures cited in this chapter are 1995 costs.

chemical analysis required on each sample are very important. Also the method of taking the samples can make a great difference. Drilling wells will cost between $50 and $75 per foot. Alternatives are pushing a probe into the aquifer to take a sample and then removing the probe. No further sampling is required and the cost is about 25% of drilling. Another option is digging a hole with a backhoe to test groundwater. It is much less expensive than drilling. Also, the geotechnical analysis is more complete and easier as it comes from visual observation of the complete area rather than visual observation of core samples.

If there is a well on the site, the regulator will want it sampled, generally on a quarterly basis and occasionally on a monthly basis. If there is no well, no sampling can occur.

The cost to perform an analysis of volatile materials such as benzene, ethyl benzene, toluene, and xylenes is about $100. The cost to test eight metals normally required for RCRA investigations is about $100. These two samples will tell about all that is needed initially for hydrocarbon contamination. The agency people are told in the guidance documents to require a complete Appendix IX chemical analysis; this includes about 222 chemicals that EPA has designated for testing in groundwater monitoring. This will cost about $1,000 per sample. If the requirement from the agency also requires *toxicity characteristic leaching procedure* analysis of the soils, the total cost will be about $2,000 per sample.

With this in mind, the number of sampling sites, the method of obtaining the samples, the types of chemical analysis required, and the type and frequency of sampling over the term of the study and remediation are very important. It is not uncommon for the agencies to approve a lighter sampling and analysis plan after the initial sampling is done. For example, they may require an Appendix IX analysis for the first sampling event and allow tests the next three quarters for only those contaminants found in the first round or expected because of the nature of the process causing the contamination.

The agency may require the complete chemical analysis again at the start of the second year. Following these sampling events, they may allow analyzing only for the contaminants found for the rest of the sampling and analysis work. This approach can reduce the cost of analytical work to 25% of the cost recommended in the guidance manuals.

G. Writing Reports

Historically the reports sent to the agencies have been very large and very involved—and thus very expensive. The agency is no more excited about reading these large documents than the author is about writing them. In many instances, the original plan of action includes much background information on the site and the contamination. This background information may cover geological information; surface water and groundwater location and use; location of receptors and pathways to reach them; previous work at the site; potential to emit contaminants; and general information.

The guidance manuals suggest that even though this information does not necessarily change, it should be included in each subsequent report. Discussions with the agency people during the acceptance of the plan of action can often lead to reductions in report size by referencing the previous work. This is especially true with data that has been generated and reported previously.

The final report for a major study can be an accumulation of reports written after each phase of the study. Normally the agency will require interim reports outlining the progress of the work. These interim documents should be written so they become various chapters of the final report thus reducing dramatically the cost to assemble the final report.

The use of electronic methods to report data is becoming more widespread. Baseline data used in assessing cleanup or in establishing permit conditions can be reported by disc rather than written page. Eventually, extensive volumes of data will be handled by computer disc.

H. Writing Permits

The same comments apply to permit applications as to report writing. The guidance documents require many sections that give the background and previous history and data of the site to be included in the permit application. Discussions with the agencies can lead to allowing references to documents rather than reproducing them again in the permit application. This can reduce a six-volume permit application to one or two volumes.

I. Providing Financial Assurance

Providing financial assurance for the cleanup or permitting activities is a major task that few people realize must be completed. The financial implications of the financial assurance requirements are very large. The Environmental Protection Agency uses multipliers in determining how much financial assurance is necessary. Normally the cost of the study, cost of remediation, and cost of postclosure care are required to be assured by some financial instrument. The Environmental Protection Agency uses a multiplier of the actual cost to determine the amount of the instrument. This multiplier can be six times actual cost. This multiplier is to cover the guarantees that fail or to cover costs that were underestimated.

In many instances, the consultant will establish the cost to complete a study or to clean up a site based on literature data or rules of thumb. These costs are always high in order to cover the difficult conditions that can occur. It is critical to review the cost estimates proposed and reduce them where ever possible as the savings will be multiplied.

IV. ESTIMATING THE COST OF REMEDIATION

The cost to perform remediation of a hazardous release is extremely site-specific. It depends on the quality of the negotiations completed during the assessment phases of the project, specifically during risk assessment. While conserving costs during the assessment phase is important, an incomplete survey and analysis can lead to major unnecessary work with its attendant costs.

For example, an oil company determined the extent of contamination at a site and estimated cleanup at $3,000,000. During the cleanup work, they discovered another area, larger than the original area that had to be cleaned up. The company had already agreed with the state agency to excavate and remove the contamination. When they completed the total excavation the cost had escalated to $17,000,000. Had they known the full extent of contamination at first, they would have chosen an in-situ stabilization approach or a no-action approach.

A. Cost Estimates for Simple Sites (Less than One Acre in Size)

Simple sites can be defined as those areas smaller than one acre. These sites may be service stations or underground storage tanks. Such sites tend to be similar and not very complicated. In these simple sites, there may be two to three levels of assessment. Assessment costs generally run $2,000–5,000 for the initial assessment and $20,000–25,000 for the verification assessment (which uses 4–6 ground monitoring wells). If additional *extent of contamination* work is needed, it may require additional wells and will likely cost between $15,000–25,000 more. By the end of the assessment phase, the best remediation method will be determined and the total cost of the cleanup can be estimated.

Wells for product recovery and/or monitoring usually cost $50–75 per linear foot of depth. Vapor extraction wells or soil venting wells will be a little more expensive, generally $60–80 per linear foot of depth. If air sparging is added to the system, more wells are needed and air sparging will generally add an additional $20 per well for existing wells. About $100 per foot is a typical cost for a new combination vapor-extraction, air-sparging system.

Analytical work for simple sites such as Leaking Underground Storage Tanks can generally be limited to those analytes anticipated either due to process knowledge or previous analytical data. Testing for any additional chemicals is expensive and adds nothing to the study. Table 7.2 gives examples of analytical costs that are usually involved in remediation activity. Installation of pump and treat systems generally costs about $16,000 per operating well. Operating costs generally run about $1,500 per well, per year.

Installation of vapor extraction with no air treatment will cost about $4,000 per well. The cost for addition of air treatment to vapor extraction will range between $40,000 and $100,000 per site. The annual operating costs for vapor extraction and air sparging are generally about the same and range between $1,200 and $1,500 per well, per year.

Cleanup costs at smaller sites will generally run from $10,000 (no remediation) to $500,000 (major remediation) and average about $150,000.

B. Cost Estimates for Sites Larger Than One Acre

Table 7.3 compares the various elements of a remediation project for medium- and large-sized sites with simple sites. The estimate should include all of the

Table 7.2 Analytical Costs

Method of analysis	$US/sample (1995)
Full TCLP	1,800
Appendix IX	1,200
Skinner's list (modified Skinner's list)	800
BETX	90
BOD/COD/TSD	50

Table 7.3 Estimating Remediation Costs ($1000 US)

Phase of remediation	Simple	Medium	Large
Assessment			
Initial	2–5	5–15	20–50
Second	20–25	50–75	100–200
Extended	15–20	40–60	100–150
Pilot test	15–20	15–20	20–40
Analytical			
Initial	1	3	5
Second	10	30	50
Extended	10	30	50

elements shown. The following rules of thumb can be helpful in estimating the cost of the remediation.

- The assessment will involve determining the extent of contamination, which will likely involve monitoring wells. Establish enough to locate all potential releases. The wells will cost the amounts shown previously.
- Analytical work will generally run about $2,000 per well for the first sampling event. Later events will be less. Knowledge of the site history and the processes on the site can help establish the possible analysis. From this cost laboratory work costs can be estimated.
- The cost of consultants is generally 50% of the total cost of the work in the assessment phase. With the number of wells estimated and the laboratory costs estimated, the total cost of the assessment phase can be determined.

C. Current Remediation Costs for Various Disposal Options

Table 7.4 is a summary of the cost of disposal in today's market. The trend in cost of disposal has been declining as the industry becomes more competitive with more disposal sites and available disposal ideas. Get price information from several sites before deciding on the contractor or the method. Most disposal systems will negotiate on price.

Once the extent of contamination and the components of the contamination are understood, the remediation methods can be evaluated and cost estimates made.

D. Selecting Disposal Sites and Methods

Of major importance is the selection of the disposal site if one is used. There are ample case histories of improper disposal at sites that required the original suppliers of the hazardous waste to clean up the site. A company in Utah was accepting waste oil, solvents, and other volatile organic compounds for disposal. The site was licensed with the state so everyone using the site felt it must be okay. The truth was that the hazardous material was not being handled according

Table 7.4 Unit Cost for Various Remediation Options

Option	1995 unit cost ($)
Incinerate:	
Liquid wastes	$0.60/lb plus shipping and analytical costs
Solid wastes	$1/lb plus shipping and analytical costs
Used as fuel for power generation	$0.22/lb plus shipping and analytical costs
Excavate and move	$35/ton for nonhazardous materials, plus shipping and analytical costs
	$100–$300/ton for hazardous materials that meet land disposal regulations, plus shipping and analytical costs
Vitrify	$1/lb plus shipping and analytical costs
Stabilize in place	$60–$100/ton plus analytical costs

to regulations. The site was finally shut down by EPA. The disposal company went broke, the owner was indicted, and there was no company money to clean up the site. The Environmental Protection Agency required the site to be cleaned up under CERCLA jurisdiction. As a result, each company that deposited anything there is legally responsible for its share of the remediation costs. With the addition of legal and attorney fees, the problem became very expensive.

Before selecting a disposal site, it is crucial to evaluate the site carefully. The evaluation should include but not be limited to the following:

- Contact the state and EPA about condition of the disposal site's license. Ask if the site has had any violations in the last 3–5 years.
- Check the financial strength and insurance of the site in case of failure to perform.
- Visit the site and perform a personal audit or evaluation to ensure that the site is well managed and in compliance.
- Negotiate a contract that gives as much protection as possible.

E. Estimating Closure Costs

Estimates of closure costs are also site-specific. Closure usually involves capping the waste in place, stabilizing in place, or a combination of the two. Table 7.4 gives the cost to stabilize. The cost of capping can be determined by classical civil engineering estimating once the permeability of the cap has been decided. Engineering costs for closure activities usually run about 20% of the total cost due to the need for certification of the closure and the extra supervision required.

F. Estimating Postclosure Costs

Postclosure care for sites that have affected the groundwater can go on for 30 years. If the closure is without impact on the ground or groundwater, then the sampling is usually completed in less than 5 years. If the site is in-between, then the monitoring can go on for several years.

The costs of postclosure are analytical, reporting, and maintenance of the site. Again, these are site-specific. Usually the analytical costs can be reduced over time as the results become more predictable. It is important to negotiate a reduced testing load as time goes on. Analytical costs generally run about $2,000 for each well at the beginning of the postclosure period. Sampling for the first year will probably be quarterly. After a site history is developed, sampling can be reduced to semi-annual and then annual periods. The analyte list can also be reduced to bring laboratory costs to less than $500 for each well.

The high cost of postclosure care is great incentive to remove the contaminated materials from the site. New waste treatment facilities should be built in such a manner as to *clean close* them. Clean-closed sites require only one year of monitoring, whereas other sites that leak contamination need to be monitored for several years.

V. CONTROLLING COST IN REMEDIATION

A. Site-Specific Risk Analysis and Cleanup Standards

Before selecting a remediation method, an understanding and agreement on the risk posed by the site should be reached with the agencies. Always insist on a site-specific risk assessment to determine the risk.

Many states do not understand risk assessment or what it means. They have adopted a single set of cleanup standards based on literature values or research done in other parts of the country which have little bearing on any specific location. This may drive the cleanup to a level higher than is necessary and on a time frame that is unnecessary.

Risk assessment is a process to review the nature of the contamination, understand its pathway, recognize the potential receptors of the contamination, and determine how much of an increase of the contaminant there is to the natural background which may increase the harm to human health and the environment.

Cleanup levels for contamination are generally driven by the potential contamination of the groundwater. Contaminated soil must be cleaned to a level that will keep the contaminant from leaching to the groundwater.

Once the water is contaminated, the cleanup level is generally driven by the use of the water. In most instances, the standard will be drinking water standards or maximum contamination levels. These are very high standards and in many instances cannot be reached very quickly or can only be reached with the help of dilution.

Therefore, cleanup standards may need to be based on what the technology can deliver rather than some other standard. One example of this approach is to treat groundwater until there is negligible improvement in the aquifer. At that time, the cleanup system can be shut down for a while, operating it again as needed until no further improvement occurs. It is important to have a way to trigger no further cleanup action required by the agency. Otherwise the cleanup can go on and on with no real improvement to the environment.

Negotiate with the agencies the point of compliance for the remediation. Try to establish the compliance point at the edge of the property so that the critical cleanup is from the property line out. If there is low risk outside of the property, the agency may allow use of nature to help without causing harm to human health or the environment. The remediation activity that occurs on the property can then be integrated with other normal operational activities and dramatically reduce the cost.

Once the above activities are complete, the remediation method for the site can be selected. The time frame required, the points of compliance, and the levels are now known. If time permits, the remediation may be implemented in phases. For example, hydrocarbons may exist as a free product and as dissolved constituents in the groundwater. If the groundwater is not moving very quickly, a vapor extraction system may be implemented first to see how it performs. If the cleanup is too slow, the system can be enhanced by introducing air sparging. If vapor extraction will do the job alone, why include air sparging or a skimming pump? The remediation method can be simple and still accomplish the task.

B. Selecting Contractors

The same principles used in selecting contractors for normal construction work apply for seeking contractors to perform studies, design and install remediation systems, and run analytical tests. In the case of environmental work, additional sets of safety issues must be addressed, especially if the hazardous materials are highly toxic. Remediation work may require special and expensive clothing, respirators, and equipment. If the contractor is not sure what will be required, the bid price may be increased to cover these unknowns. Therefore, invite the contractors that bid on the project to see the site and have a clear understanding about the work.

The whole market for environmental work has developed much slower than most expected. There are several *treatment, storage, and disposal* (TSD) facilities that were built expecting very large projects and large amounts of material to be handled. They expected their services to be very important and unique, and therefore charged high prices to cover the many unknowns they faced. Because of good management by industry, these markets have grown slowly and are very competitive. The costs to use TSD facilities continue to fall. It is important to compare options and find the best fit for the company.

The cost of professional services are more likely to be negotiated now than a few years ago. The conditions change, but by discussing the project with a contractor and then working with the consultants to define a scope of work that fits the needs, costs can be reduced. The cost can be negotiated based on the work and the time frame needed to accomplish the work.

C. Controlling Cost in Closure and Postclosure

Part of the remediation work may be closing sites that were used in the past to dispose of hazardous materials. The purpose of the closure is to stop any hazardous material that may be deposited on the site from leaching into the groundwater or from contaminating adjacent soils.

Regulations require that the cost of performing the site closure be estimated as if an outside contractor would do the work. This will ensure that if the cleanup must be managed by an agency, the agency will have adequate monies to complete the work. Unfortunately, many consultants perform the cost estimates using standard methods and rules-of-thumb rather than dealing with specific sites. This leads to excess costs for mobilization, per diem expenses, earth moving, etc. The cost hurts twice as it is used to determine the size of the financial instrument used for financial assurance.

D. Renegotiating Closure or Postclosure Scope

Postclosure care is required for most sites, even in facilities that continue to operate. If a hazardous disposal site is taken out of service, or if the agency requires a site to be closed according to regulations, a postclosure care plan will be required. Normally this requires a series of monitoring wells be placed around the closed site to determine if a release of hazardous materials occurs from the site in the future. The monitoring frequency and the chemicals monitored become critical to the cost and to the amount of financial assurance needed.

The location of the facility and the site-specific risk assessment should be central to renegotiating closure and postclosure activities. Agencies can be persuaded to follow a less aggressive program or allow leniency in the time frame to reach cleanup standards when they are convinced that the project is being well-managed and is under control. Closure and postclosure plans need to emphasize the management plan and how it will keep the site from becoming a hazard to human health and the environment in the future.

E. Renegotiating Frequency of Sampling

After the first few years of monitoring, a reduced level of sampling and a reduced chemical list can be negotiated with the agency, based on one or two years of data that show that nothing is happening. If the agency will not agree in the initial plan, sample for one or two years, gather good data, and obtain a revision of the postclosure care plan to reduce the level of analytical work required.

VI. NOMENCLATURE

Appendix IX	A list of approximately 222 chemicals designated by the Environmental Protection Agency to be tested for in groundwater monitoring
BETX	Benzene, ethyl benzene, toluene
BOD	Biological oxygen demand
CAAA	Clean Air Act Amendments
CERCLA	Comprehensive Environmental Response, Compensation, and Liability Act (Superfund)
CFC	Chlorofluorocarbon
COD	Chemical oxygen demand
CWA	Clean Water Act
EPA	Environmental Protection Agency
EPCRA	Emergency Planning and Community Right-to-Know Act
ESA	Endangered Species Act
FEPCA	Federal Environmental Pesticides Control Act
HMTA	Hazardous Materials Transportation Act

HSWA	Hazardous and Solid Waste Amendments to RECRA
LUST	Leaking underground storage tank
MCL	Maximum containment level
Modified Skinner's list	Skinner's list of chemicals reduced to less than 55 chemicals
MBTA	Migratory Bird Treaty Act
NEPA	National Environmental Policy Act
NPDES	National Pollution Discharge Elimination System
OSHA	Occupational Safety and Health Administration
OPA	Oil Pollution Act
PCB	Polychlorinated biphenyl
RCRA	Resource Conservation and Recovery Act
SARA	Superfund Amendments and Reauthorization Act
SDWA	Safe Drinking Water Act
Skinner's list	A list of 55 chemicals specific to hydrocarbon contamination
TCLP	Toxicity characteristic leaching procedure
TSCA	Toxic Substance Control Act
TSD	Treatment, storage and disposal
UST	Underground storage tank

8

Contracting—Front-End Risks, Key Contract Administration, and Cost-Schedule Considerations

I. GENERAL OBJECTIVES

It is essential that contract administration and management result in reducing risks, maximizing cost savings, minimizing claims, and improving economic return. These results can only be achieved through effectively managing contract risks by developing tough but fair contract documents, engaging in aggressive negotiating practices, and using outstanding communication skills.

The process of reaching a contract requires a specific sequence of steps. In taking these steps, the project manager must make a series of choices between priorities for project objectives, degrees of risk to be assumed by the contracting parties, control over project activities, and the cost of achieving selected goals. This process must first be fully understood by the project manager, then be tempered by experience, and finally be expanded into the ability to reach a contract through the exercise of negotiating and communications skills.

II. WHAT IS A CONTRACT?

A contract is a mutual business agreement recognized by law under which one party undertakes to do work (or provide a service) for another party, for a consideration. Contracting arrangements cover such subject as:

- contract conditions,
- commercial terms and pricing arrangements,
- scope of work (technical), and
- project execution plan.

III. WHY HAVE A CONTRACT?

A written contract is the document by which the risks, obligations, and relationships of all parties are clearly established, and which ensures performance of these elements in a disciplined manner. For the owner, the contract is the means by which the contractor can be controlled, and ensures that the work and end product satisfy the owner's requirements. For the contractor, the contract specifies risks, liabilities, and performance criteria, and outlines the terms and conditions of payment.

IV. PARTIES TO THE CONTRACT

Most projects are executed under a three–party contractual relationship among:

- the owner, who establishes the form of contract and general conditions;
- the engineer, who can have the following three roles:
    - designer: carrying out the detailed engineering work and purchasing equipment and material on the owner's behalf,
    - arbitrator: acting as the owner's agent in administering the contract, and impartially deciding on certain rights of the parties under the contract, and
    - project manager: handling design, procurement, and construction or construction management/services; and
- the contractor, who responds to the risks and liabilities of the general conditions.

The normal contractual relationship among these three parties on a single project is for the owner to have one contract with the engineer for design, procurement, and other services, and a separate contract with the contractor for the construction work. No contractual relationship exists between the engineer and the contractor. This is usually referred to as a *divided* or *split responsibility* arrangement. In an alternative arrangement called *single responsibility*, a general contractor is awarded total responsibility for engineering, procurement, and construction. This is known as an *engineering, procurement, and construction* (EPC) turnkey or *design and build* contract.

The project manager must carefully decide on a specific contracting arrangement as outlined in the "Contracting Strategy" section.

V. CONTRACT RESPONSIBILITY

The project manager should be responsible for the contract strategy that is developed as part of the project strategy. However, the proposed division of work, contracting arrangements, forms of contract, and bidder's lists should be developed in conjunction with the company's contracts department and the engineering/construction groups.

In the contracting process, dividing responsibility between the project manager and the contracts department can lead to inefficiencies, disagreements, and delays, since the organizational conflict can have a negative impact on the project cost and schedule. Close coordination and effective communications must exist among all groups to ensure complete agreement and commitment to the proposed contracting program. This is particularly important in all submissions to contract committees and/or senior management.

The project manager must obtain agreement from the company's legal, contracting, and insurance departments before committing to contractual language regarding liability, indemnity, or insurance.

VI. ASSESSING RISK AND COST LIABILITIES

Engineering and construction contracts can be drawn up in any number of formats, depending on the project objectives, the skills and resources of the company, and the skills, resources, and financial resources of the contractor. The most successful contracts have at least one fundamental in common: thoughtful and thorough preparation by the company before the contract is let. Figure 8.1 depicts a flowchart of typical contracting arrangements and illustrates the major steps in this process.

Project complexity, changing and increasingly costly legal/insurance requirements, and difficult business environments mandate a correct contracting arrangement, even though contracts must be made early in the life of a project. Yet failure to institute quality front-end planning is common and often leads to poor contracting arrangements. To develop proper contracting arrangements while simultaneously providing for the risks of uncertainties, gaining improved performance, and promoting innovation, is a major challenge for both company and contractor.

Risk is defined as the possibility of financial loss or personal injury. It can mean delay in schedule, with a resultant loss of market and poor quality of engineering and construction. Both parties must approach contracting with the aim of meeting their respective goals, recognizing the interests of the other, and allocating responsibility for risks in according with the ability of each to control or minimize those risks.

All projects have at least three goals: achieving the most economical (but not necessarily the cheapest) cost, quality, and schedule. These goals are not always fully attainable in any one contract, so compromise is usually necessary to achieve a balance. The contractor needs to make a profit on the contract, but may well have other goals, such as long-term needs for survival, growth, a greater share of the market, and to keep competitors out.

Allocation of risk must be carefully considered. It should be based on the degree of liability, the potential profit from proper risk management, incentive provisions to perform the contract more effectively, and the relative ability of the parties to protect themselves against risks. Naturally, risk is a two-way street, since exposure to economic loss is balanced by possible extra gains through proper risk management.

The underlying philosophy of any contracting program should be that the contract provides the means to manage and allocate risk. The contracting strategy needs to be built around a recognition of the relationship between contract terms and conditions, and the accompanying risks and cost impacts. The type of contract and specific language should flow from this analysis.

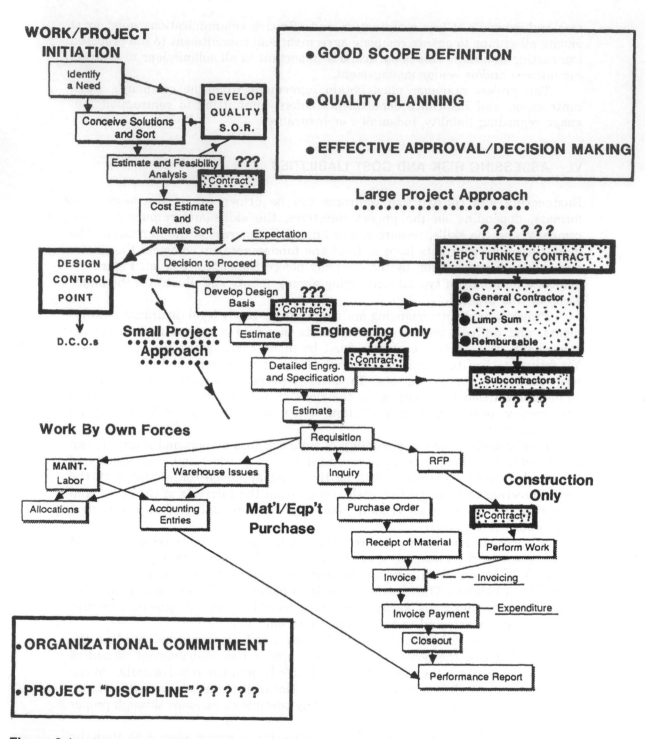

Figure 8.1 Contracting arrangements.

Failure to manage risk can result in project cost overruns of a catastrophic magnitude. One study of major U.S. owners and contractors concluded that contracting practices alone can result in cost impacts equivalent to 5% of the total project costs. These impacts can be positive or negative, and, in a buyer's market, could be even greater than 5%.

In a competitive and inflationary economy, project managers must become increasingly risk conscious in order to survive. They need to be able to recognize risk elements, understand risk accountability, know the capabilities required to manage risk, and be able to properly share risk through adequate contractual arrangements.

Since the overall financial responsibility for any project rests initially with the company, it is logical that the company take the lead in risk evaluation. This activity should cover every element of the engineering and construction process that can be characterized as a potential cost. Figure 8.2 contains examples of typical potential risks requiring contractual coverage.

Initially, all risks are the company responsibility. It is only after a contract is consummated that some of the risks become the contractor's responsibility. It is therefore important that the project manager understand how accountability varies with the type of contract. For example, the cost impact of labor productivity is the contractor's risk in a fixed-price contract, yet it is the company's risk under a reimbursable arrangement. Figure 8.2 also identifies accountability for various risk elements under fixed-price and reimbursable contract arrangements.

After identifying the risks associated with a particular project, the project manager must decide which risks the company should assume. If experienced personnel are available, the project manager may decide to use a reimbursable contract for schedule reasons. Likewise, if the workload is high or the company does not have sufficient experience or resources, the project manager may select a fixed-price contract, which will transfer the risk to the contractor and use the contractor's expertise. The main objective is to match capability and accountability with project objectives and to avoid the mistake of assuming risk accountability without the resources for effective risk management.

It is important that both company and contractor understand and accept the concept that effectively managed projects should reward both parties. Contracts that facilitate risk management benefit the company through lower costs and shorter schedules. Acceptance of accountability and superior risk management by the contractor should result in additional profit.

VII. PROJECT EXECUTION STRATEGY

An essential decision in execution strategy is whether to contract and to what extent. Before developing an execution plan, the scope of work must be established. This starts with identifying the type of facility, the required operational date, desired project life, reliability, necessary supporting facilities, a statement of the scope of work, and a preliminary work sequence can be developed for use in preparing a bid package.

A prerequisite for achieving the best contract is the project manager's commitment to thorough project planning. It is crucial that all project support groups be represented in the planning so that all constraints and trade-offs are considered.

Labor Productivity / Scope / Indirect Costs / Quality Construction

	Cost Accountability		Cost Impact to Owner	
	Fixed Price	Reimbursible	Fixed Price	Reimbursible
Labor Productivity			L	H
a. Management of workforce	C	O		
b. Timing and quality of engineering data and equipment	O/E	O/E		
c. Quality assurance	O	O		
d. Quality control	C	O		
Scope			H	L
a. Initial scope definition	O/E	O/E		
b. Changes in scope	O/E	O/E		
Indirect Costs			L	H
a. Staff	C	O		
b. Consumables	C	O		
c. Support crafts	C	O		
d. Materials management	C	O		
Quality Construction			M	M
a. Complexity of design	O/E	O/E		
b. Completeness of engineering drawings	O/E	O/E		
c. Construction procedures and methods	C	O/E		
d. Construction schedule	O/E	O/E		
e. Experience of crafts	C	O		
f. Training of crafts	C	O		
g. Supervisory personnel	C	O		
h. Construction equipment and tools	C	O		
i. Quality control procedures	C	O		

Safety / Schedule / Labor Relations / Project Management

	Cost Accountability		Cost Impact to Owner	
	Fixed Price	Reimbursible	Fixed Price	Reimbursible
Safety			M	M
a. Training	C	O		
b. Contractor's minimum standards	C	O		
c. Owner's mandatory standards	O	O		
d. Regulatory standards (OSHA, etc)	C	C		
e. Industrial hygiene	S	S		
Schedule			H	H
a. Manufacturer's promised deliveries	C	O		
b. Owner-supplied material	C	O		
c. Contractor-supplied material	C	O		
d. Personnel resource	C	O		
e. Personnel productivity	C	O		
f. Scheduling techniques	C	O		
g. Schedule duration	O	O		
h. Extended overtime or shifts	O	O		
Labor Relations			L	L
a. Jurisdictional disputes	C	O		
b. Illegal strikes and walkoffs	C	O		
c. Contract expiration strikes	C	O		
d. Jurisdictional disputes between contractors	C	O		
Project Management			L	L
a. Adequate design drawings	O/E	O/E		
b. Timely procurement and delivery of materials and equipment	O/E	O/E		
c. Limitation of number of changes and revisions to drawings and specifications	O/E	O/E		
d. Quality of fabrication of materials and equipment	O/E	O/E		

LEGEND

C - Contractor
O - Owner
E - Engineer
S - Shared
L - Low
M - Medium
H - High

Figure 8.2 Cost impact of controllable risks.

VIII. CONTRACTING STRATEGY

A. Introduction

Once the decision is made to contract, there is a wide variety of single or multiple contracts from which the project manager may select a contracting strategy. These fall into two major categories: *fixed-price*, in which the contractor has primary cost responsibility, and *cost-reimbursable*, in which the company has primary cost responsibility. Variations and combinations of these two types can be formed, depending on the degree of risk assumed by either party. The three objectives of cost, time, and quality must be placed in an appropriate priority, since tradeoffs will probably be necessary in deciding on the specific contract.

If the contractor is to assume direct cost responsibility, a fixed-price contract is appropriate. However, the total project time is usually longer with this type of contract, since the project drawings and specifications must be more complete before bids are solicited. The bidding time is longer as well.

Cost risks must be balanced against the need for speed—an increasingly important aspect if borrowed funds are being used to finance construction in an era of high interest rates. With a cost-reimbursable contract, it is more difficult to predict the final cost, but shorter construction schedules can usually be a-chieved. The schematic drawing contained in Figure 8.3 illustrates the wide range of schedule tradeoffs that project managers must consider in formulating a con-tracting strategy. The project manager can also specify the particular quality objectives or other performance goals that are desired. Under either type of contract, the level of quality must be established through the specifications.

In addition to the company's goals, the contractor's objectives should be considered. On large and long-term projects, for example, a contractor is reluctant to accept the risk of a fixed-price contract. On the other hand, if there is vigorous competition for work due to economic or other reasons, contractors may readily accept fixed-price contracts, which should result in lower bid prices.

B. Summary of Contract Strategy

As covered in the project strategy, the following are major considerations when developing a contract strategy for the project:

- When and how will the work be divided up? Should it be EPC turnkey or separate EPC?
- How will the division of work affect client/project team/main contractor/ven-dor/subcontractor interfaces? This division enables the project coordination procedures to be properly prepared.
- What type of contract should be used? Segment the project into discrete work packages to facilitate management, and subject the work packages to available resources. Consider the contract philosophy, the type of contract best suited to the project, contract interfaces, bid evaluation techniques, and bid documentation. This enables the contract strategy to be produced in liaison with the contracts department.
- What roles are licensors and consultants expected to play? This allows arrangements to be made for prequalifying suitable contractors, issuing invitations to bid, evaluating bids, and making award recommendations.

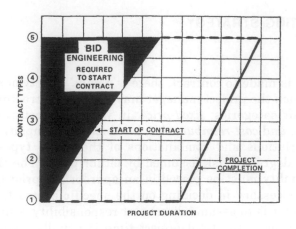

1   COST REIMBURSABLE W / % FEE
2   COST REIMBURSABLE W / FIXED FEE
3   TARGET PRICE
4   GUARANTEED MAXIMUM PRICE
5   LUMP SUM FIXED PRICE

Figure 8.3   Project schedule duration versus type of contract.

- Are there potential conflicts of interest with other owner projects in contractor offices, vendor workshops, or fabrication yards? Such conflicts can have an impact on the bidder's list.
- What is the availability of skilled labor? What is the local industrial relations climate at fabrication yards and the construction site? Lack of labor can delete a contractor from the bidder's list.
- What is the quality and availability of personnel to develop, evaluate, and administer the required type of contract/contract conditions?
- What is the total financial risk to bidder's financial strength?

IX.   CONTRACTING ARRANGEMENTS

Three principal types of contracts exist: reimbursable, measured (unit price), and lump-sum. The following forms of contract are typical of these types:

- cost-reimbursable (time and materials),
- cost-reimbursable with percentage fee,
- cost-reimbursable with fixed fee,
- cost-reimbursable plus cost/schedule bonus and/or penalties,
- measured unit price (mostly construction),
- guaranteed maximum price, and
- lump-sum/fixed-price.

The objectives of cost, time, quality, risks, and liabilities must be analyzed and prioritized, since tradeoffs will probably be necessary in deciding the type of contract to be used.

A. Reimbursable-Cost Contracts

These require little design definition but need to be drawn in a way that allows expenditures to be properly controlled. The major advantage of a reimbursable-cost contract is time, since a contract can be established during the early stages of a project. This type of contract presents a disadvantage to an owner, however, since poor performance by the contractor can result in increased costs and because final costs are the owner's responsibility. Further, the final or total investment level is not known until the work is well advanced.

Reimbursable-cost contracts can contain lump-sum elements such as the contractor's overhead charges and profit, which is usually preferable to calculating these costs on a percentage basis. Reimbursements may be applied to such items as salaries, wages, insurance and pension contributions, office rentals, and communication costs. Alternatively, reimbursement can be applied to all-inclusive hourly or daily rates for time spent by engineers on the basis that all office support costs are built into these rates. This form of contract is generally known as a fixed-fee or reimbursable-cost contract, and can be used for both engineering and other office services as well as for construction work.

Such arrangements give the owner greater control over the contractor's engineering work, but reducing the lump-sum content of the contractor's remuneration also reduces its financial incentive to complete the work economically and speedily. Further, it lessens the owner's ability to compare and evaluate competitive bids, since only a small percentage of the project cost is involved. Finally, it is possible that the best contractor may not quote the lowest prices.

1. Requirements

- competent and trustworthy contractor,
- close quality supervision and direction by the owner, and
- detailed definition of work and payment terms covered by lump sums and by all-inclusive rates.

2. Advantages

- flexibility in dealing with changes (which is very important when the job is not well defined), particularly if new technology development is proceeding concurrently with the design;
- early start can be made;
- useful where site problems such as trade union actions such as delays or disruptions may be encountered; and
- owner can control all aspects of the work.

3. Disadvantages

- final cost is unknown;
- difficulties in evaluating proposals (strict comparison of the amount quoted may not result in selecting the best contractor or in achieving the lowest project cost);
- contractor has little incentive for early completion or cost economy;

- contractor may assign its second-division personnel to the job, make excessive use of agency personnel, or use the job as a training vehicle for new personnel;
- owner carries most of the risks and faces the difficult decisions; and
- biased bidding of fixed-fee and reimbursable rates may not be detected.

B. Target Contracts (Cost And Schedule)

Target contracts are intended to provide a strong financial incentive for the contractor to complete the work at minimum cost and within minimum time. In the usual arrangement, the contractor starts work on a reimbursable-cost basis. When sufficient design is complete, the contractor produces a definitive estimate and project schedule for owner review, mutual negotiation, and agreement. After agreement is reached, these become targets. At the end of the job, the contractor's reimbursable costs are compared with the target and any saving or overrun is shared between the owner and the contractor on a pre-arranged basis. Similarly, the contractor qualifies for additional payment if the contractor completes the work ahead of the agreed-upon schedule. The main appeal this form of contract has to the contractor is that it does not involve competitive bidding for the target cost and schedule provisions.

1. Requirements

- competent and trustworthy contractor,
- quality technical and financial supervision by the owner, and
- competent estimating ability by the owner.

2. Advantages

- flexibility in controlling the work;
- almost immediate start on the work, even without a scope definition;
- economic and speedy completion (up to a point) encouraged; and
- contractor is rewarded for superior performance.

3. Disadvantages

- final cost initially unknown;
- no opportunity to competitively bid the targets;
- difficulty in agreeing on an effective target for superior performance;
- variations are difficult and costly once the target has been established (contractors tend to inflate the cost of all variations so as to increase profit potential with easy targets); and
- if the contractor fails to achieve the targets, it may attempt to prove that this was due to owner interference or to factors outside the contractor's control; hence, effective control and reporting are essential.

C. Measured (Unit Price) Contracts

These require sufficient design definition or experience in order to estimate the unit/quantities for the work. Contractors then bid fixed prices for each unit of work. The time and cost risk is shared, with the owner responsible for total quantities, and the contractor assuming the risk of a fixed unit price. A quantity increase greater than 10% can lead to increases in the unit prices.

1. Requirements

- adequate breakdown and definition of the measured units of work,
- good quantity surveying/reporting system,
- adequate drawings and/or substantial experience for developing the bill of quantities,
- financial/payment terms that are properly tied to the measured work and to partial completion of the work,
- owner-supplied drawings and materials must arrive on time,
- quantity-sensitivity analysis of unit prices to evaluate total bid price for potential quantity variations,
- ability to detect biased bidding and/or front-end loading, and
- contractor experience with this contracting arrangement.

2. Advantages

- good design definition is not essential (typical drawings can be used for the bidding process);
- very suitable for competitive bidding and relatively easy contractor selection subject to sensitivity evaluation;
- bidding is speedy and inexpensive, and an early start is possible; and
- flexibility (depending on the contract conditions, the scope and quantity of work can be varied).

3. Disadvantages

- final cost is not known at the outset, since the bills of quantities have been estimated on incomplete engineering;
- additional site staff are needed to measure, control, and report on the cost and status of the work; and
- biased bidding and front-end loading may not be detected.

D. Lump-Sum/Fixed-Price Contracts

In this type of contract, the contractor is generally free to employ whatever methods and resources it chooses in order to complete the work. The contractor carries total responsibility for proper performance of the work, although approval of design drawings and the placement of purchase orders and subcontracts can be monitored by the owner to ensure compliance with the specification. The work to be performed must be closely defined. Since the contractor will not perform unspecified work without requiring additional payment, a fully developed specification is vitally important. The work has to be done within a specified period of time, and the owner can monitor status/progress to ensure that completion meets the contractual requirements.

The lump-sum/fixed-price contract presents a low financial risk to the owner, and the required investment level can be established at an early date. This type of contract allows a higher return to the contractor for superior performance. A good design definition is essential, even though this may be time-consuming. Further, the bidding time can be twice as long as that for a reimbursable contract bid. For contractors, the cost of bidding and the high financial risk are factors in determining the lump-sum approach.

1. Requirements

- good definition and stable project conditions,
- effective competition in a buyer's marker,
- several months for bidding and appraisal, and
- minimum scope changes.

2. Advantages

- low financial risk to owner since maximum risk is on the contractor;
- cost and project viability are known before a commitment is made;
- minimal owner supervision (mostly quality assurance and schedule monitoring);
- contractor will usually assign its best personnel to the work;
- maximum financial motivation of contractor (maximum incentive for the contractor to achieve early completion at superior performance levels);
- contractor has to solve its own problems and do so quickly; and
- contractor selection by competitive bidding is fairly easy, apart from deliberate low price.

3. Disadvantages

- variations are difficult and costly (the contractor, having quoted keenly when bidding, will try to make as much as possible on extras);
- an early start is not possible because of the time taken for bidding and for developing a good design basis;
- contractor will tend to choose the cheapest and quickest solutions, making technical monitoring and strict quality control by the owner essential; schedule monitoring is also advisable;
- contractor has a short-term interest in completing the job, and may cause long-term damage to local union relationships by doing such things as setting poor precedents/union agreements;
- bidding is expensive for the contractor, so the bid invitation list will be short; technical appraisal of bids by the owner may require considerable effort;
- contractors will usually include allowances for contingencies in the bid price and they might be high; and
- bidding time can be twice that required for other types of contracts.

E. Conditions of the Contract

While the same risks/liabilities can be established for most forms of contract, the price for those risks/liabilities can vary significantly, depending on contracting skills and the business environment/market place.

F. Typical Forms of Contract

1. United Kingdom

Institution of Civil Engineers (ICE). Mainly for civil and construction-only contracts.

Federation Internationale des Ingenieurs-Conseils (FIDIC). Primarily for offshore and overseas work.

Institution of Mechanical Engineers (IMechE). Primarily for design and erection of mechanical plants).

2. United States

American Institute of Architects (AIA). Mainly for engineering work and project/ construction management; the architect/engineer usually functions as the owner's agent on a fee/reimbursable basis;

Associated General Contractors (AGC). Mainly for construction work and construction management; the contractor usually functions as an independent contractor on a lump-sum/fixed-price basis;

The Engineers Joint Contract Documents Committee (EJCDC) Standard Agreement Contract Documents. Issued jointly by the National Society of Professional Engineers, American Consulting Engineers Council, American Society of Civil Engineers, and Construction Specifications Institute, Inc., and approved by the Associated General Contractors; often used by many engineering firms. In addition, it is becoming more prevalent for an owner to develop a form of contract that is specifically customized to fit its particular needs. Similarly, an engineering/construction contractor may develop its own form of contract for use on projects where it acts as the construction/project manager for the owner. At least two basic options exist:

- use one of the standard contracts and customize it to fit a particular project; and
- use the boiler plate or front-ends developed by the engineer/contractor for use on projects where it is responsible for preparing the bidding documents and where the owner does not have its own form.

G. Summary of Contracting Arrangements

It is possible to devise a form of contract with appropriate terms and conditions to suit many different circumstances. Some basic considerations leading to the best choice are listed below.

- There must be a clear definition of each party's contractual responsibilities. Shared responsibilities are unsatisfactory, although they are unavoidable in some circumstances.
- A lump-sum form of contract provides the best financial risk for the owner, gives the contractor the maximum incentive for early completion, and produces the greatest benefit of competitive bidding. Conversely, reimbursable contracts provide no such incentives. It is dangerous, however, to attempt to use a lump-sum contract if the essential conditions are not satisfied—notably, a clear and complete definition of the scope of the work.

- The owner must have the contractual right to exercise control adequate to ensure the success of the project, but the temptation to assume excessive control should be resisted.
- Control and responsibility go together—the greater the owner's control, the less responsibility is carried by the contractor.
- Additional liability/risk should result in greater profit to the contractor.
- Finally, the form of contract must be decided early in the course of project development, and the choice must be made known to the engineers before they write the specification. Obviously, the specification will be much more precise and comprehensive if it is to be used for a lump-sum contract than would be required for a reimbursable contract.

X. CONTRACTUAL AND LEGAL REVIEW (FOR COST IMPLICATIONS)

The risks accepted by the contractor through the contractual language can be difficult to define in absolute terms. Still, the implied risks in practically any contract, if not tempered by insurance, can drive a contractor into bankruptcy if the risks become realities. A wide range of expertise in developing contracts, coupled with the many vicissitudes of give and take in contract negotiations, can yield contract provisions that range from minimal impact to possibly catastrophic consequences for contractors. Thus, an owner's project manager can unwittingly increase the project costs by being overly protective when using contract language to require a contractor to assume risks over which the contractor has only minimal control at best. Apart from refusing to bid, the contractor's only defense is to carefully evaluate the risk and potential cost of the over-protection, and to increase the cost of the proposed services accordingly. The market place, of course, has an important bearing on risk and liability. When work is in short supply, it is possible to get contractors to accept greater risks due to a very competitive buyer's market. This also means that the buyer (owner) can dictate the contractual conditions to a much greater degree than usual.

Development of the form of contract and the contractual language may be the direct responsibility of the contracts department, working under direction of the project manager. In many cases, bid proposals will be reviewed by two groups of company personnel—the project team and contract/legal personnel. The project team has total evaluation and selection responsibility, but concentrates on the contractor's project execution capability. This leave contracts/legal personnel to advise on contract conditions, language, liability, guarantees, and other pertinent areas.

A. Contract And Legal Review

Assuming the bidders are allowed to take exceptions and/or offer alternatives, this essentially covers the following two objectives:

- detecting any deviations from the prescribed bid basis that would require resolution before award of the contract, and
- analyzing the contractor's exceptions to the proposed contract that could affect the selection.

The review will involve detailed analysis of the proposed exceptions/alternatives and their associated impact on risk, liability, and cost. It is possible that significant deficiencies and/or exceptions could disqualify a contractor.

Since many business and contractual conditions can have cost implications, it is important that close cooperation exist between the contract and project cost personnel.

B. Cost Aspects Of A Reimbursable-Cost Agreement (EPC Contract)

The allocation of costs to the fixed-fee and reimbursable categories can vary in a contractor's proposals. Even though the bid documents specify the required division of costs, it is very difficult to cover every detailed cost element. Contractors will therefore provide terms and conditions that are clearly different or ambiguous so as to allocate items as reimbursable when, in fact, they were intended by the company to be part of the fixed fee. Following are examples of areas in which this can occur.

1. Home Office Services

- Contractor personnel policies should be thoroughly reviewed in relation to the conditions outlined in the contract agreement. Are training and recruiting reimbursable, or are they part of the fee? Are contractor-quoted salary ranges fixed for the duration of the project, or are salary increases automatically passed on to the company? Are travel expenses and proposed trips subject to company approval before the fact or afterward? Are transfer and assignment of overseas personnel subject to approval by the company?
- A common point of contention is payment for home office engineers temporarily transferred to the job site. The overhead on field personnel is generally half that of home office personnel. What rate should apply?
- Is there adequate identification of the personnel and services that are included in the fee and those that are reimbursable?
- Purchasing policies, vendor service personnel, insurance, duties, inspection, and expediting services are costly items. Are adequate procedures contained in the contract agreement?
- Payment for reproduction and computer services should be thoroughly reviewed. Payment of costs for resident company personnel should be clearly outlined. These would include offices, equipment, services, secretarial, and typing assistance.

2. Construction Services

- Personnel policies for construction staff are quite different from those for home office personnel and need to be clearly stated in the contract agreement. Such topics as jobsite allowances, home leave, completion bonus, family and bachelor status, relocation, and replacement should be clearly outlined, particularly on overseas projects.
- Payment conditions for construction equipment, maintenance, small tools, and consumables should be clear and not left open to interpretation, since they can become points of contention. What is the relationship between

construction equipment rental rates and transportation costs? Does rental start on dispatch from the storage yard or at the date received on the job site? Are there adequate buy-out clauses? Are the terms for major and minor maintenance clearly stated?

- Payment for construction services is often fixed at a unit cost per field job-hour. Again, what field jobhour? Direct and indirect, direct hire, subcontract and field staff? The costs on differing jobhour bases can be considerable.
- Contractual conditions should be specific. If they are not, contract and cost engineers should estimate potential costs against each item and collate the information on the bid evaluation form. Subsequent negotiations will then determine the status of all exceptions, alternatives, and anomalies.

XI. COST ASPECTS OF A LUMP-SUM AGREEMENT (EPC CONTRACT)

A. Quality Owner Contract Package

There must be a complete definition of facilities, site working conditions, and contract scope in order for bidders to accurately program the project and estimate its cost. There are four main sections in owner contract/bid packages and contractor proposals:

- design basis/services,
- commercial terms,
- contract conditions, and
- execution program requirements.

For an EPC turnkey contract, the owner's scope of work or design basis is the key element. A poor design package can only result in poorly drawn proposals from contractors.

B. Proper Risk Analysis

Both parties need to carefully develop, evaluate, and agree on all risks that are required and contained in the proposed contract terms and conditions. From the owner's viewpoint, the major thrust will be to develop contract risks and liability for poor performance/schedule slippage. As detailed in the contractor's proposal section, the contractor must carefully review all documentation to fully understand the scope of work, services to be provided, working conditions, and the proposed reporting requirements/owner involvement. There must be strong competitive market conditions for EPC turnkey contracts, which requires that a sufficient number of qualified and competitive contractors be willing to bid. Beyond these, the activity level in the relevant contract market must be such that contractors need more work than is available to them under other, lower-risk contracts. Then there is the people risk consideration: it is one thing to identify the risks inherent in a specific project, but it is quite another matter to properly analyze the competence of company personnel to properly handle those risks. Recognizing personnel capabilities and limitations is essential with an EPC turnkey contract, especially from the contractor's viewpoint.

Project size must be acceptable so that contractors have the financial strength to bear the risks. There must be stable economic, political, and social conditions at the project location, as well as for the relevant materials, equipment, and labor market conditions. This is required in order to minimize the need for bidders to include large risk contingencies.

C. Estimating Capability

The ability to produce a ±10 to 15% estimate is essential. This requires historical experience/data, a quality estimating program, and experienced estimators. Owners will use the estimate to evaluate the contractor's proposed price, and contractors will use the estimate to establish the price.

An appropriate, detailed execution program is essential in developing a detailed estimate. The anticipated working conditions should be carefully evaluated, and organization requirements and schedules should be developed. The estimate should be based on quotations for all major equipment. Bulk materials should be estimated with quantity takeoffs from plot plans, P&IDs, layout drawings, and specifications. Engineering jobhours and costs should be based on numbers of drawings/documents. Direct construction labor should be estimated from equipment/material quantities multiplied by unit jobhours multiplied by jobhour labor rates. Construction indirect such as construction equipment, supervisory/administration staff, and field office expenses, should be individually estimated.

D. Cost of Bidding

Engineering, procurement, and construction turnkey bids are expensive to produce. A contractor has to carry out engineering in order to develop bulk quantity takeoffs, seek quotations for all major equipment, obtain subcontract bids for portions or all of the construction work, develop schedules, define organizational requirements and staffing levels, and develop a detailed execution plan.

Experience shows that with established technology, EPC turnkey proposals cost about 3–5% of the total project cost. Where new technology is significant or the bidder's experience is minimal, the proposal cost could rise to 5–7% of the total project cost. Therefore, a serious consideration in seeking or entering into EPC turnkey contracts is the question of companies being able to bear the cost of bidding. Failure to gain profitable work can lead a contractor into financial difficulty if there is no offset against proposal costs.

E. Key Contractor Considerations

A buyer's market or a soft market can expose a contractor to difficult conditions and risks such as:

- breakeven or below-cost pricing levels,
- buying the job in order to maintain/employ resources,
- overprotection of contract conditions by owner,
- higher risk conditions of lump-sum/turnkey contracts,
- schedule penalty arrangements,
- poor owner-contractor relationship,

- reduced or poor negotiating position,
- establishing an aggressive claims operation, and
- going bankrupt or out of business.

From the legal viewpoint, the most appropriate method of surviving is to carefully review all owner requirements (as expressed in the owner contract documents) and attempt to negotiate away those requirements as necessary. The owner-contractor relationship is not (or should not be) an adversary one. To the extent that, at the commencement of a project, the parties can fully understand and agree on the things that are really required and what the contractor's price does cover, the more likely the relationship will remain as intended. *This understanding/agreement is established through the process of certainty.*

Experience suggests that the best approach to achieving the desired result is for the contractor to assemble as much in-house expertise as possible (such as project management, engineering, procurement, estimating, and cost control) and to thoroughly review the owner's requirements. This is important because it enhances everyone's understanding of the requirements: not just those stated in the contract, but those contained in the specifications, drawings, and project procedures as well.

The contractor team should, in essence, educate itself to the point that all involved realize the impact on price and schedule of owner inspection, reporting, and other requirements. The key with an EPC turnkey project is to get all requirements defined as specifically as possible. Otherwise, a contractor can find itself in the unenviable position of being selected as the winning bidder and realizing early on during the course of the project that there is very little opportunity to make money.

With the advantage of a buyer's market, owners should nevertheless construct contract conditions (risks) and negotiate in a responsible manner. In the long run, contractor financial hardship/bankruptcy is not in the owner's best interests. The rationale by some owners that "it's now our turn" is a poor excuse for good business practice.

XII. PAYMENT OF THE FIXED/TURNKEY PRICE OR THE FIXED FEE (OF A REIMBURSABLE CONTRACT)

Payment should never be based on a fixed, periodic payment schedule. In the past, this was a common practice, but it often happened that all payments had been made long before a project was complete. Thus, payments should be based directly on physical completion of the work.

Progress values for engineering, material commitment, and construction should be developed, and these should then be weighted together to develop an overall progress figure. This will be the percentage figure, less retention, that will be applied to the fixed price for payment. Engineering and construction progress can be measured according to a physical measurement system. Material commitment progress could be based on the number of purchase orders or requisitions where a weighted progress is determined as work is completed in stages: engineering issue, inquiry issue, purchase order issue, vendor fabrication, and jobsite receipt. Planned and actual progress curves can be developed as an owner requirement to provide project status information as well as the system for payment.

Since these systems can sometimes take many months before they become effective, it is possible that a certain percentage of the lump-sum price or fixed fee could be made available at contract award. The contractor may propose the figure and provide an explanation to justify it.

XIII. BIASED BIDDING

Biased bidding is a deliberate pricing strategy used by a bidder. It poses a significant cost threat to the purchaser and must be detected as soon as it is practiced. The following are examples of biased bidding.

A. Fixed-Fee Reimbursable Contracts

The majority of these contracting arrangements are based on a minimal design package, so it is possible that the final scope will be greater due to changes and poor scope definition. An experienced bidder can recognize these increases, and may take profit out of the fixed fee and add it to the reimbursable jobhour rates. This results in greater profit if the scope of work increases as anticipated by the bidder. A negotiation of this bias can result in a proper distribution of profit to fee and rates, as specified.

B. Unit-Price Construction Contracts

The most efficient use of this contracting arrangement is in time management, with the work quantities (unit prices) being developed by the purchaser on an estimated basis. As the size of the contract increases, so does the difficulty of developing accurate estimated quantities increases. Significant errors, usually underestimating, can occur on individual work items. An experienced bidder can detect such errors, taking profit out of correctly estimated items and adding it to underestimated items. As the work quantities increase, the resulting extra profit can be considerable. As previously outlined, a quantity sensitivity analysis can identify such a situation, and negotiation to rebalance this practice can reduce and/or eliminate this profit enhancement technique.

C. Front-End Loading

This practice applies to unit-price contracts where the purchaser adds more profit to the early work activities by reducing the profit of later work items. As the work starts and progresses, the bidder earns additional profit and greatly improves its cash flow. There is also an added work risk to the purchaser, in that if the bidder defaults on the contract, the remaining funds are insufficient to complete the work. Again, negotiation should attempt to reduce or eliminate this practice.

D. The Deliberate Low Bid (Buying the Job)

There are two forms of low bidding: the deliberate low bid, and a low bid produced by lack of experience and/or a poor purchaser bid package (see next section). By definition, a low bid is less than the purchaser estimate by 20% or more. Since it is assumed that the purchaser has developed a reasonable or good quality estimate, a deliberate low bid is thus a premeditated pricing strategy. This strategy is often referred to as buying the job, and may be employed for any number of the following reasons:

Survival. In times of low business activity and intense competition, a purchaser can decide to bid for work at a low price, with little or no profit built in, so as to keep its existing staff employed;

Reducing or eliminating competition. Low bidding can be used to maintain a purchaser's market share and/or to reduce competition by other companies;

Breaking into a new market. Without a proven track record in a specific industry or category of work, the major way a company can secure work in that industry is to attract a new purchaser by submitting a low price. It is assumed that a general technical capability and company resources are available and that a company can get onto one of the purchaser's existing bidder's lists; and

Detecting significant errors in the purchaser's request for quote. If the purchaser's request for quote contains significant errors, there is a chance for the contractor to earn additional revenue. The contractor must first, however, be successful in getting the job through a low bid and an accompanying claims program.

Each instance of possible deliberate low bidding has to be evaluated carefully to prevent companies from developing poor financial, quality, and scheduling results. The greatest risk can be a company defaulting on the contract.

E. The Low Bid Through Lack of Experience

Apart from performance problems, there is the legal risk of the process of certainty, which requires that the purchaser make certain that the other party fully understands the scope of work. When there is a lack of experience, lack of certainty can be a direct result. Thus, clarification meetings and further technical analysis of the proposed bidder's programs are essential. If lack of certainty is present, the award of a low bid can be rescinded or can lead to successful claims by the bidder.

9

Cost and Schedule Trend Analysis—
Forecasting of Baselines

I. DEVELOPING REAL COST CONSCIOUSNESS—NO SMALL TASK

A. Project Manager Responsibility

The project manager is directly responsible for creating an environment that will enable project control to be properly exercised. Thus the project manager must seek counsel, accept sound advice, and stretch cost/schedule personnel to the extent of their capability. Team building and team stretching are key elements of successful project management.

On smaller projects, where the project manager is also the project control engineer, it is essential that the project manager possess project control skills and/or motivate the supporting/service groups to provide the quality information that is needed for creative analysis and effective decision making.

B. Establishing Project/Cost Consciousness

Effective project control requires the timely evaluation of potential cost and schedule hazards and the presentation of recommended solutions to project management. Thus the cost/schedule specialist must be a skilled technician and also be able to effectively communicate at the management level. Sometimes, an experienced project control engineer's performance is not adequate because of poor communication skills. Technical expertise will rarely compensate for this lack. As in all staff functions, the ability to sell a service can be as important as the ability to perform the service. On larger projects, project teams are usually brought together from a variety of melting pots, and the difficulty of establishing effective and appropriate communications at all levels should not be underestimated. In such cases, the project manager must quickly establish a positive

193

PROJECT COST CONSCIOUSNESS
A People & Team Responsibility

Figure 9.1   The Bean-Counter syndrome.

working environment where the separate functions of design, procurement, construction, and project control are welded into a unified, cost-conscious team.

C.   The Bean-Counter Syndrome

This is a widespread practice, where effective cost control is absent or greatly diminished. This practice has two major contributing factors. Firstly, the project manager does not want an aggressive, creative, analytical function for the cost engineer and, therefore, relegates the work to a retroactive, record-keeping function (hence the term *bean-counter*). Secondly, the cost engineer can be directly responsible for this practice, as the individual may not possess the essential analytical skills or may not believe in an aggressive trending approach and/or may not possess the essential people/communication skills. The cost engineer may, in fact, be content with a bean-counting role. Figure 9.1 illustrates the bean-counting function. There is a much wider acceptance today of the need for dynamic, proactive cost engineering-trending, and it is to be hoped that the function will become a pivotal project role, as effective cost trending is essential for project success.

After a lifetime of project work, the author has learned this fundamental truth:

> Projects are designed and built by people, not companies. People do it
> singly or in multiple groups, and if there are good people relationships,

there is a chance of success. If the relationships are poor, there is little chance of success.

D. Project Control Defined

Effective project control consists of a process that:

- identifies potential hazards well in advance of their occurrence;
- evaluates the impact of such hazards and, where possible, proposes actions to alleviate the situation; and
- provides constant surveillance of project conditions to effectively and economically create a *no surprises* condition, except for force majeure situations.

E. A Cost-Effective Program

All project control programs must live up to their own principles and be cost effective. Many large and allegedly sophisticated architect/engineering firms and contractors over-control and over-report. This occurs because of the ever-present tendency, as projects become larger or more complex, to create additional levels of control, reporting, and personnel. Yet more is not necessarily equivalent to better. Project managers and project control supervisors must carefully evaluate their company's program against project needs, eliminating all instances of over-control and over-reporting.

II. BUSINESS DECISION MAKING VERSUS TECHNICAL DECISION MAKING

Three important steps must be taken to ensure effective project control.

A. Business Decision Making

The initial step is to ensure that key decisions makers—the project manager, the engineering manager(s), the procurement/contract manager, and the construction manager—base their decisions on sound business practice. When these people are not motivated by a solid business ethic, cost overruns and schedule delays become both common and inevitable.

B. Business Information/Analysis

The second step is to ensure that all information needed for making sound decisions is available at the right time and in the right place. All too often, the information necessary for analyzing options and alternatives is not available or simply hasn't been developed. Yet much of this information can be found in a firm technical scope of work, a quality estimate, and a good schedule. When these items are not available, decision-making may be flawed or ineffective.

The early stage of a project is often the time when significant business decisions have to be made; consequently, having quality information is vital at that point. Yet firm information may not be available at this point: technical options are still being considered, and execution plans and contracting strategies are still being developed. Creative analysis and experienced judgment is, therefore, essential to bridging this gap, sometimes referred to as the *blackout period*. This is discussed at greater length later.

C. Communication Channels

The third step involves the project manager's responsibility for actively promoting *project consciousness* with all involved departments and key personnel. This ensures that the project team and/or service groups work toward the agreed project objectives and execution plan. Active and open communication channels are also critical to properly coordinating and interfacing with the client to establish a positive atmosphere for making decisions and obtaining approvals. Limits of authority, lines of communication, degrees of responsibility, and approval requirements must be clearly established in the project coordination procedures. Personnel motivation and leadership skills are essential qualities in a project manager if project consciousness and effective communication channels are to be established.

III. TIMELY COST ACCOUNTING/REPORTING

A. Typical Program

An accounting program is an integral part of cost control, as it specifies levels of approval, billing and payment procedures, banking arrangements, accounts numbering systems, and all functions necessary for project financial reporting. Tracking the status of financial objectives and providing a base for developing cost predictions is essential to project success. Thus, the key element in cost control is timely and accurate reporting of commitments and expenditures.

Delayed payments or lost invoices inevitably cause poor cost predictions. Further, many control techniques are based on unit costs, which can be greatly distorted if the cost-to-date is inaccurately reported or if the information is reported late. On smaller projects, an accurate accounting/cost report should be available no more than two days after the cutoff date; on larger projects, availability in no more than five days is essential.

B. Project Accounting Group

As most operating company accounting systems are commodity based, they are unsuitable for *project accounting*. Therefore, an effective organizational arrangement is to establish a project cost accounting group within the general accounting department but with a cost coding system that is *project driven*. The group concentrates solely on accounting for capital projects and reports directly to project management for day-to-day activities. It is, however, often difficult to establish such a group, as accounting managers, fearful of losing authority, may resist the project approach. A more common and less efficient approach is to have the project group develop a second set of books and associated coding system, which is then maintained by project/cost personnel. It is to be hoped that such management failures/abuses will be corrected as TQM cultures become common-place in the construction industry.

IV. EFFECTIVE TRENDING SYSTEM

A. The Blackout Period

A difficult time for cost control occurs during the transition from the feasibility estimate to the *full funding* estimate. This is often called the blackout period, during which time the estimate is upgraded to a higher quality (but is still a conceptual estimate) and on large projects this period can last six months or longer. During this time, many engineering decisions are made without full recognition of their cost impact. Accordingly, having an effective and cost-conscious project team is absolutely vital, so that full and proper trending of project-design development and project change is generated by all to all. Cost-consciousness, especially on large and complex projects, can only be achieved by direct leadership of the project manager, in conjunction with a highly effective *team building program*. The weekly trend meeting then builds on this program with detailed development and analysis of all change.

B. Owner Trending System Versus Contractor Trending System

On larger reimbursable-type projects, it is common for an owner to maintain an independent trending program. In contracts with a small owner project task force, a large contractor's operations and organization can be cumbersome, inflexible, and slow to respond. In such cases, an owner trending program can be more current, accurate, and responsive to changing circumstances. It should be a beneficial, not adversarial, program that works positively with a contractor from the same project data-experience and shares information as trends are detected. There is also a question of owner oversight, especially on reimbursable projects, and also there can be a lack of contractor skills that make owner independent analysis essential.

C. Typical Trending Situations

On reimbursable-type projects the risk of design change is higher than on lump-sum projects, and it is common for a greater proportion of key engineering decisions to be initiated by owner-engineers. In addition, design and project discipline is less prevalent with owner engineering personnel. It is, therefore, vital that cost engineering change control ensures that communication channels are properly developed with both owner and contractor engineers to constantly provide an accurate assessment of the developing design. General design speci-fications and equipment specifications should be monitored for conflict, preference (that adds cost), and "gold plating". All changes to engineering specifications, scope, procurement, subcontracts, etc. should be recorded, via a *design log pro-gram*, as they occur or as they are being considered by the engineering groups. Figure 9.2 illustrates typical trending situations, and Figure 9.3 is an example of a trend report.

1. **ALL SCOPE CHANGES**

2. **DESIGN CHANGES / DEVELOPMENT**
   - Especially During Early Design Phase
   - Owner Engineers
   - Contractor Engineers
   - Design Change Log
   - Cost / Schedule Consciousness
   - "Gold Plating"

3. **SCOPE REDUCTION PROGRAMS**

4. **PROJECT / CONTRACTUAL CONDITIONS**
   - Plant Operations
   - Breach of Contract
   - Site Conditions

5. **EXECUTION PLAN CHANGES**
   - Priority of Project Objectives
   - Schedule Acceleration

6. **APPROVED TRENDS**
   - Formally Authorized By Project Manager

7. **POTENTIAL TRENDS**
   - Early Identification Can Prevent, OR
   - Assessment of Cost Increase / Schedule Slippage

Figure 9.2   Typical trending situations.

Changes to the project execution plan—whether they are contractual, environmental, regulatory, or schedule oriented—should be included as well. Potential and approved trends should be reported, and preparation of related cost estimates should be routine. Approved trends are those that have been formally authorized by the project manager. Potential trends include items verbally approved and items that the cost engineer, through discussions with task force personnel, believes are likely to occur. Potential changes should be shown separately from approved trends.

D.   The Weekly Trend Meeting

Of the many meetings held during the execution of a project, the weekly trend meeting is probably the most important. This is not a decision-making meeting but a time when information is gathered and shared by key technical/services specialists. The project manager generally leads the meeting, and the project cost engineer often serves as secretary. All current and potential influences, changes, extras, and trends are reviewed and discussed. The key meeting objective is the

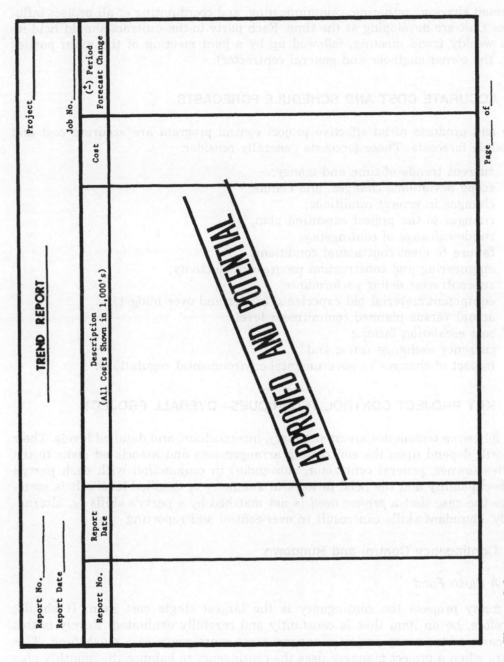

Figure 9.3 Trend report example.

common sharing, gathering, communicating, and coordinating of all project influences that are developing at the time. Each party to the contract should hold its own weekly trend meeting, followed up by a joint meeting of the main parties (i.e., the owner-engineer and general contractor).

V.   ACCURATE COST AND SCHEDULE FORECASTS

The end products of an effective project control program are accurate cost and schedule forecasts. These forecasts generally consider:

- current trends of time and money;
- scope deviations, changes, and claims;
- changes in project conditions;
- changes to the project execution plan;
- rundown/usage of contingency;
- failure to meet contractual conditions;
- engineering and construction progress/productivity;
- subcontractor dollar performance;
- equipment/material bid experience (under and over budget);
- actual versus planned commitment levels;
- cost escalation factors;
- currency exchange rates; and
- impact of changes in governmental-environmental regulations.

VI.   KEY PROJECT CONTROL TECHNIQUES—OVERALL PROJECT

The following techniques are at summary, intermediate, and detailed levels. Their use will depend upon the contracting arrangements and associated risks to the parties (owner, general contractor, subvendor) in conjunction with each party's skills-capability and the need to work at summary or detailed levels. It is sometimes the case that a *project need* is not matched by a party's skills or, alternatively, abundant skills can result in over-control and reporting.

A.   Contingency Control and Rundown

1.   A Slush Fund

On many projects the contingency is the largest single cost item. It should, therefore, be an item that is constantly and carefully evaluated. Rarely is this the case, as too many project managers treat contingency as a *slush fund*. This occurs when a project manager uses the contingency to balance the monthly *plus trends*, without properly evaluating the plus trends with a good risk analysis program. To reduce the contingency by the amount of the plus trends in order to maintain the previous monthly cost forecast can be a dangerous technique. In such cases, it is quite common for the contingency to be spent well before the project ends. Contingency is, essentially, to be used for *unknowns*, and these unknowns only become known when commitments are made. It is then that the *cost reality* and validity of the estimate-project budget becomes apparent.

The calculation process shows a rundown which is calculated on the basis of uncommitted and unspent costs and is computed as follows:

Major Cost Center	Cost Prediction Uncommitted (%)	Committed but Unspent (%)
a) Material and Equipment	10	5
b) Labor	20	15
c) Labor Subcontracts	20	15
d) Material and Labor Subcontracts	15	10
e) Home Office, Engineering and Fee	10	5
f) Field Indirects and Temporaries	10	5
g) All other costs	10	5
h) Owner Costs	10	5

This simple but practical method covers risk on work yet to be committed and work committed but not yet paid for.

Figure 9.4   Contingency control—simple method.

2. Rundown Routine

Figure 9.4 represents a simple but effective calculation routine to run down the contingency over the life of a project. The calculation process shows a rundown that is calculated on the basis of risks inherent in uncommitted and unspent costs. The computations are made as shown in the figure.

This simple and practical method covers risk on work yet to be committed and work committed but not yet paid for. The basic presumption is that, until the last and final invoice has been paid, there is still a risk that cost increases will occur, quite often due to late, missing, or lost invoices. Construction labor and subcontracts are usually the most volatile items and, as shown in Figure 9.4, carry the greatest contingency percentage. The maximum contingency that this set of numbers will develop is 13%, and such a low percentage is too low for controlling a conceptual estimate/project budget. Alternatively, if the project had a detailed estimate with a 10% contingency, then the illustrated percentages should be reduced. This set of numbers, which is based on historical data, is appropriate for use in a project control estimate where the contingency is generally in the 12–15% range.

3. Startup/Commissioning

A further *risk consideration* is the amount of contingency required at mechanical completion to ensure that funds are available for late changes, commissioning accidents, material startup requirements, and final invoices or claims that were not anticipated earlier. If adequate funds have already been allowed in the estimate for these considerations, then further hold-back would be unnecessary.

B. Cash Flow Evaluation/Control

1. General Objectives

The major objective of cash advances and the bank handling procedures is to ensure that the project is funded by the responsible party in accordance with the contract. On reimbursable-type contracts, a cash flow forecast is usually prepared for a two-month period and presented on a biweekly basis for cash advances. A good forecast should generate no more than a 5% excess requirement. Payments terms for a lump-sum contract usually require an initial payment (approximately 10% of the contract value) at contract award, followed by monthly payments directly tied to the physical progress of the work. Progress is mutually verified and agreed upon.

2. Documentation (Reimbursable-Type Contract)

As soon as practical after the last day of each month, the contractor should prepare a complete statement of the amounts actually paid. This is supported by copies of invoices, payrolls, bank statements, and properly executed waivers of lien for each contractor and subcontractor.

3. Banking Arrangements (Reimbursable-Type Contract)

A commonly used banking arrangement is the *zero bank account system*. This assumes that the contractors do not finance the project and that owners are required to advance sufficient funds (in accordance with the cash flow forecast) to the contractor's bank. Interest accruals will be credited to the owner. This procedure should form part of the contractor's proposal and should specify penalties to the owner for failure to provide timely funding.

C. Cash Flow Curve

Figure 9.5 depicts a typical format for a project total cash flow curve, showing planned and actual cumulative curves on a monthly basis. The planned curve should be evaluated every month to ensure accuracy of the short-term (three-month) forecast compared with the overall budget. This curve, plus appropriate backup, should be presented for approval on an agreed periodic basis.

D. Earned Value System for Fixed Budgets

1. Overview

When overall percent complete for a combination of unlike work tasks or for an entire project must be determined, an essential technique called the *earned value system* (EVS) is used. (The terms *achieved value*, *accomplished value*, or *physical quantity measurement* can also be used). A project's budget is expressed in both jobhours and dollars, and earned value is keyed to the project budget. Most projects are constrained by fixed budgets; others have floating, or variable, bud-

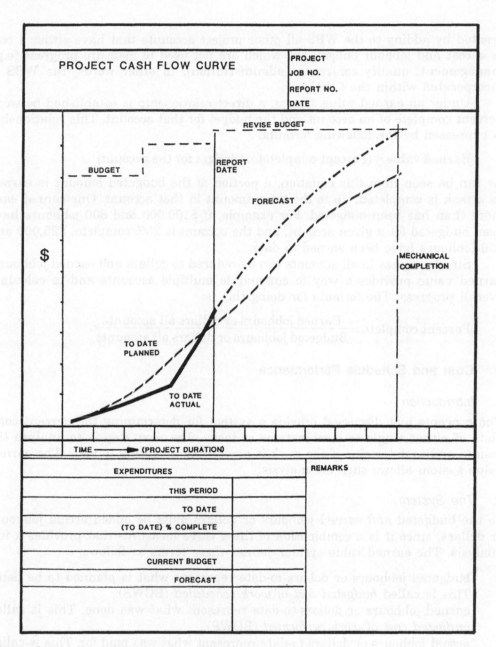

Figure 9.5 Project cash flow curve format.

gets. Earned value techniques can be applied in both situations, although differences exist in the detail of application.

2. The System

When developing a control system, a project should be segmented into its controllable parts. A work breakdown structure (WBS) is developed, which includes all work tasks to be used in determining project progress. Each task should have its own dollar and jobhour budget. A project cost breakdown structure (CBS) is created by adding to the WBS all other project accounts that have either a cost or a cost and jobhour budget, but which are not used to measure progress (e.g., management, quality control, or administration). In other words, the WBS is incorporated within the CBS.

Under an earned value system, a direct relationship is established between percent complete of an account and the budget for that account. This relationship is expressed by the following formula:

$$\text{Earned value} = (\text{Percent complete}) \times (\text{Budget for the account})$$

As can be seen from this equation, a portion of the budgeted amount is earned as a task is completed, up to the total amount in that account. One cannot earn more than has been budgeted. For example, if \$100,000 and 600 jobhours have been budgeted for a given account, and the account is 25% complete, \$25,000 and 150 jobhours have been earned to date.

Since progress in all accounts can be reduced to dollars and earned jobhours, earned value provides a way to summarize multiple accounts and to calculate overall progress. The formula for doing this is:

$$\text{Percent complete} = \frac{\text{Earned jobhours or dollars all accounts}}{\text{Budgeted jobhours or dollars all accounts}}$$

E. Cost and Schedule Performance

1. Introduction

The concepts just discussed provide a system for determining the percent complete of either single or combinations of tasks. The next step is to analyze the results and to determine if the work is proceeding according to plan. The earned value system allows such an analysis.

2. The System

To the budgeted and earned jobhours or dollars must be added actual jobhours or dollars, since it is a combination of these three measures that provides a full analysis. The earned value system defines these terms as follows:

- budgeted jobhours or dollars-to-date represent what is planned to be done. This is called *budgeted cost of work scheduled* (BCWS).
- earned jobhours or dollars-to-date represent what was done. This is called *budgeted cost of work performed* (BCWP).
- actual jobhours or dollars-to-date represent what was paid for. This is called *actual cost of work performed* (ACWP).

Schedule performance is a comparison of what was planned to what was done. In other words, jobhours were budgeted and earned. If the budgeted jobhours are less than the earned jobhours, more was done than planned, and the project is ahead of schedule. The reverse would place the project behind schedule.

Performance against budget is measured by comparing what was done to what was paid for. To do this, earned jobhours are compared to actual jobhours. If more was paid for than was done, the project is over budget.

The following relationships can be expressed as formulas:

$$\text{Schedule variance (SV)} = \frac{\text{Earned jobhours or dollars}}{\text{Budgeted jobhours or dollars}}$$

or $= \text{BCWP} - \text{BCWS}$

$$\text{Cost variance (CV)} = \frac{\text{Earned jobhours or dollars}}{\text{Actual jobhours or dollars}}$$

or $= \text{BCWP} - \text{ACWP}$

$$\text{Cost performance index (CPI)} = \frac{\text{Earned jobhours or dollars-to-date}}{\text{Actual jobhours or dollars-to-date}}$$

or $= \text{BCWP} + \text{ACWP}$

A positive variance and an index of 1.0 or greater denotes favorable performance. Figure 9.6 a plot shows the relationships between BCWS, BCWP, and ACWP. Other earned value systems are covered later in this chapter.

F. Design or Development Allowance

This allowance, identified as a separate line item in the estimate, is not part of contingency. It is a known condition that is very common on *fast-track* projects and represents money that will be used to cover design changes after an equipment purchase has been made. The fast-track approach requires early placement of all critical equipment, even before the design has been completed. As the design advances to completion, changes can and do occur to the already-committed equipment. This, then, leads to change order requests from the equipment vendors. The design allowance is used to cover these added costs.

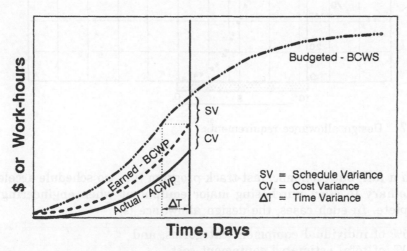

Figure 9.6   Relationships between BCWS, BCWP, and ACWP.

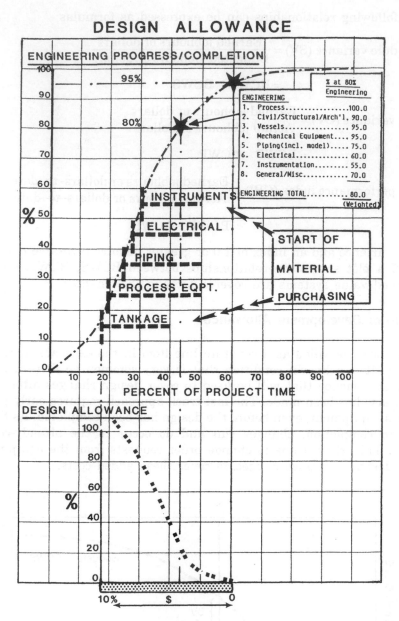

Figure 9.7 Design allowance requirements.

With a full, but economic, fast-track program (i.e., no schedule acceleration), it is customary to start purchasing major equipment when engineering is only 20% complete. In such cases, the design allowance is:

- 5–15% of individual equipment categories, and
- 8–10% of total estimated equipment cost.

The amount of design allowance directly depends on the degree of engineering completion and the associated risk of design change. Figure 9.7 shows two curves for a full, economic fast-track program. The top curve shows the historical engineering progress curve, with separate material category *bars* for the start of purchasing. The lower curve is a guide to indicate the percent of design allowance that should be available as engineering advances. The example assumes that design allowance has been estimated at 10% (of total equipment), and the curve shows the allowance being run down to zero at 95% completion of engineering. At that stage, no risk of design changes should exist. The control of design allowance is by both procurement and engineering. This technique is also covered in Chapter 3 as it relates to estimating.

G. Project Rundown Control (Reimbursable Contract)

1. Purpose

As construction approaches mechanical completion, an efficient rundown can be difficult to achieve, especially if a contractor has low workload and does not want to lay off staff. On large projects, this situation can be particularly serious and lead to increased costs. Moreover, the earned value system is ineffective at this point, since the weight of accumulated performance now obscures incremental performance. As punchlists and completion lists are generated, it is difficult to ascribe a measurable scope of work to a multitude of different items, many having top priority for completion. One effective technique for reducing these problems is shown in Figure 9.8. It involves preparing a detailed estimate of the remaining work items, with work activities/punchlists, and then individually controlling, by item, on a daily/weekly basis. The figure graphically depicts a rundown program for direct labor on a reimbursable project. In addition to field labor, other costs must be considered as well.

2. Material

At this stage, field-purchased material usually reaches a peak, due mainly to design changes and rework. An estimate of these expenditures should be made, with planned dollar expenditure curves drawn for major categories and actual expenditures plotted.

3. Subcontracts

Direct-hire forces can be replaced at this time, and the remaining work completed under subcontracts. The subcontracts are often awarded on a time-and-material basis, making dollar expenditure curves the only way to retain proper control. There is also the need for closing out the regular construction subcontracts, many of which may require settlement of claims and extras. Resolution of these items requires time and attention.

4. Design Changes and Startup Problems

At the commissioning stage there can be significant design changes, due, in part, to the insistence by the maintenance and/or operations departments that their requested changes are necessary. It is probable that most of these changes are preference items, and the project manager can take a hard position and refuse the changes, especially if the work is to the approved drawings and specifications. This situation, of course, depends on the power of the project manager and the

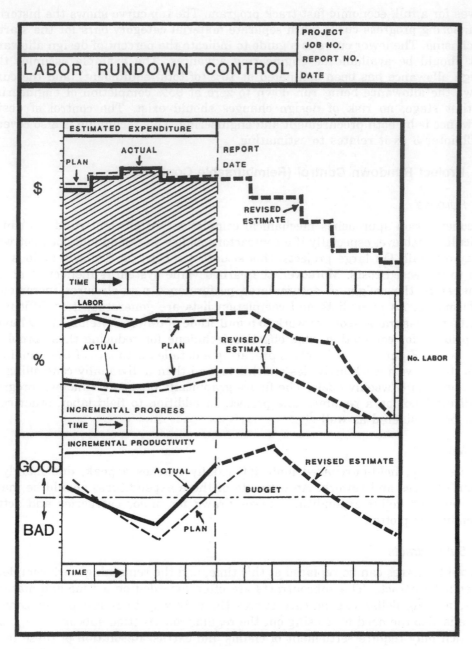

Figure 9.8 Labor rundown control program.

project discipline of the maintenance and operations departments. In practice it should be a give-and-take situation, and it is the wise project manager who tries to accommodate the maintenance and operations departments and at the same time observe project fiscal responsibility.

A field change log can help identify the scope of this work, and designation of the initiator can assist in a proper allocation of costs. Thus, on large projects, it is vital that personnel, jobhours, and costs be tightly controlled through the rundown stage.

The same rundown control system can also be effectively used to close out engineering.

H. Monthly Cost Report

Figure 9.9 depicts a typical monthly cost report. This is a summary report and should list major accounts or cost centers. Columns 1 and 2 contain committed and expended costs. Columns 3, 4, and 5 cover the original budget, updated with approved changes. Columns 6 and 7 carry the forecast and variance of forecast to the current budget. The same format can be used for detailed as well as summary-level reports. The monthly cost report is a key management document and, as such, should be timely, accurate, and readily communicate an overall cost evaluation.

I. Commitment and Expenditure Report

Figure 9.10 contains an example of commitment and expenditure curves. Curve formats are far more conducive to developing forecasts than a tabular report. A curve shows the trend of performance to date and can, with a good database and good judgment, be developed to a final prediction. This format is particularly efficient, as it shows the development of the control budget and forecast against time.

The example in Figure 9.10 shows an original control budget of $80 million developed in mid-1990 and, with scope changes, increasing to $94 million by 1994. The forecast starts at $90 million in 1991 and rises to approximately $120 million by 1994. With a good database, it would be fairly obvious in mid-1991 that with commitments of $45 million and a mechanical completion in late 1994, a forecast of $90 million would be unlikely. Historical experience would indicate a straight-line profile to end in 1992, followed by a gradual rundown to mechanical completion. The status is reported as of April 1994, with commitments of $113 million and a forecast of $118 million.

This technique can be used for detailed elements as well as for the overall project.

J. Home Office Jobhour Expenditure Curve

Figure 9.11 shows incremental and cumulative curves for overall home office jobhours. Planned curves should be developed, and actual experience plotted on a monthly basis. This example shows a significant overrun against the plan as of month nine. Evaluations of individual elements of the home office confirm that the overrun is irrecoverable; therefore, a final overrun is forecast. With a curve format, this trend was really discernible in month six. Depending on the accuracy

PROJECT COST REPORT

PROJECT
JOB No
PERIOD
REPORT No
SHEET ___ OF ___

ALL FIGURES =

CODE	ITEM / DESCRIPTION	1 COMMITTED TO DATE	2 EXPENDED TO DATE	CONTROL BUDGET			6 CURRENT FORECAST	7 VARIANCE (6)−(5)
				3 INITIAL(AFE) BUDGET	4 APPROVED CHANGES	5 REVISED BUDGET		
1								
2								
3								
4								
5								
6								
7								

Figure 9.9 Project cost report.

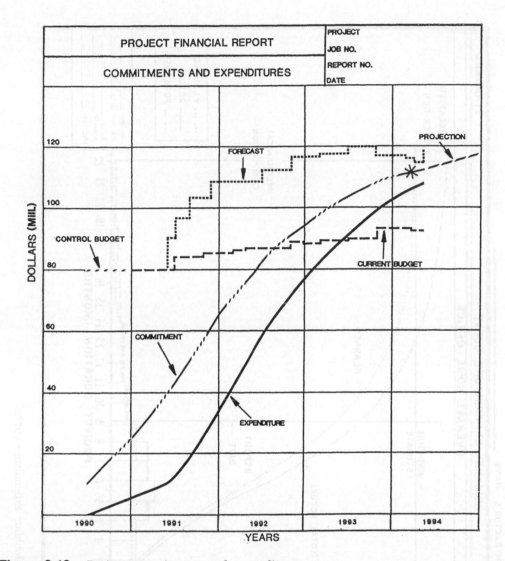

Figure 9.10 Project commitment and expenditure curves.

of the planned curve, a potential overrun, evident in month six, becomes a reality by month nine.

Nevertheless, in this particular example, a jobhour overrun does not necessarily mean a cost overrun. It is possible (though not probable) for an underrun in expense budgets and personnel costs to compensate for an overrun in jobhours. Consequently, it is also important to monitor jobhour costs as well as jobhours.

K. Home Office Costs and Expenditure Curve

Figure 9.12 shows planned and actual expenditure curves of home office costs. Planned curves are developed by judgment and historical experience and can be compiled from profiles of more detailed elements.

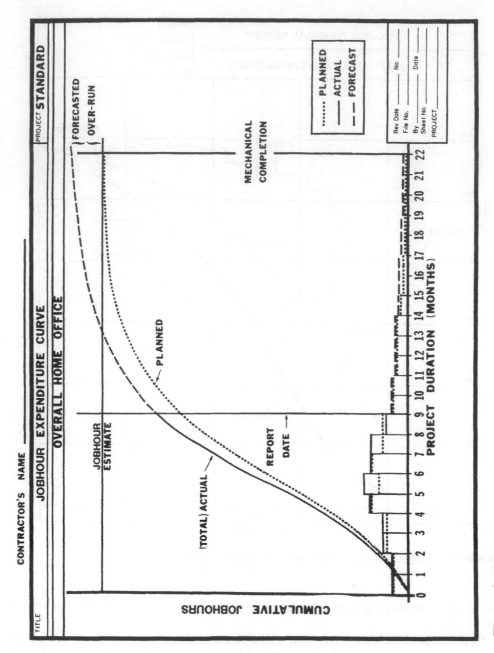

Figure 9.11 Jobhour expenditure curve.

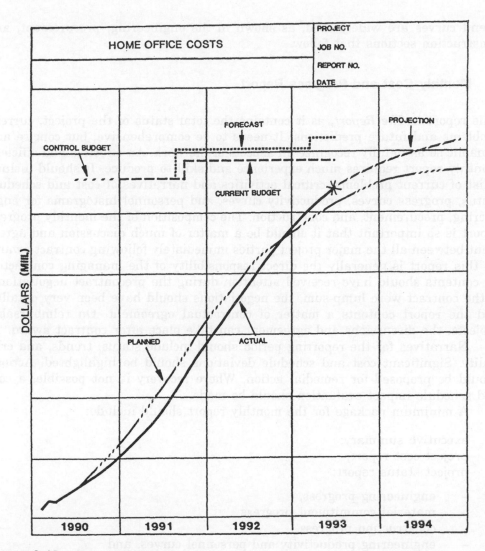

Figure 9.12 Home office cost curves.

This example shows the actual expenditure curve consistently underrunning the planned until the end of 1992 where it continues straight on instead of commencing to run down. A trend was discernible in mid-1992 and, in fact, a forecasted overrun had been made in the third quarter of 1991. A further overrun was made in the first quarter of 1993 due to failure of the curve to commence a rundown.

Home office costs typically consist of:

- project management;
- project control;
- procurement services;
- computer, reproduction, and communication;
- home office construction; and
- accounting, clerical, and administrative.

Trend curves are widely used, as shown in the engineering, procurement, and construction sections that follow.

L. Monthly Cost and Progress Report

This report is *The Report*, as it contains the total status of the project, current problems and future predictions. It needs to be comprehensive, but concise and dynamic so as to fully recognize all aspects of the work. An effective and efficient monthly report requires much experience and skill to produce. It should include a list of current problems, critical activities and narratives of cost and schedule status, progress curves, productivity curves, and personnel histograms for engineering, procurement, and construction. The composition of the monthly progress report is so important that it should be a matter of much discussion and agreement between all the major project parties immediately following contract award. As this report is generally the direct responsibility of the managing contractor, its contents should have received attention during the precontract negotiations. If the contract were lump-sum, the negotiations should have been very detailed and the report contents a matter of contractual agreement. On reimbursable projects, the discussions and agreement can take place after contract award.

Narratives for the reporting period should include status, trends, and criticality. Significant cost and schedule deviations should be highlighted. Actions should be proposed for remedial action. Where recovery is not possible, a cost and schedule impact evaluation should be made.

A minimum package for the monthly report should include:

- executive summary;
- project cost report;
- project status report:

  - engineering progress,
  - material commitment progress,
  - construction progress,
  - engineering productivity and personnel curves, and
  - construction productivity and personnel curves;

- project master schedule;
- subcontracts report (if the project has major subcontract elements);
- trend report; and
- contingency rundown curve.

A brief narrative should:

- outline the progress for the period, the cost status, deviations, and trends;
- highlight critical activities; and
- provide a forecast of cost and probable schedule completion.

The *executive summary* is the key to a quality report; it must be short, sharp, and attention-getting. Also, maximizing graphical content of the monthly cost and progress reports will enable management to more quickly absorb the status, trends, and forecast for the project.

M. Monthly Executive Progress Report

In addition to a weekly activity report (major and critical items) and the detailed monthly cost and progress report, a brief monthly report should be produced for executive management. This is a vital document as it is probably the only project document that is studied by senior management and it can, therefore, obtain or lose executive management attention and support. Figure 9.13 illustrates a suitable one-page report. The report should be written in a terse style and should include:

- concise, but complete, financial statement;
- brief report on project progress versus predicted schedule; and
- description of any anticipated delays.

- If appropriate, the executive summary of the monthly cost and progress report can be added to this report.

Each monthly report should reflect a complete statement of project status and should not rely on previous reports for a detailed picture of the financial and schedule status. Actual versus planned progress curves for engineering and construction should be known.

N. Backcharge Register

Charges against vendors or contractors for extra work can be significant. Backcharges result from poor vendor-contractor workmanship, schedule delays, lack of resources, failure to clean up, etc. Accurate record-keeping and documentation are vital.

Several important points should be observed regarding backcharge administration:

Documentation. All pertinent costs associated with backcharges should be recorded and updated regularly;

Early notification. The vendor or contractor should be advised in writing of a backcharge, including an estimate of the associated cost, as soon as possible; and

Cost forecast. The actual cost should be discounted when taking credit for potential recoverable backcharges in a cost forecast, since full recovery is very unlikely, often falling in the range of 20–40%.

Figure 9.14 illustrates a typical backcharge register. The data compiled on this form can be used for:

- assessing the financial impact on project costs as backcharges are settled with vendors, and
- highlighting problem vendors and contractors for quality control purposes.

VII. KEY PROJECT CONTROL TECHNIQUES FOR DESIGN ENGINEERING

A. Engineering Cost Control—Overall

The greatest proportion of design changes (in-scope and out-of-scope) will occur during the engineering phase. These changes can impact on all or many phases

MONTHLY EXECUTIVE PROGRESS REPORT		
PROJECT			

LOCATION	JOB NO.	DATE

SCOPE (Brief Description)

FINANCIAL

AFE ESTIMATE – _____ DATE _____

REVISED AFE – _____ DATE _____

REVISED AFE – _____ DATE _____

SCHEDULE

ORIGINAL AFE SCHEDULE _____ REVISED TO _____ DATE _____

REVISED TO _____ DATE _____ REVISED TO _____ DATE _____

CONTRACTOR

CONTRACT BASIS

PROJECT STATUS–COMMENTS

	ENGINEERING	CONSTRUCTION
100		
90		
80		
70		
% 60		
50		
40		
30		
20		
10		

	CURRENT STATUS/FORECAST	ORIGINAL PLAN
MECHANICAL COMPLETION DATE		
OPERATIONAL COMPLETION DATE		
ENGINEERING % COMPLETE		
PROCUREMENT % COMPLETE		
CONSTRUCTION % COMPLETE		

PROJECT COST	AFE	COMMITED TO DATE	FORECAST LAST PERIOD	FORECAST THIS PERIOD

OWNER PROJECT MANAGER	CONTRACTOR PROJECT MANAGER

Figure 9.13 Monthly executive progress report form.

Figure 9.14 Backcharge register.

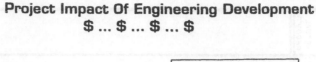

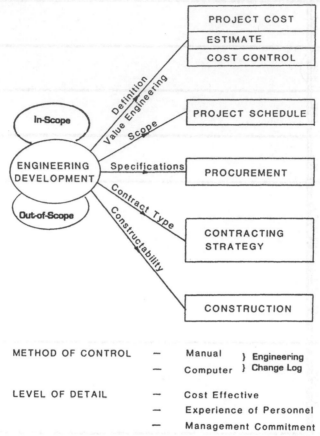

Figure 9.15   Engineering development flowchart.

of the work, such as reduced engineering productivity, increased material costs, different contracting arrangements, additional construction hours, more field resources, etc. Figure 9.15 is a flowchart that illustrates these elements.

It is, therefore, vital that these changes/development be identified and tracked from the earliest possible moment. The tracking can be accomplished with a trending/design change log procedure. The bases for control purposes are the drawing and requisition registers or indexes, where the first listings are developed through the experience and estimates of the lead design engineers. The quality of these estimates can be enhanced with good historical databases. As the drawings and requisitions indexes are refined, so too is the quality of control refined. In addition, the use of an earned value system, with the monitoring of engineering productivity analysis of actual against budgeted levels, can show deviations of jobhours and costs.

Figure 9.16 shows a flowchart of an engineering cost control system and displays typical control methods. If applicable (depending on the contractual basis), owner design

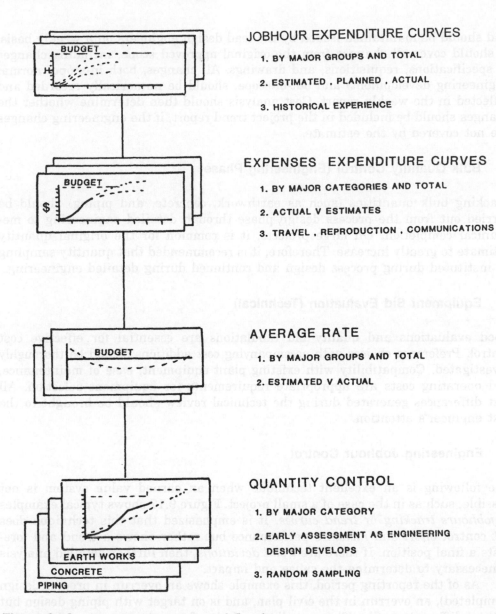

Figure 9.16 Engineering cost control flowchart.

specifications should be carefully evaluated for conflicts with contractor specifications. During the early engineering phase of the project, cost engineers should ensure that effective communication channels are set up with design personnel, provide prompt and accurate cost estimates, and develop cost consciousness within all groups. This requires people skills in addition to technical expertise.

B. Engineering Change Log

The objective of the engineering change log (see Fig. 9.15) is to track changes and their cost impact. This log is a vital part of the project trending program and should be maintained-issued by the lead design engineers on a weekly basis. It should cover all changes from the original approved scope, including changes to specifications, requisitions, and drawings. All changes, both in-scope (normal engineering development) and out-of-scope, should be numerically recorded and reflected in the weekly report. Cost analysis should then determine whether the changes should be included in the project trend report, if the engineering changes are not covered by the estimate.

C. Bulk Quantity Control (Engineering Phase)

Tracking bulk quantities (such as earthwork, concrete, and piping) should be carried out from the process design phase through detailed engineering to mechanical completion. On large projects, it is common for the original quantity estimate to greatly increase. Therefore, it is recommended that quantity sampling be instituted during process design and continued during detailed engineering.

D. Equipment Bid Evaluation (Technical)

Good evaluations and quality bid tabulations are essential for effective cost control. Preference choices with accompanying cost additions should be thoroughly investigated. Compatibility with existing plant equipment, ease of maintenance, and operating costs are appropriate requirements for equipment selection. All cost differences generated during the technical review should be brought to the cost engineer's attention.

E. Engineering Jobhour Control

The following is an excellent technique when an earned value system is not possible, such as in the case of a small project. Figure 9.17 shows typical examples of *jobhours tracking* or *trend curves*. It is emphasized that this technique does not control effort and measure performance but rather shows a trend and forecasts a final position. If there are *trend deviations*, then further detailed analysis is necessary to determine the cause and impact.

As of the reporting period, this example shows an overrun in process design (completed), an overrun in the civil plan, and is on target with piping design but with a different profile. The overall cumulative forecast is overrun. Projecting forward at a rundown profile similar to the plan will give a forecasted overrun at completion. However, it could be that this *reported overrun* is because work is ahead of the plan and is not, in fact a *cost overrun*. For appropriate size projects, target curves should be developed for individual engineering disciplines (such as process, civil and structural design, equipment and mechanical, piping, electrical, and instrumentation).

Historical data can assist in developing accurate planned curves. However, development will require skill and experience. In practice (especially on large, reimbursable-type projects), recycling of engineering, causing increases in jobhours, can be due to owner changes. Such situations can be controlled with the jobhour tracking curves and the following jobhour rate profiles.

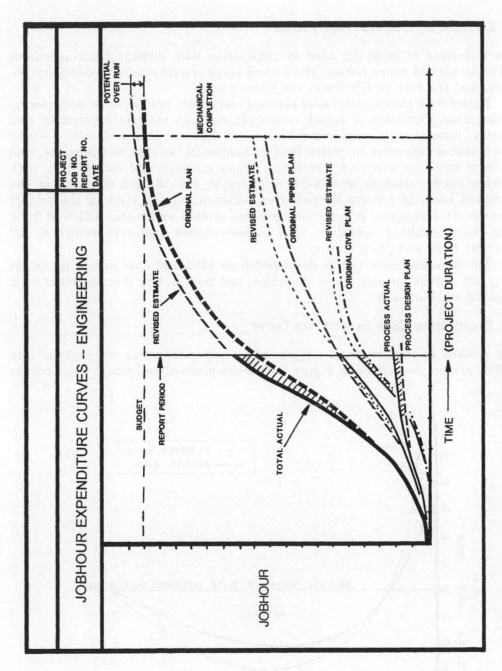

Figure 9.17 Jobhour expenditure curves—engineering.

F. Engineering Jobhour Rate Profile

This technique is generally used in conjunction with jobhour tracking curves and/or an earned value system. The earned value system tracks productivity/jobhours, and the rate profile tracks the jobhour cost.

Figure 9.18 shows cumulative planned and actual profiles of an engineering jobhour rate. Deviation of actual versus planned can indicate a potential cost overrun. Rate overruns can occur because of different mixes of salary levels, unanticipated increases in salary levels, changes in benefits and burdens, and premium costs for overtime. The figure shows an estimated rate of $20, with tracking curves starting at $23–26, dropping to $17–18, and finishing at the estimated level. It reflects high-salaried personnel at the start of the project (department managers, project management, process engineers), followed by a lower-priced drafting operation, with higher-salaried engineers returning for punchlist work and plant startup.

The planned curve can be developed from historical data and/or by evaluating rate curves for each major discipline, and then adding them together on a weighted jobhour basis.

G. Engineering Cost Expenditure Curve

This method is used on small projects where separate curves and jobhour rate profiles are not cost effective. Figure 9.19 shows planned and actual trend curves

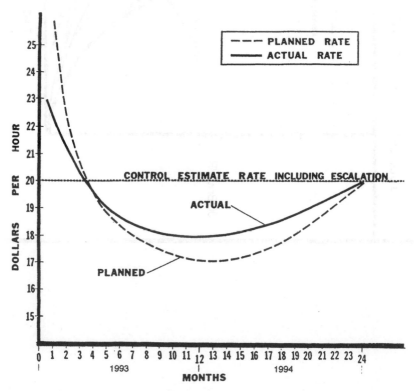

Figure 9.18   Engineering jobhour rate curve.

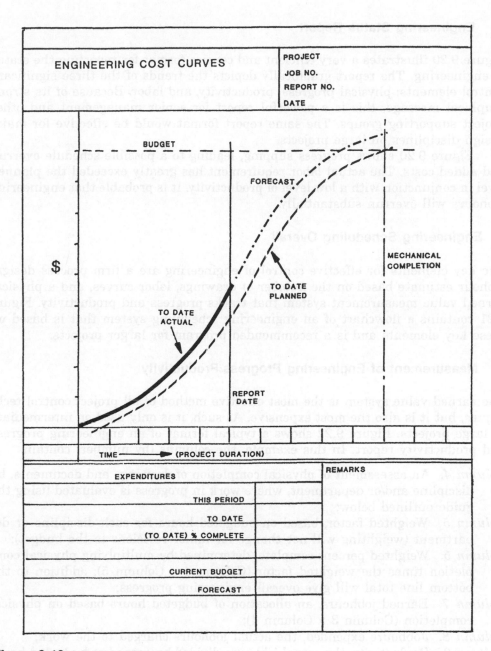

The figure contains the following labels:

ENGINEERING COST CURVES

PROJECT
JOB NO.
REPORT NO.
DATE

BUDGET

FORECAST

MECHANICAL
COMPLETION

$

TO DATE
PLANNED

TO DATE
ACTUAL

REPORT
DATE

TIME ⟶ (PROJECT DURATION)

EXPENDITURES		REMARKS
THIS PERIOD		
TO DATE		
(TO DATE) % COMPLETE		
CURRENT BUDGET		
FORECAST		

Figure 9.19   Engineering cost curves.

for total engineering costs. A planned curve should be developed with judgment, using the engineering schedule and related historical data.

This example shows actual expenditures following a similar profile as planned but consistently at a higher level. It is, therefore, reasonable to predict a final overrun with a projection following the same rundown profile but starting from the actual value (overrun) at the month twelve reporting date.

H.   Engineering Status Report

Figure 9.20 illustrates a very efficient and concise format for reporting the status of engineering. The report graphically depicts the trends of the three significant control elements: physical progress, productivity, and labor. Because of its strong graphical message, this is a powerful report for senior management and other project supporting groups. The same report format would be effective for major design disciplines on large projects.

Figure 9.20 shows progress slipping, leading to a possible schedule overrun and added costs. The actual labor requirement has greatly exceeded the planned level in conjunction with a low level of productivity. It is probable that engineering jobhours will overrun substantially.

I.   Engineering Scheduling Overall

The key elements for effective control of engineering are a firm process design, jobhour estimate based on the number of drawings, labor curves, and a physical earned value measurement system that tracks progress and productivity. Figure 9.21 contains a flowchart of an engineering scheduling system that is based on these key elements and is a recommended program for larger projects.

J.   Measurement of Engineering Progress/Productivity

The earned value system is the most effective method of all project control techniques, but it is also the most expensive. As such, it is only used on intermediate or large projects. Figure 9.22 shows a typical format of an engineering progress and productivity report. In this example, the columns (by number) contain:

Column 4. An assessment of physical completion of drawings and documents, by discipline and/or department, where work in progress is evaluated using the guide outlined below;

Column 5. Weighted factor, based on budgeted hours for each discipline or department (weighting will not change for minor revisions to the budget);

Column 6. Weighted percent complete, determined by multiplying physical completion times the weighted factor (Column 4 × Column 5); addition to the bottom line total will give overall engineering progress;

Column 7. Earned jobhours, an allocation of budgeted hours based on physical completion (Column 3 × Column 4);

Column 8. Jobhours expended, the actual jobhours charged to the work;

Column 9. Productivity, the actual jobhours divided by earned or budgeted hours (Column 8 ÷ Column 7); and

Columns 10, 11, and 12. Used for historical records if necessary.

K.   Guide to Completion of Design and Drafting

The stages of completion given in Table 9.1 are a guide to determining the percent complete of specifications or drawings. These percentages are determined by discipline supervisors. The guide covers only drawings and specifications. Other items (such as coordination and supervision) within the engineering scope are

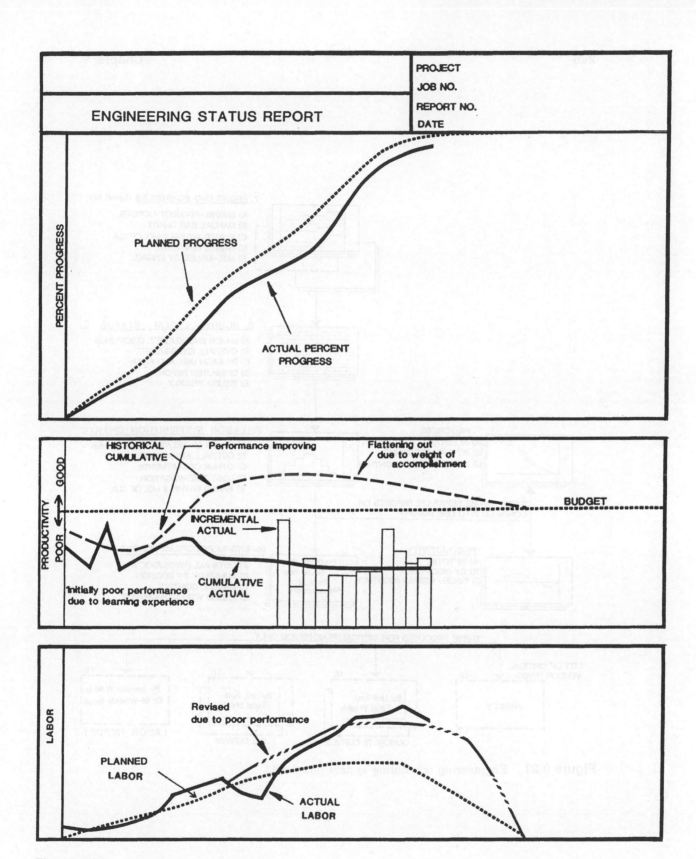

Figure 9.20 Engineering status report.

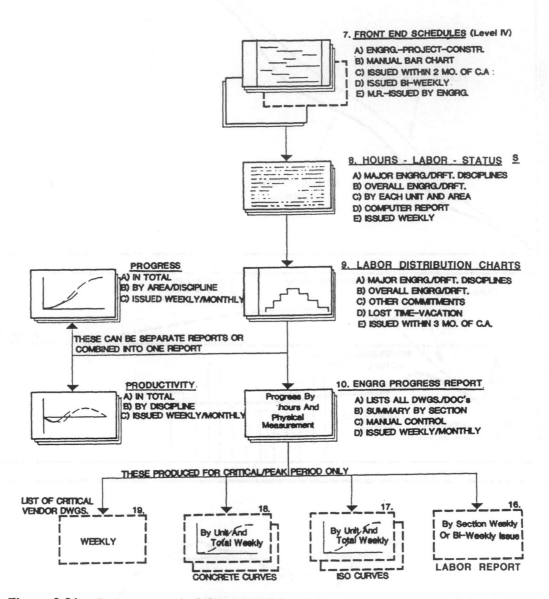

Figure 9.21 Engineering scheduling system flowchart.

ENGINEERING PROGRESS & PRODUCTIVITY REPORT									PERIOD		PROJECT		
1	2	3	4		5	6		7		8		9		10	11	12	
A/C	SECTION	BUDG. HRS.	%		WTD. FACTOR	WEIGHTED-%		EARNED HOURS		ACTUAL HOURS		PRODUCTIVITY		NO. OF DWGS.	BUDGET HOURS PER DWG.	ACTUAL HRS. PER DWG.
			INC.	CUM.		INC.	CUM.	INC.	CUM.	INC.	CUM.	INC.	CUM.			INC.	CUM.
TOTAL																	

Figure 9.22 Engineering progress and productivity report.

Table 9.1 Stages of Completion

	% complete
Specifications	
Complete draft	20
Write specification	70
Check specification	85
Issue for approval	85
Issue for construction	85
Issue revisions as required	95–100
Architectural drawings	
Complete sketches and general arrangements (GA)	15
Issue sketches and GA for approval as required	25
Complete drawing	75
Issue for approval	85
Issue for construction	90
Issue revisions as required	95–100
Civil drawings	
Complete preliminary site plan (building locations and site elevation)	10
Issue preliminary site plan for approval as required	25
Complete design calculations	30
Complete drawing	80
Issue for approval	85
Issue for construction	95
Issue revisions as required	95–100
Concrete and foundation drawings	
Complete design calculations	25
Complete drawing	60
Check drawing	85
Issue for approval	90
Issue for construction	95
Issue revisions as required	95–100
Steel and superstructure drawings	
Complete design calculations	25
Complete drawing	60
Check drawing	85
Issue for approval	90
Issue for construction	95
Issue revisions as required	95–100

Table 9.1 (Continued)

	% complete
Electrical drawings	
Complete design calculations	15
Complete drawing	75
Check drawing	85
Issue for approval	90
Issue for construction	95
Issue revisions as required	95–100
Instrumentation drawings	
Complete design calculations	50
Complete drawing	70
Check drawing	85
Issue for approval	90
Issue for construction	95
Issue revisions as required	95–100
Mechanical general arrangement drawings	
Complete design calculations	15
Complete preliminary GA drawings	20
Issue preliminary GA drawings for approval	40
Complete drawing	65
Issue for approval	75
Issue for construction	80
Issue revisions as required	95–100
Mechanical and piping drawings	
Complete design calculations	10
Complete drawing	65
Check drawing	85
Issue for approval	90
Issue for construction	95
Issue revisions as required	95–100
Flow sheet drawings	
Complete design calculations	40
Complete drawing	60
Check drawing	70
Issue for approval	80
Issue for construction	95
Issue revisions as required	95–100

not so easily quantified. These should be treated as below the line items and given the same measure of completion as the quantified work.

L.   Engineering/Drawing Status Report

Figure 9.23 shows a detailed and efficient format for reporting the status of engineering/drawings. The report covers progress, schedule, and jobhours. It provides source information for the project report documents and for evaluating engineering progress. Individual drawing and document status is evaluated by discipline supervisors. Due to schedule pressure and over-optimism, it is common for engineering supervision to overstate the progress of the work, therefore it is essential that the numbers be verified.

M.   Engineering Progress (Curves)

The curves shown in Figure 9.24 monitor overall cumulative engineering progress. Actual and forecasted progress are shown relative to the original or revised schedule curves. Overall progress is compiled from the individual curves developed for each engineering discipline. Progress should be physical measurement based on completion of drawings and engineering documents, in accordance with the techniques outlined above.

Work unit tracking curves for production drawings provide a substantial part of the data needed to assess overall engineering progress. Jobhour expenditure can be a poor basis for reporting progress; it is, however, a temporary alternative that can be used until drawing and document quantities are known.

N.   Engineering Productivity (Curves)

An estimated productivity profile should be developed as early as possible. It should be based on historical data and specific project conditions. Such a profile can provide an early warning of potential overruns if actual productivity varies significantly from the anticipated curve.

1.   Engineering Productivity Profile or Index

Figure 9.25 shows a typical cumulative productivity profile. Productivity is assessed by comparing earned value, which is based on percent of budgeted jobhours complete, against the number of actual jobhours expended:

$$\text{Productivity} = \frac{\text{Physical completion} \times \text{budget jobhours}}{\text{Actual jobhours}}$$

This example shows productivity during the initial stage of engineering to be unusually low, due to unproductive jobhours expended at kickoff meetings, setting up project controls, recycling of the process design, abortive layout work, etc. Productivity should improve and stabilize during production engineering, and gradually decrease over the latter phases of the project when more jobhours are required for engineering rework and closeout. For ease of reporting, progress and productivity curves may be combined on a single report.

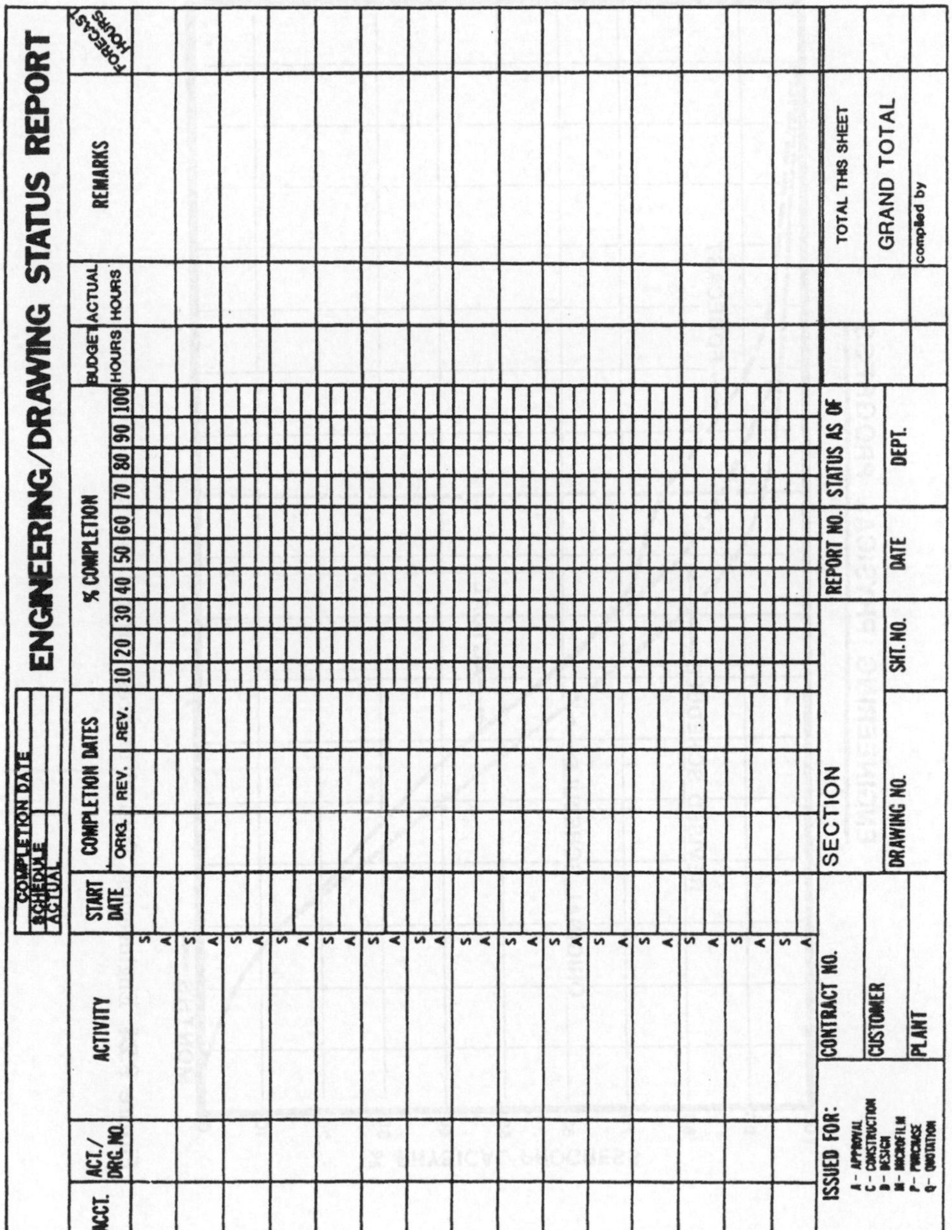

Figure 9.23 Engineering/drawing status report.

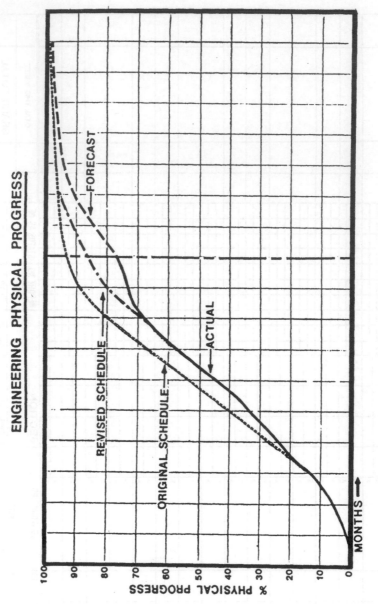

Figure 9.24 Engineering physical progress curves.

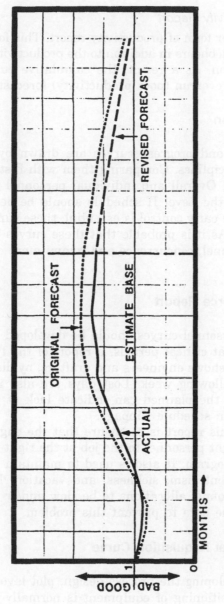

Figure 9.25 Productivity index profile.

2. Engineering Productivity Report

Figure 9.26 shows another form of productivity report. This format covers performance data and forecasted jobhours in addition to the productivity curve. The example shows sharp deterioration. As a result, the cumulative curve drops below the historical curve, and an overrun (poor productivity) forecast is made.

O. Personnel Allocation

The planned personnel and progress curves are drawn by weighting together curves for individual disciplines. Comparing them with historical standards can validate the assessment. Overall and individual personnel curves based on the engineering budget and the Level II schedule should be developed. Figure 9.27 shows such curves. Since early control is essential, these curves should be drawn up as soon as possible. As it is probable that these curves will be developed by project/scheduling personnel, concurrence and commitment of engineering supervision must be obtained.

P. Engineering Workforce Report

As stated previously, personnel curves should be developed as soon as practical. In the early stages and at critical periods, a report of the type shown in Figure 9.28 is recommended; it shows engineers and drafters, by discipline, assigned for the week and for the following week. Lost days are also reported. Differences between the actual and the planned can indicate lack of personnel or lack of work, which can result in schedule slippage.

The major use of this report is to ensure that the engineering department properly allocates the right personnel to the job at the right time and, thereafter, maintains the correct program. It also is used to maintain a check on lost time allowances, such as absenteeism, sickness, and vacation. It is common at the start of projects for personnel allocations to be slow and/or the wrong mix. This technique can provide the data to prevent this problem.

Q. Engineering Material Requisition Curve

In conjunction with developing the process design, plot layouts, and engineering specifications, the requisitioning of equipment is normally on the critical path. Thus, plotting the actual issuance of requisitions against time can provide an excellent overview and control of the work.

Figure 9.29 shows an engineering material requisition control curve. Tabulated data can be detailed further by showing material categories if desired. This example shows that actual issues are behind the planned curve. If the planned curve has been drawn from the detailed engineering schedule, the situation could be serious and require further investigation at the detailed level.

The curves represent the planned and actual issues of material requisitions from the engineering department to the procurement department. The procurement department will then require another two to four weeks to add the commercial sections and issue the completed inquiry package. A curve representing the later issue is shown in the procurement section of this chapter.

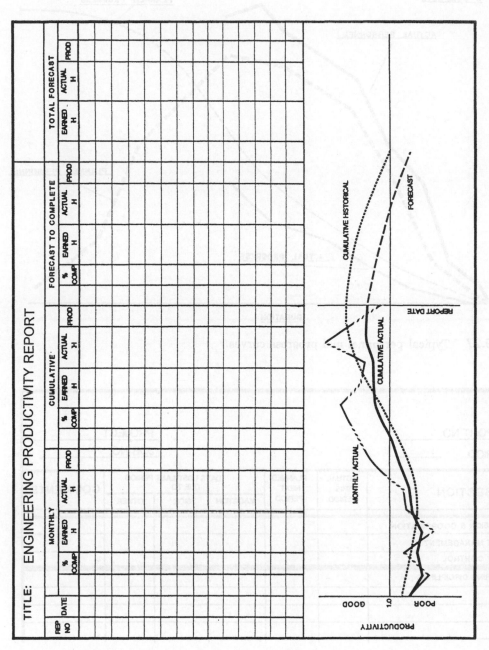

Figure 9.26 Engineering productivity report form.

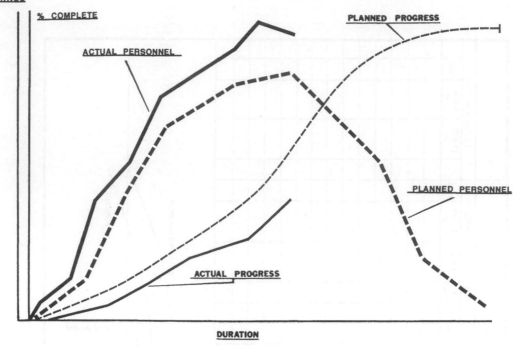

Figure 9.27 Typical personnel and progress curves.

<table>

| REPORT NO. ____ | | | | PROJECT ____ | | |
| PERIOD ____ | | | | JOB NO. ____ | | |
</table>

SECTION	ACTUAL THIS PERIOD	PLANNED NEXT PERIOD	DAYS LOST LAST PERIOD DUE TO			COMMENTS
			VACATION	SICK	OTHER	
SUPERVISION & COORDINATION						
PROJECT MANAGEMENT						
PROJECT CONTROL						
ENGINEERING DISCIPLINES .						

TOTAL						
CUMULATIVE TOTAL						

Figure 9.28 Engineering—force report.

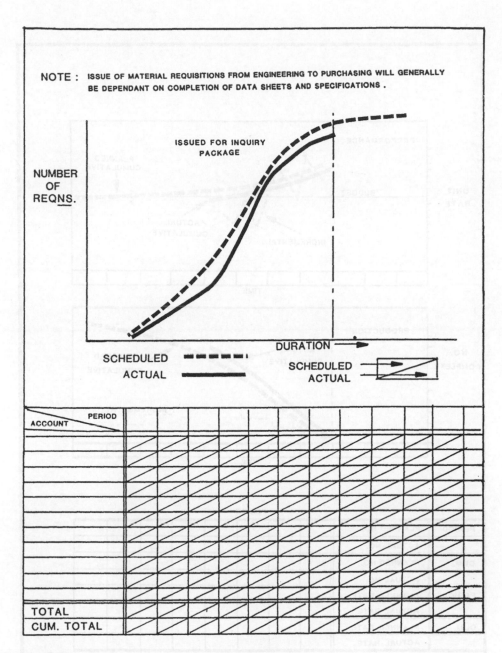

Figure 9.29 Engineering material requisition curve.

The number of requisitions approved by engineering for purchase may also be tracked, with the curve normally being three to four months later than the inquiry curve.

The total number of requisitions should be estimated as early as possible and should be updated as engineering progresses. The curves are key techniques for use in evaluating the overall progress of the engineering-purchasing cycle. Delays in issuing material requisitions can result in late ordering and corre-

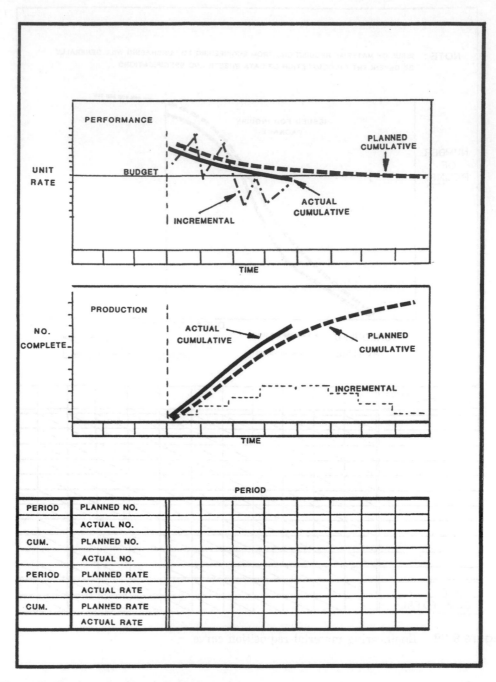

Figure 9.30 Engineering work unit tracking curves.

spondingly late delivery of materials and equipment to the job site. Feedback of the vendor's information to engineering is also affected, which in turn can delay the engineering program. For these reasons, careful monitoring and control of the material requisition program is essential.

R. Engineering Tracking Curves

Production of concrete foundation drawings and piping isometrics usually require careful monitoring. Both disciplines use a high proportion of engineering jobhours and are usually critical to the project schedule. Curves such as those shown in Figure 9.30 can be used to monitor the work.

Productivity curves (shown in the top diagram of Figure 9.30) record monthly and cumulative values for jobhours spent per drawing. The histograms (shown in the center diagram) indicate planned and actual numbers of drawings issued each month. Cumulative totals are also plotted. Statistical data is tabulated at the bottom of the form. This technique can be used for any significant element of work. During critical periods, the frequency of monitoring would be increased from monthly to bi-weekly or even weekly if necessary.

S. Piping Design Activities and Progress Curves

Since piping design is often the critical path, detailed schedules and programs should be developed for this work. Figure 9.31 provides an example of a curve for this work. It shows a typical relationship among major piping design activities for a single unit and weights them together for overall progress. Personnel resources should be applied to the individual activities.

Projects can vary considerably in their process design, plot layouts and detailed engineering philosophies, so this typical schedule should be adjusted for project-specific circumstances. Project-specific schedules should be drawn against a calendar base. The typical schedule shown is based on return data and can therefore be used to develop a project-specific schedule.

T. Vendor Drawing Control

An engineering schedule can be affected by the lack of vendor drawings. Foundation design and piping design depend on vendor drawings for details of foundation bolts, bearing loads, nozzle orientation, pressure, and capacity data. Thus an effective vendor drawing control system is essential for planning and scheduling engineering work. A bid tab evaluation should cover the issue of vendor drawings, and the successful vendor should be monitored thereafter to ensure compliance with promised or need dates. This is essential in the case of critical vendor drawings. A vendor material expediting program should also cover the supply and issue of vendor drawings. Schedule engineers should coordinate this effort with the expediting department.

Figure 9.32 shows a form that can be used to control critical vendor drawings. The same format can be used for all vendor drawings; however, critical drawings should be identified separately.

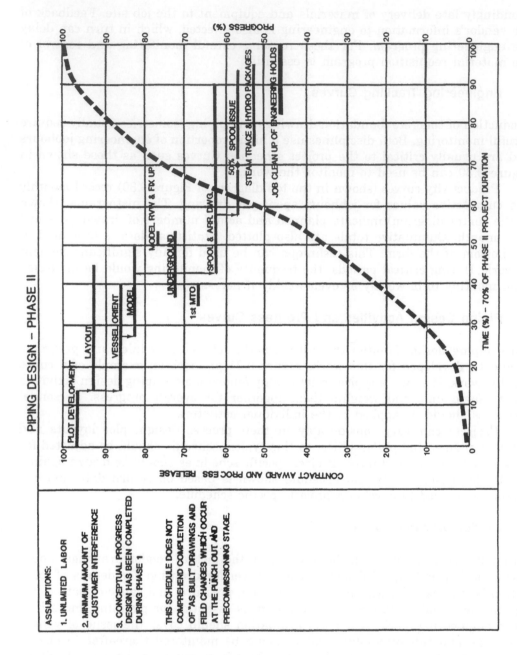

Figure 9.31 Piping design progress curve.

DATE				PROJECT						
				FOR APPROVAL			FINAL			
P.O. NO	EQUIPMENT	VENDOR	VENDOR DRAWING	REQ'D	REC'D	RET'D	REQ'D	RET'D	REQ'D	REC'D

CRITICAL VENDOR DRAWING LIST

Figure 9.32 Critical vendor drawing list.

VIII. KEY PROJECT CONTROL TECHNIQUES FOR PROCUREMENT

A. Procurement Cost Control—Overall

Project purchasing strategy should specify a proposed split of domestic and international purchasing. In the control budget, escalation rates and design allowances should be established by prime account. Use and coordination of overseas purchasing offices should be specified. Project conditions such as schedule acceleration, national purchasing preferences, and special owner requirements need to be evaluated for their cost impact.

Figure 9.33 displays a flowchart of a purchasing cost control system and shows methods by which cost control can be exercised. A purchase order commit-

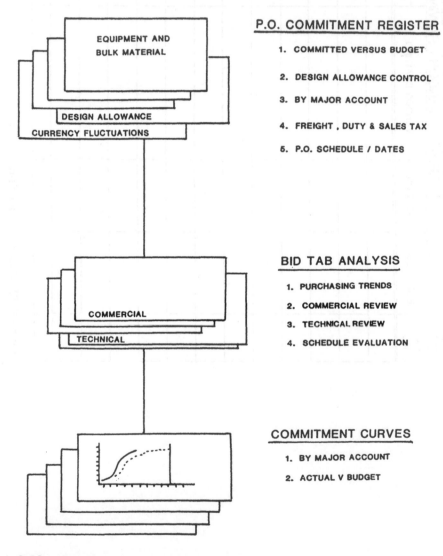

Figure 9.33   Procurement cost control flowsheet.

ment register should be maintained by prime account and separated for the following major categories:

- firm-price equipment orders,
- open-ended bulk material orders, and
- field-purchased materials.

Curves showing actual versus planned commitments should be considered for these categories, for larger projects and where schedule/progress is critical.

Generally, *bid groupings* exceeding the control budget by 20% should be thoroughly investigated. Such deviations could signal a fat design, poor bid documents resulting in a high bid, tight market conditions, or a poor estimate.

Several major considerations must be taken into account when planning for procurement cost control:

- optimize commercial terms by taking advantage of cash discounts for quantity and by making prompt payments;
- on large projects, maximize discount terms by ordering similar items on a single purchase order;
- minimize import duties and taxes by requesting special waivers from host countries or by reclassifying items to lower duty categories;
- minimize freight charges by careful selection of carriers; and
- after evaluating potential quantity increases, make a sensitivity analysis of unit price bids to verify that an apparent low bidder is in fact still lowest.

B. Purchasing Plan for Critical Material

Material on a critical path could require premiums for vendor drawings and shorter delivery. Alternatively, special expediting arrangements, such as teams resident in vendor shops, can be required, as can air freight and special handling. Such requirements often lead to additional costs. Refer to the section later in this chapter on procurement scheduling for additional information on the purchasing plan for critical material.

C. Bid Tabulation Procedure

The bid tabulation procedure should cover preparation, control, review, and approval requirements of bid tabulation documents. Levels of approval should be defined, and instructions for opening of tenders and evaluations of tender documents should be given. Budget comparisons should be made, and a negotiations philosophy stated. Bid tabulations should also contain information based on knowledge of or experience with specific bidders if this may have a bearing on their selection.

D. Purchase Order Commitment Register

Figure 9.34 shows a purchase order commitment register listing purchase orders placed, date placed, vendor, value, currency, and delivery date. Design allowances should be carefully reviewed against the latest supplement to a purchase order.

1	2	3	4	5	6	7	8	9	10	11	12
REQN. NO.	P.O. NO.	DATE P.O.	REQD. P.O. DATE	ITEM	VENDOR	COMMITTED	EXPENDED	DESIGN ALLOW. BUDGET	BUDGET	FORECAST	VARIANCE FORCAST BUDGET

PURCHASE ORDER COMMITMENT REGISTER

PROJECT
JOB NO.
REPORT NO.
DATE

Figure 9.34 Purchase order commitment register.

As engineering advances, so the need for design allowance reduces. Freight costs, duties, and taxes should be segregated and monitored.

E. Cost Forecasts of Major Purchase Orders

The purchase order commitment register just discussed may become voluminous. For this reason, *it is recommended that the major cost purchase orders be segregated and reviewed each month for a final cost forecast.* Design allowances for fast-track projects are of particular importance, and currency fluctuations, escalation, and delivery conditions should all be carefully considered.

F. Equipment Cost Curves

Figure 9.35 illustrates a monetary commitment curve for equipment. The curve tracks actual cumulative commitments relative to a planned profile, but only indicates a trend of total commitments. The curve can be used:

- as a graphical representation of work status, and
- as a guide for procurement progress, assuming the planned curve reflects the current schedule.

Significant deviations of actual versus planned values can provide a trend for schedule and cost deviations. Separate curves should be drawn for equipment, bulk material, and major package units that are *supply and install*.

Figure 9.35 shows an overrun of commitments versus the plan, although the overrun could simply be that commitments have been made ahead of the plan and not that costs are overrunning. Alternatively, a schedule check can reveal that the work is on or behind the plan, which would indicate a potential cost overrun. As shown, there is a revised budget (increase), indicating that cost was a problem.

G. Procurement Scheduling—Overall

Major elements for effectively scheduling purchasing work are an accurate material requisition list, a purchase order schedule, and a viable purchasing strategy. Figure 9.36 illustrates key techniques and shows methods by which effective control can be exercised:

- equipment purchase schedules (level IV),
- material requisition curves for inquiry and purchase,
- short-range purchase plan,
- material status report,
- list of critical vendor drawings, and
- material logistics schedule.

The key activities are developing and using:

- a purchasing plan for critical material;
- a project buying strategy and purchasing program (i.e., quality worldwide purchasing and the use of overseas satellite offices);
- effective vendor lists;

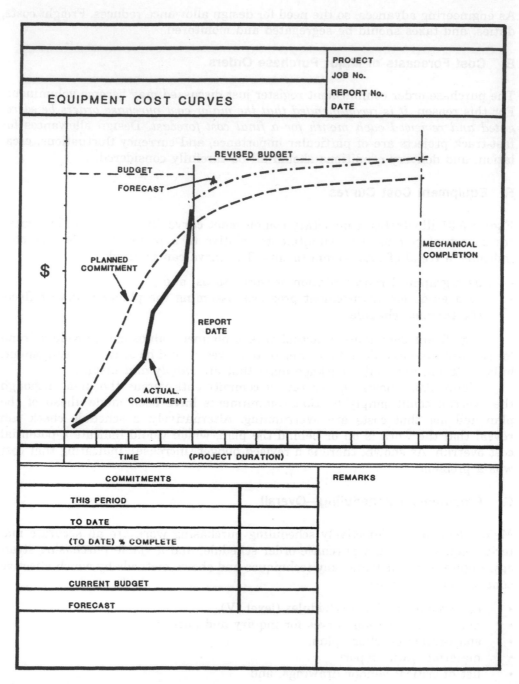

Figure 9.35 Equipment cost curves.

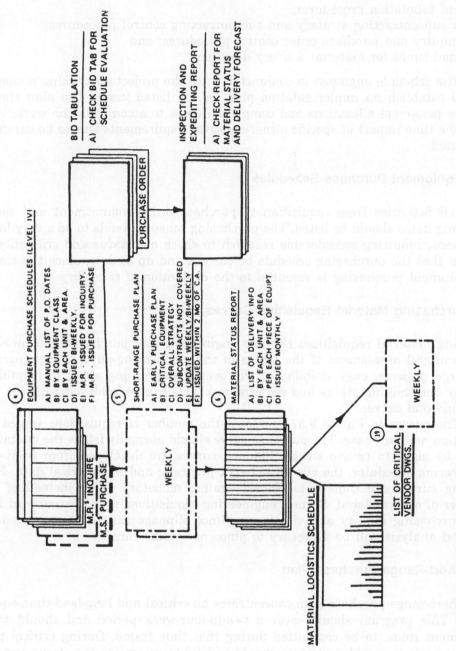

Figure 9.36 Procurement scheduling flowsheet.

- bid tabulation procedures;
- a subcontracting strategy and subcontracting control procedures;
- inquiry and purchase order control procedures; and
- lead times for material delivery durations.

The schedule engineer, in conjunction with the project purchasing manager, should establish an implementation plan for the listed items. The plan should include personnel allocations and completion dates to accomplish the work. Further, the time impact of specific owner-approval requirements should be carefully evaluated.

H. Equipment Purchase Schedules

Detailed activities from requisition to purchase order commitment and corresponding dates should be listed. The purchasing schedule tends to be a very large document, requiring considerable research to check out status and criticality. It is vital that the purchasing schedule be current and up to date, since the status of equipment purchasing is essential to the evaluation of criticality.

I. Purchasing Material Requisition Curves

Tracking material requisitions through engineering and purchasing can provide a meaningful assessment of the status of the engineering-purchasing program. On large projects, easy visibility of the overall purchasing effort is helpful in quickly determining status and performance. This can be achieved with the use of requisition curves.

The curves in Figure 9.37 represent the number of requisitions issued for quotation and purchase. The purchase curve should normally follow the quotation curve by eight to twelve weeks. Planned curves are developed from front-end engineering schedules, the overall network program, and/or historical data. Purchasing curves are sometimes plotted against monetary value instead of the number of orders placed. Several engineering requisitions may be combined into one purchasing inquiry and vice versa. Since slippage can have many causes, detailed analysis will be necessary to pinpoint specific causes.

J. Short-Range Purchase Plan

The short-range purchase plan concentrates on critical and long-lead-time equipment. This program should cover a two-to-four-week period and should show equipment items to be committed during that time frame. During critical purchasing periods, weekly meetings should be held to determine the status and the need for any remedial action. Project managers should attend these meetings.

K. Material Status Report

Like the purchasing schedule, the material status report is voluminous. Again, special abstracts and summary information are essential in achieving effective control. The most important information contained in the material status report is the anticipated date equipment will arrive at the job site.

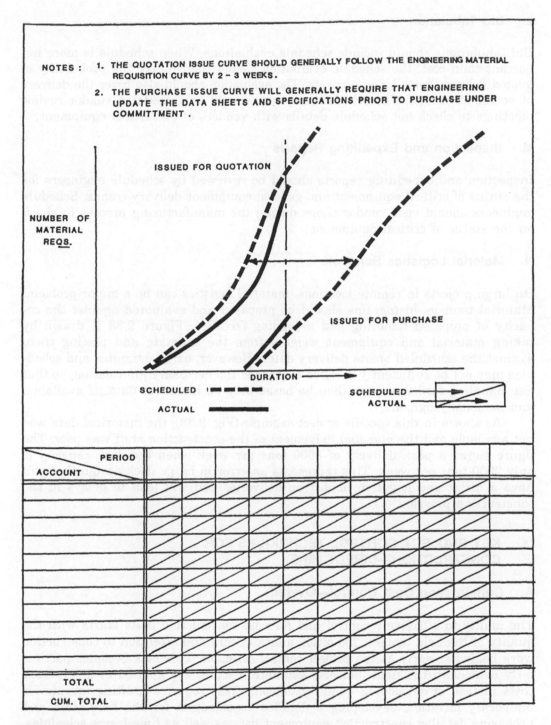

Figure 9.37 Purchasing—material requisition curves.

L. Bid Tabulation

Bid tabulations should include schedule evaluations. When schedule is more important than cost, the schedule evaluation should then reflect the probability of quoted delivery dates being accurate. The evaluation should also cover the delivery of vendor drawings. The schedule engineer should participate in vendor review meetings to check out schedule details with vendors of all critical equipment.

M. Inspection and Expediting Reports

Inspection and expediting reports should be reviewed by schedule engineers for the status of critical equipment and general equipment delivery trends. Schedule engineers should visit vendor shops during the manufacturing process to check on the status of critical equipment.

N. Material Logistics Schedule

On large projects in remote locations, material logistics can be a major problem. Material tonnage histograms should be prepared and evaluated against the capacity of proposed handling and unloading facilities. Figure 9.38 is drawn by taking material and equipment weights from the estimate and placing them against the scheduled onsite delivery dates. However, early estimates and schedules may not be sufficiently detailed to provide the necessary information, so that site handling facilities must then be based only on historical data (if available) and personal judgment.

As shown in this specific project example (Fig. 9.38), the historical data was not available and the personal judgement of the construction staff was poor. The figure shows a peak delivery of 4000 tons per week when the dock capacity is only 2000 tons per week. This represents an error in barge dock design that will have a direct and negative impact on deliveries and will lead to delays in the construction program.

IX. KEY PROJECT CONTROL TECHNIQUES FOR CONSTRUCTION—DIRECT HIRE

A. Construction Cost Control—Overall

The ability to effectively control and forecast construction costs starts with the quality of the field budget. Too often, a lack of care and attention to construction work begins as early as the project control estimate. This lack becomes particularly evident in the field indirects estimate. Factors are applied to direct-labor costs instead of developing quantity takeoffs. This requires drafting layouts for temporary facilities, developing a detailed organization for the field staff, and preparing detailed construction equipment lists as well as time-frame schedules. Many projects experience overruns in field indirects due to a poor estimate rather than a lack of control and poor site management.

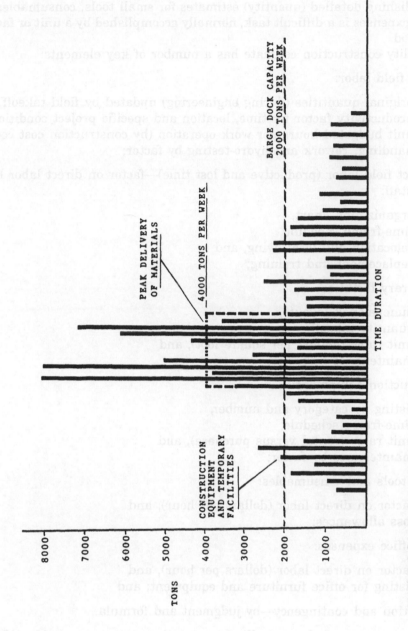

Figure 9.38 Material logistics schedule.

Establishing detailed (quantity) estimates for small tools, consumables, and field office expenses is a difficult task, normally accomplished by a unit or factored basis method.

A quality construction estimate has a number of key elements:

- direct field labor:

    - original quantities (during engineering) updated by field takeoff,
    - productivity factor for time, location and specific project conditions,
    - unit budgeted hours per work operation (by construction cost codes),
    - handling, rework and hydro-testing by factor;

- indirect field labor (productive and lost time)—factor on direct labor hours;
- field staff:

    - organization chart,
    - time-frame schedule,
    - relocation and local living, and
    - replacement and training;

- temporary facilities:

    - dimensioned layouts,
    - quantity takeoff,
    - unit rates (dollars per square foot), and
    - maintenance by factor;

- construction equipment:

    - listing by category and number,
    - time-frame schedule,
    - unit rates (rental versus purchase), and
    - maintenance by factor;

- small tools and consumables:

    - factor on direct labor (dollars per hour), and
    - loss allowances;

- field office expenses:

    - factor on direct labor (dollars per hour), and
    - listing for office furniture and equipment; and

- escalation and contingency—by judgment and formula.

The major elements for controlling construction costs are early identification of quantity variances, labor productivity, and craft rates, and continuous evaluation of field indirects. A quantity field budget, showing clearly defined units of work, is essential. On large, reimbursable-type projects, it is recommended that a separate labor rundown control system be instituted for the last 20% of the project. Figure 9.39 is a flowchart of a cost control system and shows key cost control techniques.

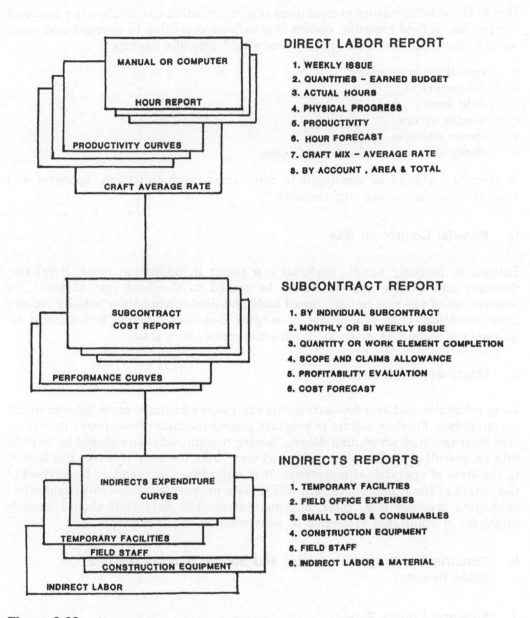

DIRECT LABOR REPORT

1. WEEKLY ISSUE
2. QUANTITIES – EARNED BUDGET
3. ACTUAL HOURS
4. PHYSICAL PROGRESS
5. PRODUCTIVITY
6. HOUR FORECAST
7. CRAFT MIX – AVERAGE RATE
8. BY ACCOUNT , AREA & TOTAL

SUBCONTRACT REPORT

1. BY INDIVIDUAL SUBCONTRACT
2. MONTHLY OR BI WEEKLY ISSUE
3. QUANTITY OR WORK ELEMENT COMPLETION
4. SCOPE AND CLAIMS ALLOWANCE
5. PROFITABILITY EVALUATION
6. COST FORECAST

INDIRECTS REPORTS

1. TEMPORARY FACILITIES
2. FIELD OFFICE EXPENSES
3. SMALL TOOLS & CONSUMABLES
4. CONSTRUCTION EQUIPMENT
5. FIELD STAFF
6. INDIRECT LABOR & MATERIAL

Figure 9.39   Construction cost control flowsheet.

B. Recording and Reporting Extra Work

Due to the volatile nature of conditions at a construction site, change is a constant companion. A field trending system that reflects costs due to changes and extra work is therefore essential. The system would typically capture:

- specification changes,
- design errors,
- field errors,
- vendor errors,
- owner changes, and
- changed or unusual site conditions.

A procedure should be developed to cover extra work initiation, approval and authorization, reporting, and closeout.

C. Material Control on Site

Failure to properly handle material can result in additional costs. Breakage, damage, and trouble reports should be routed to the field cost engineer for assessment of the cost impact. Small tools and field consumables usually require close attention. Cost evaluations of surplus materials, spares, and startup requirements should be made as construction draws to a close.

D. Machinery Protection

Long schedules and schedule extensions can cause additional costs for equipment maintenance. Further, failure to properly protect machinery can result in serious cost increases and scheduling delays. Vendor recommendations should be written into an overall procedure to protect machinery from the time it leaves the factory to the time of operational acceptance. It is important that vendors be advised of the length of time their equipment will remain unoperated, since this could affect packaging specifications, costs, and guarantees. The field staff should identify instances of additional protection or winterization requirements.

E. Construction Progress/Jobhour and Budget Reports (Earned Value System)

1. Progress/Jobhour Report

Figure 9.40 depicts a format for reporting actual jobhours and earned budget value in the earned value system (EVS). Progress is measured by budget jobhours. A jobhour prediction is based on a judgment analysis of productivity versus actual hours. Productivity is derived from actual hours expended versus budget hours earned. Indirect jobhours should be shown separately. Indirect jobhours are not quantity based. It is recommended that an indirects budget or equivalent earned value be based on direct work progress and judgment for each major indirect jobhour account.

SUMMARY REPORT (PRIME ACCOUNT)									JOB NO: REPORT NO: UNIT: PERIOD:		
		HOURS SPENT		BDGT, FOR WORK DONE		TOTAL	%	PRODUCTIVITY		TOTAL	
DESCRIPTION	CLASS	THIS WEEK	TO DATE	THIS WEEK	TO DATE	BUDGET	COMPL.	THIS WEEK	TO DATE	PREDICTED HOURS	
EARTHWORK											
FOUNDATIONS											
HEATERS											
↓											
SUBTOTAL DIRECT											
TEMPORARY FACILITIES											
TOOLS & EQUIPMENT											
UNALLOCABLE											
SUBTOTAL INDIRECT											

Figure 9.40   Jobhour summary report.

The earned value system is not an absolute productivity measure, since it is based on the estimate. A poor estimate means a poor productivity measurement, although progress is not similarly affected.

2. Budget Report

Figure 9.41, an earned budget report, collects quantities as the work is completed and predicts total quantities based on a field takeoff made from construction issue drawings. These quantities are then converted into an earned value of budget hours, which are subsequently entered on the jobhour report as previously illustrated.

F. Construction Productivity Report (Direct Labor)

Accurate measurement of overall productivity is essential to good craft labor planning and jobhour-cost forecasting. One widely used method, called the *earned budget hour*, or *earned value*, approach, is illustrated in Figure 9.42. This method requires:

- budgeted construction hours,
- actual jobhours expended (monthly and cumulative), and
- physical progress (percent complete, monthly, and cumulative) or earned budget hours (monthly and cumulative).

Productivity is simply a ratio or yardstick to measure performance by using actual jobhours versus budgeted or estimated hours. Poor performance can be

Figure 9.41 Earned budget report.

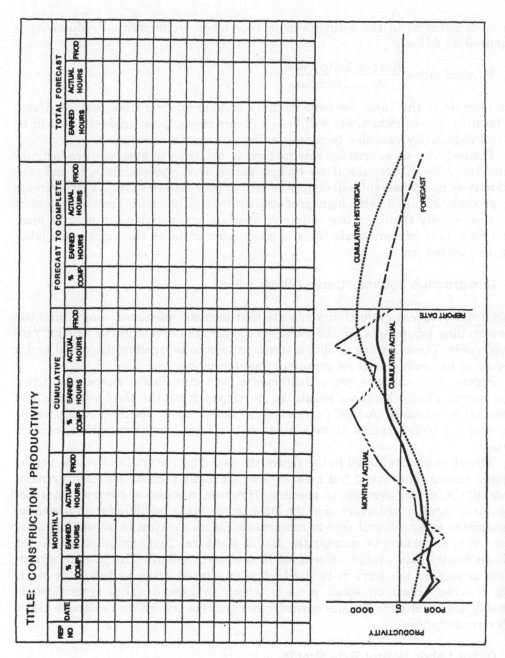

Figure 9.42 Construction productivity form.

due to deficiencies in the estimate or to poor labor productivity. Productivity is computed as follows:

$$\text{Productivity} = \frac{\text{Earned budget hours}}{\text{Actual jobhours}}$$

This formula is the same for both monthly and cumulative time periods. Using the formula, good productivity will be > 1.0; conversely, poor productivity will be < 1.0. Productivity can also be measured by an inverse calculation.

Productivity measured by this method is not true or absolute productivity of the labor force, because if no budget exists or if the estimate is bad, the productivity measurement will simply be zero or poor. Alternatively, a fat estimate will produce an artificially high productivity that is no more representative of true productivity than is a low estimate. Absolute productivity can only be measured on a unit-jobhour basis. This is often referred to as the work unit, labor unit, or jobhour unit.

G.   Construction Jobhour Curve (Direct Labor)

Apart from progress-productivity reports that measure jobhours, major methods for controlling labor costs are the jobhour expenditure curve and the jobhour rate curve/profile. These curves enable a trend pattern to be readily discerned and a forecast to be made based on cumulative performance.

Figure 9.43 shows a set of incremental and cumulative curves for direct labor hours. Planned curves should be developed from the field estimate and construction schedule. Actual experience is plotted on a weekly-monthly basis. The actual jobhour status is tabulated on the form, together with a current forecast.

The example contained in the figure shows a slight overrun of actual versus planned values. This trend has been evident for many months and improvement is not likely, so an overrun is possible. However, a jobhour overrun does not necessarily mean a cost overrun of the labor budget. It is possible for an underrun in estimated labor jobhour cost to compensate for an overrun in actual jobhours. Hence, it is important to monitor the cost of jobhours. This can be accomplished with an hourly rate profile, (discussed in the next section). The jobhour expenditure curve does not have to be used when an earned value system is in place, so it is widely used on small projects where an earned value system would probably not be employed (since earned value systems are not cost-effective under such circumstances).

H.   Direct Labor Hourly Rate Profile

Figure 9.44 shows cumulative planned and actual profiles of the direct-labor hourly rate. This rate represents total labor costs divided by total direct hours. The monthly status is tabulated as shown. Significant deviation of actual versus planned values can indicate a potential overrun. Rate overruns can occur because of different craft mixes, union contract changes, governmental regulation changes, and excess premium costs for overtime and shift work. This example shows the curves tracking closely for the first six months, then an underrun is apparent. If this trend should continue, a significant cost savings would be possible.

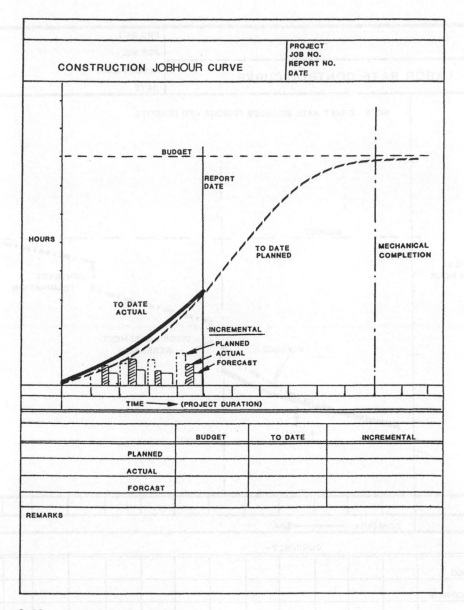

Figure 9.43 Construction jobhour curve.

Abrupt changes in a cumulative profile do not generally occur when the job is well advanced, since there is too much weight of past performance to allow instant or abrupt change. A turndown or underrun can indicate the use of lower-paid crafts, escalation lower than anticipated, or failure by trade unions to achieve pay demands. The following elements are typical for profiles of process projects:

- low at first due to low-skill civil workers (laborers);
- increasing to peak due to highly paid equipment operators, millwrights, pipe fitters, and electricians; and
- a slight downturn due to lower-paid insulators and painters.

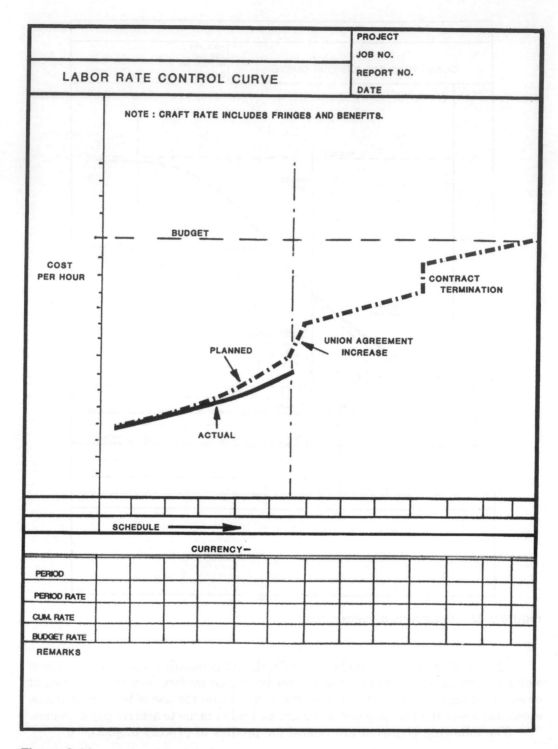

Figure 9.44   Labor rate control curve.

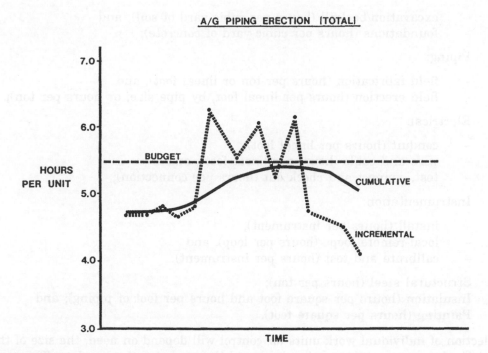

Figure 9.45 Typical work unit tracking curve.

Jobhour expenditure and rate profile curves could also be used to monitor indirect labor costs.

I. Work Unit Tracking Curves

Another strongly recommended approach to measuring productivity is recording actual direct hours (incremental and cumulative) against physical units of completed work (e.g., lineal feet of piping, cubic yards of concrete, tons of steel, etc.). These direct hours per unit are an absolute measure of productivity in contract to the relative measure previously outlined.

Figure 9.45 shows typical tracking curves for above-ground pipe erection. Incremental and cumulative performance is monitored. This example shows a budgeted level of 5.4 direct hours/unit and a current cumulative rate of 5.1—currently a good performance. Based on this performance and judgment of future conditions, a good final forecast could be made. As above-ground piping is the major element of a field budget, piping tracking curves are important control tools. As indicated by the figure, the incremental rate can vary widely; thus the focus should be on the cumulative number.

When work unit tracking curves are needed, they should be prepared early in the construction phase. Significant cumulative deviations (e.g., ±10%) from the budget should be investigated.

The following list represents major work categories and associated work units:

Civil:

- excavation/backfill (hours per cubic yard of soil), and
- foundations (hours per cubic yard of concrete);

- Piping:

  - field fabrication (hours per ton or lineal foot), and
  - field erection (hours per lineal foot, by pipe size, or hours per ton);

- Electrical:

  - conduit (hours per lineal foot),
  - wire and cable (hours per lineal foot), and
  - test, connect and check out (hours per connection);

- Instrumentation:

  - install (hours per instrument),
  - local-remote loops (hours per loop), and
  - calibrate and test (hours per instrument).

- Structural steel (hours per ton);
- Insulation (hours per square foot and hours per foot of piping); and
- Painting (hours per square foot).

Selection of individual work units for control will depend on need, the size of the project, the amount of money involved, and the ease of gathering data. This type of data is also very useful for historical purposes.

J. Indirect Labor Jobhour Curves

As indirect labor budgets are often poorly estimated, tracking curves should be developed for major indirect categories. The technique described in this section would be appropriate for use on larger projects. Figure 9.46 shows a typical indirect jobhour tracking curve. Hours and hourly costs should be tracked separately. A separate curve for average hourly rate can be drawn, or, alternatively, the hourly rate can be entered on the jobhour chart. The monthly status is tabulated as shown, and forecasts are shown. Forecasted hours multiplied by the anticipated hourly rate will give forecasted costs.

As shown by the example, and as is often the case, a significant jobhour overrun exists. The overrun started in month ten and escalated further in month thirteen. A significant overrun was evident at that time. The hourly rate is also shown to be overrunning.

Indirect labor covers hourly paid labor not directly involved in the construction of permanent facilities; indirect activities can comprise as much as 30% or more of direct labor, and can include:

- erection of temporary buildings, roads, etc.;
- erection and maintenance of temporary utility systems;
- site cleanup during and after construction;
- materials handling and preservation (warehouse operation);
- scaffolding;
- equipment maintenance;
- lost time (weather, union allowances, training, etc.);

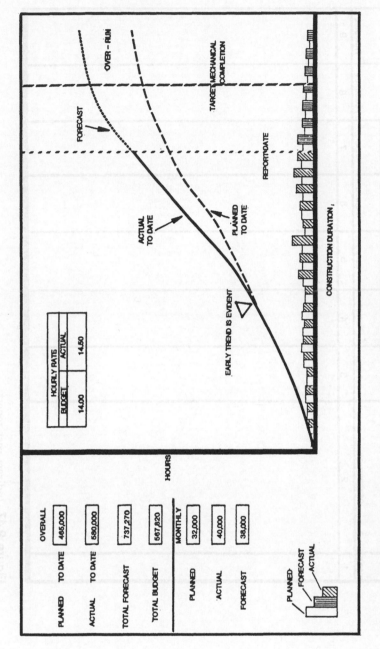

Figure 9.46 Typical construction indirect labor jobhours tracking curve.

INDIRECT COST REPORT			COMMITTED		EXPENDED				PROJECT JOB NO. REPORT NO. DATE
CODE	DESCRIPTION	BUDGET	PERIOD	CUM.	PERIOD	CUM.	FORECAST	VARIANCE	
1	2	3	4	5	6	7	8	9	

Figure 9.47 Indirect cost report.

- coffee breaks, walking time, etc.; and
- winterization.

K. Indirect Cost Report (Material)

Indirect material costs generally do not require curve techniques and can be monitored with a monthly status report. Figure 9.47 illustrates a typical monthly status report of field indirect material costs. Individual items are identified by account code; commitments, expenditures, the budget, forecasts, and variances are reported.

L. Field Staff Control

In preparing a detailed control estimate, the construction department should provide a field staff organization chart showing all necessary job functions. A listing should then be prepared, showing field staff positions, both permanent and local hire, and indicating planned arrival and release dates as well as budget hours or months. Figure 9.48 is such a typical listing and is a basic control document for allocation of field supervision. This document should be constantly updated. The contract agreement will outline the basis of charging for the construction staff.

M. Construction Equipment

1. Equipment Cost Report

On large projects, the cost of construction equipment is substantial. The report shown in Figure 9.49 serves much the same function as the construction supervision listing. It defines the plan for assigning equipment and provide for the monitoring and control of that plan during construction. The list should show actual versus planned arrival and release dates, and the rental rate (or purchase price, if bought), and should compare the total forecast equipment cost versus the budget. Timely arrival and release of equipment is of schedule and cost benefit. Thus, major deviations from the original (i.e., budgeted) plan warrant close scrutiny.

A detailed evaluation should be made of the construction equipment rental agreement to ensure that terms and conditions are economically acceptable. In the United States, rental rates are usually stated as a percentage of a nationally accepted price list. Factors to investigate are buyout conditions for equipment, terms for regular maintenance versus heavy repair, arrangements and costs for transportation from a particular area, and the date on which the rental will commence.

2. Equipment Utilization

An effective technique for controlling construction equipment costs is a comparison of actual costs per labor direct hour against budget and/or a historical profile. This technique would be appropriate for use on a large project. Figure 9.50 shows a typical format for control profiles.

FIELD SUPERVISION LISTING & SCHEDULE

PROJECT
JOB NO.
REPORT NO.
DATE

POSITION	NAME	SCHEDULE	BUDGET $	FORECAST HOURS	HOURLY RATE	FORECAST $

Figure 9.48 Field supervision listing and schedule.

EQUIPMENT	SUPPLIER	MOBILN. COSTS	MONTHLY RATE	BUDGET		FORECAST			COST	
				NO. OF MONTHS	COST	NO. OF MONTHS	COST		PERIOD	CUM.
1	2	3	4	5	6	7	8		9	10

CONSTRUCTION EQUIPMENT COST REPORT

PROJECT

JOB NO.

PROJECT NO.

DATE

Figure 9.49 Construction equipment cost report.

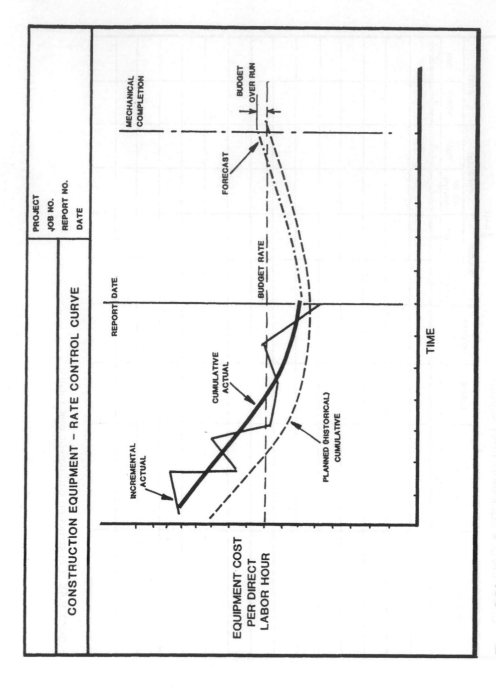

Figure 9.50 Typical construction equipment—rate control curve.

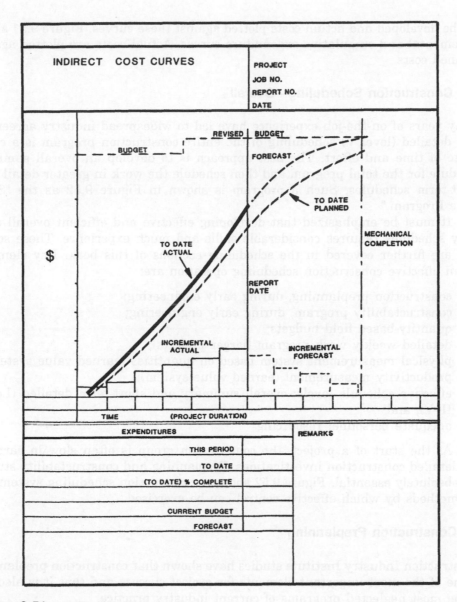

Figure 9.51 Indirect cost curves.

This record of monthly and cumulative rental costs against direct labor hours could give an evaluation of equipment commitments in relation to labor buildup. This could indicate that equipment was brought to the site too early or that a projected labor buildup was not achieved. In either case, equipment was underused.

N. Indirect Cost Curves

For cost elements that are not quantifiable (such as temporary facilities, small tools, consumables, and field office expenses), financial expenditure curves can provide meaningful evaluations. Planned curves, based on history or judgment, can be developed and actual costs plotted against these curves. Figure 9.51 shows incremental and cumulative expenditure curves. Actual costs are plotted against planned costs.

O. Construction Scheduling—Overall

Many years of on-the-job experience have led to widespread industry agreement that detailed (level V) scheduling of the entire construction program is a costly waste of time and effort. The best approach is to develop an overall summary schedule for the total program and then schedule the work in greater detail with short-term schedules. Such a program is shown in Figure 9.52 as the "Short Cycle Program."

It must be emphasized that developing effective and efficient overall summary schedules requires considerable skills and much experience. These schedules are further covered in the scheduling sections of this book. Key elements for an effective construction scheduling operation are:

*   construction preplanning, during early engineering;
*   constructability program, during early engineering;
*   quantity-based field budgets;
*   detailed weekly work program (larger projects);
*   physical measurement system based on quantities (earned value system);
*   productivity measurement (earned value system);
*   effective schedule levels: overall summary and short-term detailed (Levels II–V); and
*   adequate personnel resourcing.

At the start of a project, the construction group is often slow in carrying out detailed construction investigations. Preplanning and constructability studies are absolutely essential. Figure 9.52 shows a construction scheduling system and the methods by which effective control can be exercised.

P. Construction Preplanning

Construction Industry Institute studies have shown that construction preplanning is one of the most important elements for project success and that it is also one of the most neglected programs of current industry practice.

Preplanning for construction at the early stages of a project is absolutely essential. At an early stage, detailed planning is restricted by lack of scope definition. However, there are areas where preplanning can be reasonably definite, such as work accessibility, traffic patterns, laydown areas, rigging studies, preassembly and modularization, material selection, temporary facilities, and overall personnel resources.

A *path of construction* should be developed at this early stage, outlining the major flow of work from site preparation, to foundations, steel work, equipment installation, and pipe erection on an area and subarea basis. The schedule engineer, in conjunction with construction personnel, should establish priorities for the early and critical construction work, such as:

*   path of construction (earthwork/concrete/mechanical/piping),

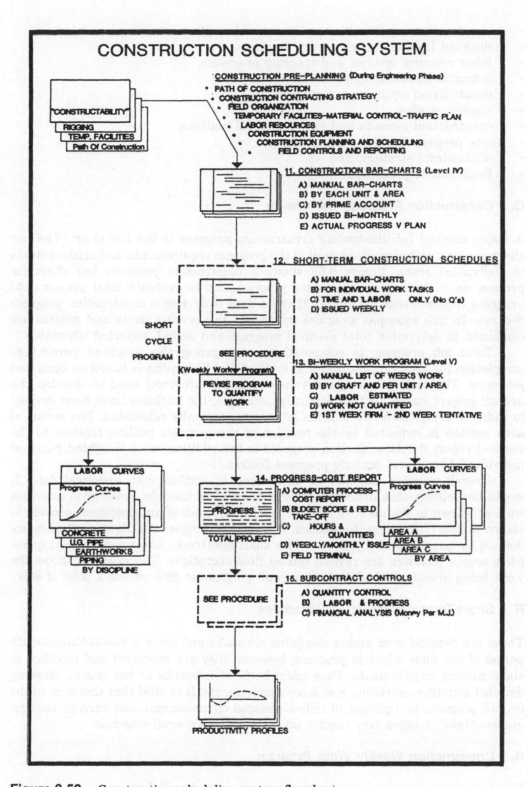

Figure 9.52 Construction scheduling system flowchart.

- construction organization and personnel assignments,
- detailed layouts for temporary facilities,
- labor resource studies and training programs,
- material handling/logistics studies,
- construction equipment requirements,
- rigging studies,
- construction permits and environmental matters,
- site preparation and early fieldwork,
- subcontract strategy, and
- field control procedures.

Q. Construction Progress Bar Charts

A major method for illustrating construction progress is the bar chart. The bar chart is also a technique for totalling the progress requirements and achievements in individual areas. Figure 9.53 shows a construction progress bar chart for process units. It is used to monitor progress and to evaluate total project (job) progress requirements, which can then be used to develop a construction progress S-curve. In this example, progress figures for two process units and offsites are combined to determine total planned progress and actual, reported job status.

Total job progress is computed by aggregating the weighted percentage completion for each process unit and total offsite. Weighting is based on budgeted jobhours. The total job progress values obtained are then used to develop the overall project curve. In this particular example, the barlines have been revised to indicate completion two months later than originally scheduled. The status of each section is indicated by the position of the progress barline relative to the vertical report dateline. Section progress is the addition, on a weighted basis, of individual lower-level activity progress figures.

Overall progress can also be determined or verified with historical data. It would be unlikely that historical experience could be used for lower-level activities such as shown in Figure 9.54. The mechanical erection of an equipment account is shown in this illustration. In addition to a relative weight breakdown, activities are defined by quantities and jobhours. This chart also tracks scope definition as quantities and/or jobhours are revised during field operations. The example shows the work being ahead of schedule, with overall progress at 38% versus a plan of 30%.

R. Short-Term Construction Schedules

These are *detailed area* and/or *discipline schedules* and cover a two-to-three-month period of the total schedule program; however, they are expanded and modified to show current requirements. They could be CPM networks or bar charts, showing detailed activities, duration, and labor requirements. It is vital that this slice of the overall program be updated to reflect changed circumstances and current requirements. Major changes may require an update of the overall schedule.

S. Construction Weekly Work Program

The technique described in this section is a very detailed program and is only cost effective on larger projects. It is the most effective of all detailed work

CONSTRUCTION PROGRESS BAR CHART - PROCESS UNITS

ITEM			JAN	FEB	MAR	APR	MAY	JUN	JUL	AUG	SEP	OCT	NOV	DEC	JAN	FEB	MAR	APR	MAY	JUN	JUL	AUG
TOTAL JOB	100	O	31.9	39.4	47.2	53.6	60.4	67.2	73.8	80.3	86	91.1	94.7	97.1	98.6	99.4	99.9	100				
		R	20.2	25.5	31.9	39.4	47.2	53.6	60.4	67.2	73.8	80.3	86	91.1	94.7	97.1	98.6	99.4	99.9	100		
		A																				
		N	21.5	25.8	30.9	36.9																
DESCRIPTION	WEIGHT	O																				
		R																				
		A																				
		N																				
VACUUM UNIT	35	O	29.7	36.3	44	51.9	60.7	69.4	77.9	85.6	92.2	97	99.3	100								
		R	18.9	24.0	29.7	36.3	44	51.9	60.7	69.4	77.9	85.6	92.2	97	99.3	100						
		A																				
		N	19.0	24.0	30.0	36.3																
FCC UNIT	50	O	34.6	42.3	49.8	55.3	61.1	66.7	72.1	77.6	82.7	87.6	91.6	95.0	97.4	98.9	99.9	100				
		R	22.3	28.2	34.6	42.3	49.8	55.3	61.1	66.7	72.1	77.6	82.7	87.6	91.6	95.0	97.4	98.9	99.9	100		
		A																				
		N	23.0	26.0	30	34.6																
OFFSITES	15	O	28.1	36.8	45.8	51.6	57.5	63.4	69.9	77.1	83.1	89.1	94.4	97.3	99.4	99.8	100					
		R	16.3	20.1	28.1	36.8	45.8	51.6	57.5	63.4	69.9	77.1	83.1	89.1	94.4	97.3	99.4	99.8	100			
		A																				
		N	22.0	29.0	36.0	45.8																
		O																				
		R																				
		A																				
		N																				

Figure 9.53 Typical construction progress bar chart.

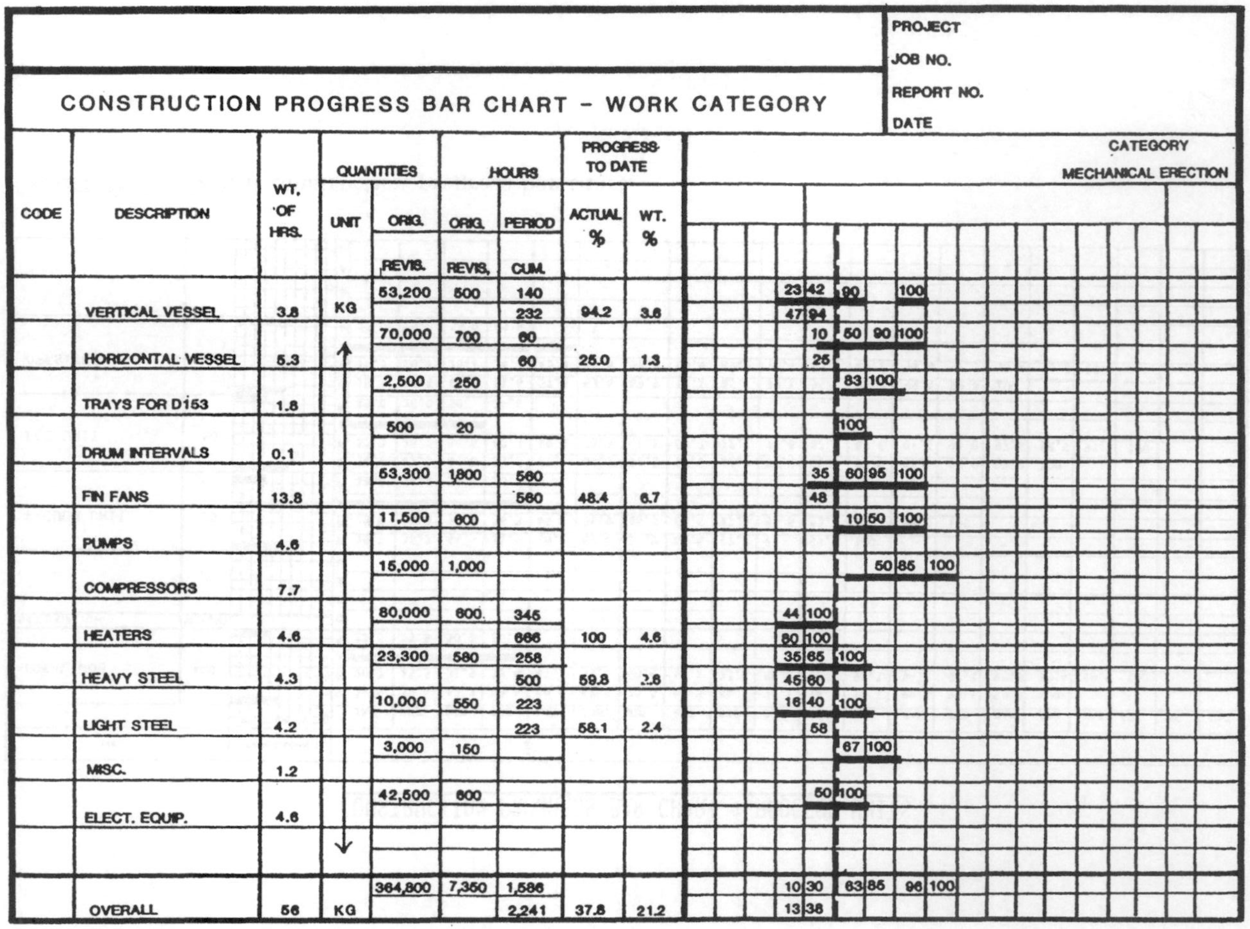

CONSTRUCTION PROGRESS BAR CHART – WORK CATEGORY

PROJECT

JOB NO.

REPORT NO.

DATE

CATEGORY: MECHANICAL ERECTION

CODE	DESCRIPTION	WT. OF HRS.	QUANTITIES REVD. ORIG.	UNIT	HOURS REVD. ORIG.	HOURS PERIOD CUM.	PROGRESS TO DATE ACTUAL %	WT. %
	VERTICAL VESSEL	3.8	53,200	KG	500	140 / 232	94.2	3.6
	HORIZONTAL VESSEL	5.3	70,000		700	60 / 60	25.0	1.3
	TRAYS FOR D163	1.6	2,500		250			
	DRUM INTERVALS	0.1	500		20			
	FIN FANS	13.8	53,300		1,800	560 / 560	48.4	6.7
	PUMPS	4.6	11,500		600			
	COMPRESSORS	7.7	15,000		1,000			
	HEATERS	4.6	80,000		800	345 / 666	100	4.6
	HEAVY STEEL	4.3	23,300		580	258 / 500	59.8	2.6
	LIGHT STEEL	4.2	10,000		550	223 / 223	56.1	2.4
	MISC.	1.2	3,000		150			
	ELECT. EQUIP.	4.6	42,500		600			
	OVERALL	56	364,800	KG	7,360	1,596 / 2,241	37.8	21.2

Figure 9.54 Typical lower-level activity construction progress bar chart.

programs, but requires considerable skills and full commitment and discipline from the construction supervision. In practice, it can use a computer program.

The weekly work program is a two-to-four-week forecast, supported by an overall schedule and detailed estimate. With proper supervisory support, it is a very efficient method of achieving control. It readily adapts to a dynamic construction environment where priorities, material deliveries, craft hour content, site conditions, resources, and weather can change very rapidly.

The essential requirement is identifying and itemizing specific pieces of work through the use of:

- quantities of measurable units,
- application of budgeted unit hourly rates,
- adjustment for current productivity,
- craft labor assessment,
- planning of extra work and rework (hours and craft labor), and
- reconciliation of weekly program to overall plan.

The flowchart contained in Figure 9.55 shows the elements of the weekly work program. It also shows the flow of information, including the overall schedule for the month's work, the budget estimate for the work in labor-unit direct hours, together with quantity takeoff sheets for detailed scope. This is for direct-hire work; subcontractor work is shown separately.

Work lists are then drawn up, usually by construction personnel, (assisted by the field planning group) who then ensure that material and engineering drawings are available. The look-ahead list will be in less detail than the current week's program, and a good system will define 80% of the work for the second week, 60% of the work for the third week, and 40% of the work for the fourth week. Control of subcontractor work depends on the contractual agreements and the capability of the subcontractors. Subcontractors can be controlled on a milestone schedule and craft labor basis alone, or by full incorporation into the weekly work program. The field planning group should work with the work list to evaluate required hours and craft labor and to coordinate direct-hire work and subcontract work.

Figure 9.56 shows a typical report format for the weekly work program. The following explanations refer to the use and function of each numbered column:

1. Check that all engineering information is available;
2. Check that all material is available;
3. Abbreviated description of work items;
4. Total quantities of measurable units for work item;
5. Budget direct hour rate for this piece of work;
6. Quantities of measurable units for work planned for week;
7. Budgeted hours for the planned weekly work (Column 6 + Column 7);
8. Extra work and/or rework (rework many times has no measurable quantities or hourly rates; in such cases the work should be estimated and entered in Column 8);
9. Remarks on priority, constraints, resource requirements, and crew sizes are appropriate;
10. Planning reconciliation (when the weekly program has been calculated and finalized, it is necessary to compare the projected output of the weekly program against the overall progress requirement, both weekly and cumulative);

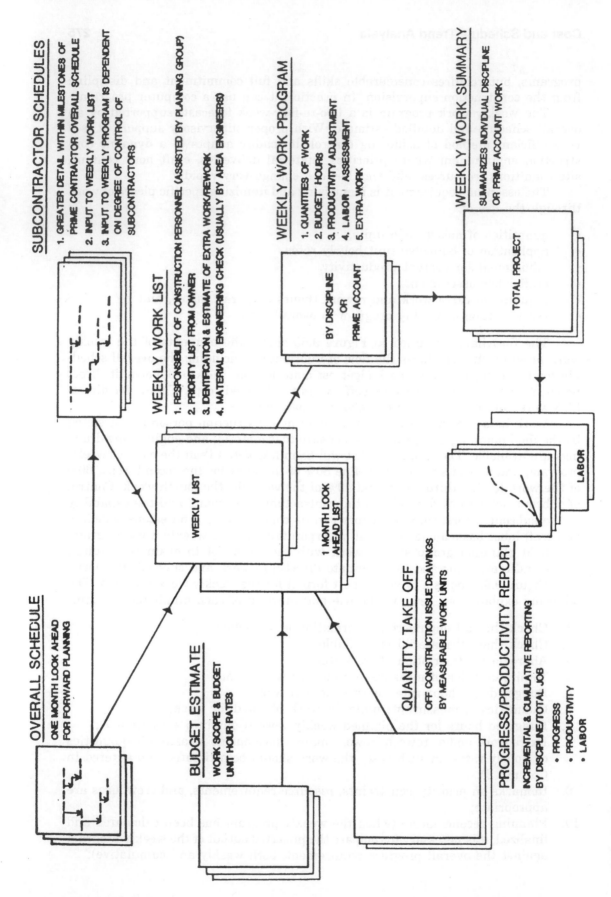

SUBCONTRACTOR SCHEDULES

1. GREATER DETAIL WITHIN MILESTONES OF PRIME CONTRACTOR OVERALL SCHEDULE
2. INPUT TO WEEKLY WORK LIST
3. INPUT TO WEEKLY PROGRAM IS DEPENDENT ON DEGREE OF CONTROL OF SUBCONTRACTORS

WEEKLY WORK LIST

1. RESPONSIBILITY OF CONSTRUCTION PERSONNEL (ASSISTED BY PLANNING GROUP)
2. PRIORITY LIST FROM OWNER
3. IDENTIFICATION & ESTIMATE OF EXTRA WORK/REWORK
4. MATERIAL & ENGINEERING CHECK (USUALLY BY AREA ENGINEERS)

WEEKLY WORK PROGRAM

1. QUANTITIES OF WORK
2. BUDGET HOURS
3. PRODUCTIVITY ADJUSTMENT
4. LABOR / ASSESSMENT
5. EXTRA WORK

WEEKLY WORK SUMMARY

SUMMARIZES INDIVIDUAL DISCIPLINE OR PRIME ACCOUNT WORK

BY DISCIPLINE OR PRIME ACCOUNT

TOTAL PROJECT

LABOR

WEEKLY LIST

1 MONTH LOOK AHEAD LIST

OVERALL SCHEDULE

ONE MONTH LOOK AHEAD FOR FORWARD PLANNING

BUDGET ESTIMATE

WORK SCOPE & BUDGET
UNIT HOUR RATES

QUANTITY TAKE OFF

OFF CONSTRUCTION ISSUE DRAWINGS BY MEASURABLE WORK UNITS

PROGRESS/PRODUCTIVITY REPORT

INCREMENTAL & CUMULATIVE REPORTING BY DISCIPLINE/TOTAL JOB

• PROGRESS
• PRODUCTIVITY
• LABOR

Figure 9.55 Construction weekly work program flowchart.

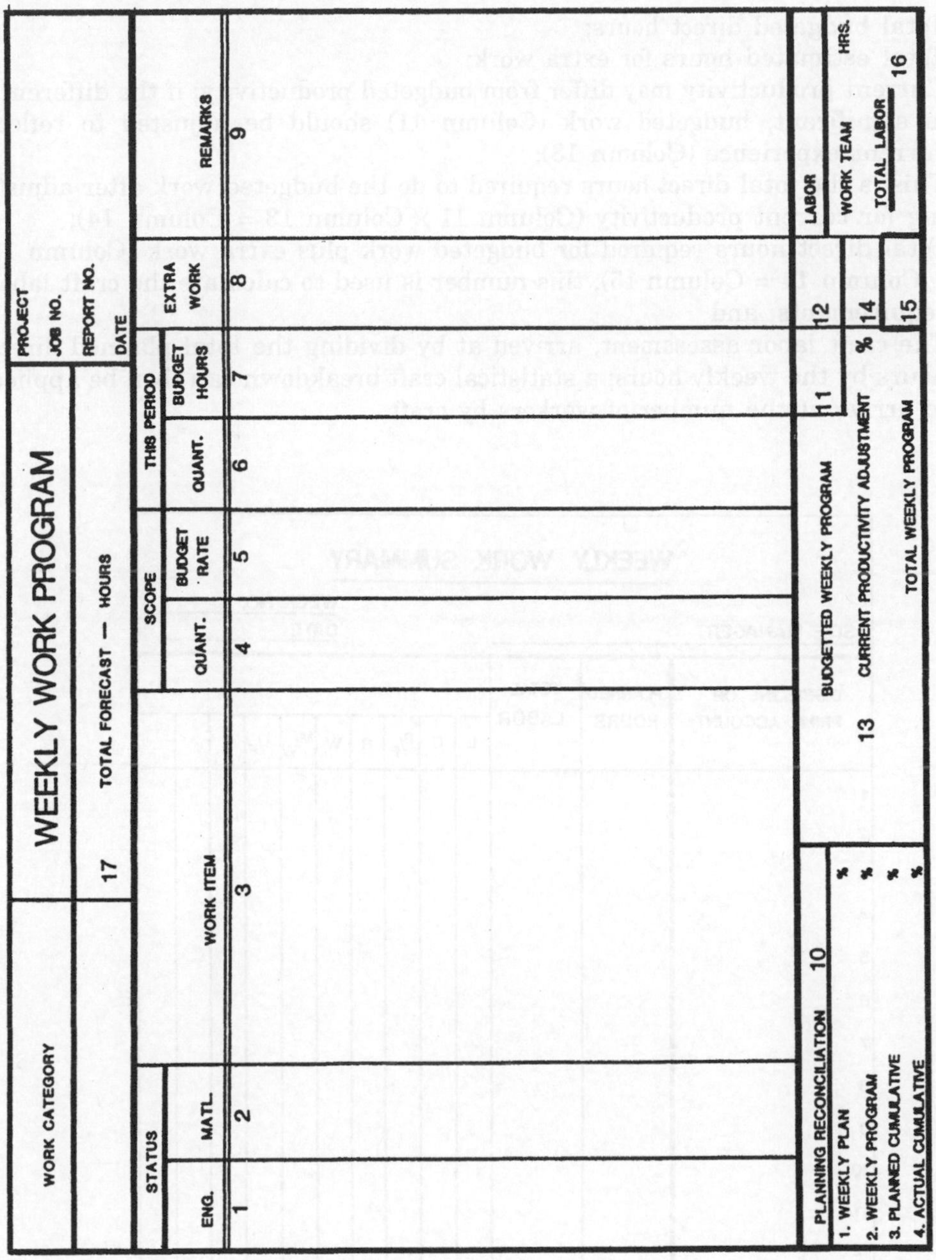

Figure 9.56 Weekly work program form.

11. Total budgeted direct hours;
12. Total estimated hours for extra work;
13. Current productivity may differ from budgeted productivity; if the difference is significant, budgeted work (Column 11) should be adjusted to reflect current experience (Column 13);
14. This is the total direct hours required to do the budgeted work after adjusting for current productivity (Column 11 × Column 13 = Column 14);
15. Total direct hours required for budgeted work plus extra work (Column 12 + Column 14 = Column 15); this number is used to calculate the craft labor requirements; and
16. The craft labor assessment, arrived at by dividing the total planned direct hours by the weekly hours; a statistical craft breakdown can then be applied to arrive at the number of workers by craft.

WEEKLY WORK SUMMARY

SITE MANAGER _____

WEEK NO. _____
DATE ___ ___ ___

DISCIPLINE OR PRIME ACCOUNT	PLANNED HOURS	TOTAL LABOR	L	C	P/F	R	W	M/W	I/F	E		
1												
2												
3												
4												
5												
6												
7												
8												
9												
10												
11												
12												
TOTAL												

PLANNED RECONCILATION   TOTAL FORECASTED MANHOUR

1] WEEKLY PLAN   %   3] PLANNED CUMULATIVE   %
2] WEEKLY PROGRAM   %   4] ACTUAL CUMULATIVE   %
                                     (LAST PERIOD)

REMARKS:

Figure 9.57 Weekly work summary form.

Note that when craft labor is a restraint, either not available or fixed due to a lack of accommodations (as in offshore work) or saturation, the calculation process is worked backward.

Figure 9.57 is a weekly work summary report. This document summarizes individual discipline programs. Planned direct hours and total craft labor are the same numbers shown on individual work programs marked in Columns 15 and 16. The craft breakdown is derived by applying a historical labor standard for appropriate disciplines.

The planning reconciliation provides for evaluation of the weekly program against the overall program.

In summary, this weekly work program is the most effective of all detailed work programs. When properly developed and operated, it will be a major contributor to the success of the construction effort. However, it requires a high degree of construction-scheduling skills and the full and constant support of construction supervision. In fact, construction supervision must lead and direct the program.

T. Construction Progress and Productivity (EVS)

S-curves (Fig. 9.58) can provide more meaningful analysis of progress than bar charts, since deviations are more easily recognized and trends or recovery plans can be more easily developed. The planned curve can be determined by using judgment, historical data, and the bar chart evaluation.

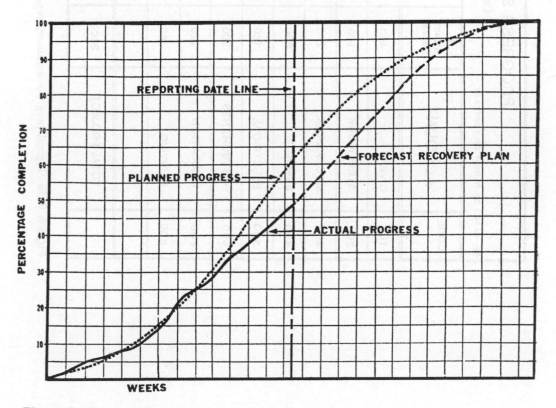

Figure 9.58 Typical construction progress S-curve.

PROGRESS SUMMARY REPORT
(× 1000)

* BASED ON BUDGET OR FORECAST

PROJECT
JOB NO.
REPORT NO.
PERIOD

CODE	ACCOUNT	BUDGET HOURS *	WEIGHT %	% COMPLETE. ACCOUNT	% COMPLETE. PROJECT (4÷5)	EARNED HOURS (3÷5)	ACTUAL HOURS	PROD. $\frac{7}{8} \times 100$	FORECAST $\frac{3}{9}$
1		3		5	6	7	8	9	10
100	CIVIL	440	20	50	10	220	200	110	400
200	STEELWORK	100	4	20	0.8	20	16	125	80
300	MECHANICAL	200	11	10	1.1	20	22	90.9	220
400	PIPING	600	33	5	1.6	30	34	88.2	680
500	ELECTRICAL	160	7	–					140
600	INSTRUMENTS	200	10	–					200
700	INSULATION	220	11	–					220
800	PAINTING	80	4	–					80
	TOTAL PROJECT	2000	100	–	13.5	290	272	106.6	2040

Figure 9.59 Typical progress summary report.

Table 9.2 Example of Percent Complete Calculation Based on Estimated Jobhours

Discipline 1000 – civil	Cubic yards (1)	Weight % (2)	% Complete (3)	Weighted % complete (4)
Foundation for C102	8	20	100	20
Foundation for D104	4	10	50	5
Foundation for P101	2	5	100	5
Foundation for P102	2	5	—	—
Foundation for P103	4	10	—	—
Pipe rack foundations	20	50	40	20
Total account				50

There are two essential requirements for effectively measuring construction progress: a quality critical path method schedule with appropriate detail, and an accurate way to measure progress. Measuring and reporting construction progress are usually variations of two basic approaches:

- measurement of physical quantities (EVS), and
- labor hours assessment.

The physical quantity measurement (EVS) is the better method, since construction progress can be expressed simply as:

$$\frac{\text{Physical quantities installed}}{\text{Total scope (qualities) of project}} \times 100\%$$

Figure 9.59 shows an overall progress report. It lists major accounts and shows labor hours, percent complete, and productivity. It also forecasts final labor hours. Each discipline is assigned a weighting (Column 4) based on the hours allocated in the construction budget (Column 3). Each discipline is further broken down into measurable quantities of work, which are then given a weight based on the estimated jobhours, as shown in Table 9.2.

During construction, the total scope forecast can change due to better definition, extra work orders, or other changes. When scope changes are sufficiently large, a reweighting of construction activities may be necessary.

The percentage completion of each discipline is calculated by totaling the weighted percentages in Column 4 of Table 9.2. The total is used in Column 5 of the summary report to calculate the overall project completion. The level of quantity measurement and progress reporting depends on the detail of the work measurement system, type of contract, and whether construction is subcontract or direct hire.

U. Guide to Field Progress Reporting

Accurate assessment of work in progress is essential, and good judgment is needed in order to assess the status of partially completed work. The guidance contained in this section covers major categories of work. Each discipline is listed, and work items are broken down into major tasks and recommended percentages for completion of the work are shown[*]:

Site preparation and earthwork: report by percent of total cubic
    yards involved
Earth tank pads

percent of compacted earth in place	85% (85)
final dressing	100% (15)

Concrete: report by percent of total cubic yards involved, with
    the following allowances:

rebar in place	20% (20)
forming complete	70% (50)
concrete poured	80% (10)
stripping complete	95% (15)
dressed and patched	100% (5)

    piles: report by number in place as percent of total required
    paving: report by square feet installed against total square
           feet required
    sewers and access holes (prefabricated)

access holes and catch basins installed	65% (65)
hookup and connections complete	90% (25)
test and checkout complete	100% (10)

Steel structures, piping supports, and miscellaneous steel

report by tons erected in place	90% (90)
bolting tension checked and completed	100% (10)

Buildings (excluding foundations)
    shelter-type (no interior work)

steel erected	50% (50)
walls and roof complete	90% (40)
checked out complete	100% (10)

    masonry-type

walls erected	30% (30)
roof framing complete	50% (20)
doors and windows installed	65% (15)
interior complete	100% (35)

Equipment installation
    columns and vessels
    shop-fabricated, no internals

set in place	60% (60)
secured and grouted	90% (30)
tested and bolted up	100% (10)

    shop-fabricated, with trays or internals

set in place	25% (25)

---

[*] The first number represents cumulative percent of the job. The number in parenthesis represents the incremental percent for the indicated step.

secured and grouted	35% (10)
internals complete	90% (55)
tested and bolted up	100% (10)

field-fabricated: report by number of prefabricated sections or rings
and internals installed; allow appropriate percent complete
for partly completed work elements

storage tanks, field-fabricated: report by base, number of rings
installed, roof, and internals from subcontractor erection
schedule; allow appropriate percent complete for partly completed work elements

exchangers

shell and tube (per unit)

set in place	60% (60)
secured and grouted	90% (30)
tested and accepted	100% (10)

fin-tube (per unit)

set in place	60% (60)
secured and grouted	90% (30)
tested and accepted	100% (10)

fin fans (per unit)

steel structure erected	20% (20)
housing erected	30% (10)
fan and driver assembled	50% (20)
coils installed	70% (20)
run-in and fan balance	90% (20)
tested and accepted	100% (10)

heaters

vertical heater (package unit)

heater set in place	50% (50)
stack erected	70% (20)
secured and grouted	90% (20)
tested and accepted	100% (10)

heater (field-assembled)

substructure complete	20% (20)
refractory installed	55% (35)
tubes installed	75% (20)
stack and breeching installed	85% (10)
burners installed	90% (5)
tested and accepted	100% (10)

pumps and drivers

pump set in place	40% (40)
aligned and grouted	90% (50)
run-in and accepted	100% (10)

compressors and drivers

package compressor (with driver)

set in place	50% (50)
secured and grouted	90% (40)

                run-in and accepted   100% (10)
            package compressor (with driver separate)
                compressor in place   25% (25)
                driver in place   50% (25)
                unit coupled and aligned   85% (35)
                secured and grouted   90% (5)
                run-in and accepted   100% (10)
Piping[*]: percent complete in this account can be reported in the
        following categories by the method indicated:
            fabricated pipe spools: as completed by count, tons, or feet
            pipe spools installed: as installed by count, tons, or feet
            straight run racked pipe: by percent of linear feet installed
            underground lines: by percent of linear feet installed
            steam tracing: by percent of linear feet installed
            hangers and supports: as completed by count or percent
                allowance
            hydrotesting: by subsystem or by holding back 10% of pipe
                spools for hydrotest and punchlist work
            handling: laydown to work area, percentage basis (by judg-
                ment)
            2.5 inch and less in diameter: by feet, screwed or socket
                weld, by size
            3–inch and more in diameter: by each fit-up and tack, by
                size, schedule, and type of material
        weld out: by cubic inches of weld plus per operation
        hydrostatic test: percentage basis or by subsystem
        punch out: percentage basis (by judgment)
        rework: percentage basis (by judgment)
        pipe fabrication, pipe supports, and hangers: unit jobhours
Electrical
        power and control equipment: as installed, by count
        lighting equipment (pole assemblies): percent installed, by count
        underground conduit and duct: percent of linear feet installed
        above-ground conduit (power): percent of linear feet installed
        above-ground conduit (lighting): percent of linear feet installed
        power and control wire: percent of total feet pulled
        power connections: percent of total complete
        grounding: percent of feet installed
        lighting wire: percent of feet installed
        push-buttons and receptacles: percent of total installed
        communications: by system complete
Instrumentation of control panels (including shop-mounted instruments)
        install panels   25% (25)
        hook up and connect   85% (60)

---

[*] This section outlines a simplified piping approach; many systems use more detailed approach
to measuring completion of pipe erection.

test and check out   100% (15)

Instruments and instrument materials
    wire and conduit: percent of linear feet installed
    pipe and tubing: percent of linear feet installed
    field-mounted instruments: percent installed, by count
    control and relief valves: percent installed, by count
    racks and supports: percent of linear feet installed
    hookups: as completed, by count
    loop check: as complete, by system

Insulation
    vessels and towers: percent of square feet installed
    piping: percent linear feet installed

Painting
    vessels, tanks, towers, and structural steel: percent of square
        feet installed
    piping: percent of linear feet covered

Good systems will be based on quantities, earned value, and a productivity assessment based on actual hours versus earned hours.

V. Construction Status Report

Figure 9.60 shows a status report by total project or unit-area. The top section of this figure shows planned and actual cumulative progress curves. Backup data for the curves would be obtained from construction progress bar charts. The center section shows planned and actual incremental and cumulative productivity curves. The bottom section shows planned and actual cumulative craft labor levels.

The report shown in Figure 9.60 enables an overall evaluation of current status and future predictions to be made quickly and accurately. It is an excellent management report, since it graphically shows the overall status of the project on a single sheet of paper. The labor histogram should be prepared from the progress curve and labor hour budget. Appropriate allowances should be made for lost time.

For areas of limited labor resources and for larger projects, detailed labor resource evaluations are essential (a part of construction preplanning). They should evaluate both other and future work and cover:

*   local labor availability;
*   travelers and imported labor;
*   local practices and regulations for labor;
*   infrastructure (housing, transportation, medical, educational, religious);
*   training requirements; and
*   wage rates and allowances.

W. Work Unit Tracking Curves

Installing concrete foundations and fabricating and erecting pipe spools are often critical areas of construction and should, therefore, be carefully monitored. Work-unit tracking curves, shown in Figure 9.61, are strongly recommended for this monitoring.

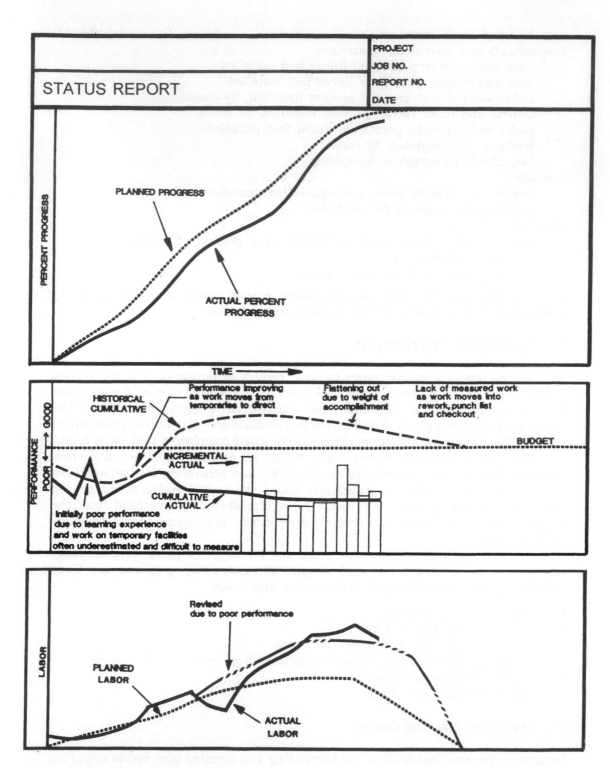

Figure 9.60 Typical status report.

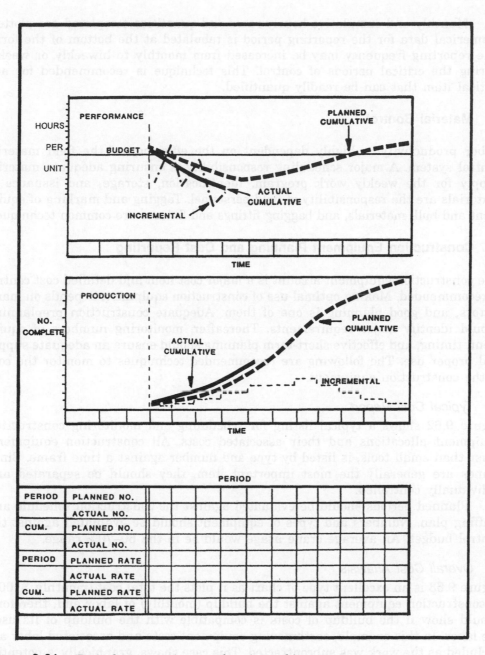

Figure 9.61 Typical construction—work unit tracking curves.

Monthly and cumulative hourly rates and quantities completed are plotted. Numerical data for the reporting period is tabulated at the bottom of the form. The reporting frequency may be increased from monthly to biweekly or weekly during the critical periods of control. This technique is recommended for any critical item that can be readily quantified.

X. Material Control

Labor productivity is highly dependent on the efficiency of the field material control system. A major scheduling responsibility is ensuring adequate material supply for the weekly work program. Identification, storage, and issuance of materials are the responsibility of field personnel. Tagging and marking of equipment and bulk materials, and bagging fittings and fixtures are common techniques.

Y. Construction Equipment Planning and Cost Reporting

The construction equipment account is a major cost item and detailed cost control is recommended. Making optimal use of construction equipment depends on many factors, and good planning is one of them. Adequate construction preplanning should identify major requirements. Thereafter, monitoring numbers of equipment, timing, and effective short-term planning should ensure an adequate supply and proper use. The following are recommended techniques to monitor the cost of the construction equipment.

1. Typical Cost Report

Figure 9.62 shows a typical listing for scheduling and monitoring construction equipment allocations and their associated costs. All construction equipment, other than small tools, is listed by type and number against a time frame. Since cranes are generally the most important item, they should be separated and individually controlled.

Planned periods should be evaluated against the construction schedule and staffing plan. Numbers and types of equipment should be evaluated against the control budget. An average crane usage would be in the 60–70% range.

2. Overall Cost Analysis

Figure 9.63 is an excellent type of chart as it plots the total cost (monthly, $1000) of construction equipment against the buildup (monthly) of labor and, therefore, should show if the buildup of costs is compatible with the buildup of its user, the labor. In this example, earthmoving equipment costs and associated labor are excluded as the work was subcontracted. This case shows, graphically, a potential loss due to construction equipment being brought on to site too far in advance of the workers; the actual requirement is not indicated by the figure. This case shows mismanagement, however, since the overall monthly equipment cost is being significantly reduced as labor is increasing. This should not occur, since equipment cost should be compatible with labor buildup. Equipment building should just lead labor so that labor is not held up and that the cost of wasted equipment is contained.

Project control personnel should ensure that the planning and scheduling of construction equipment is compatible with the labor program. This is a difficult

CONSTRUCTION EQUIPMENT COST REPORT		PROJECT JOB NO.. REPORT NO. DATE				
EQUIPMENT	SCHEDULE (PLANNED VERSES ACTUAL)	PURCHASE PRICE OR RENTAL RATE	BUDGET	TO DATE COST	FORCAST PERIOD	FORFCAST COST

Figure 9.62 Construction equipment cost report form.

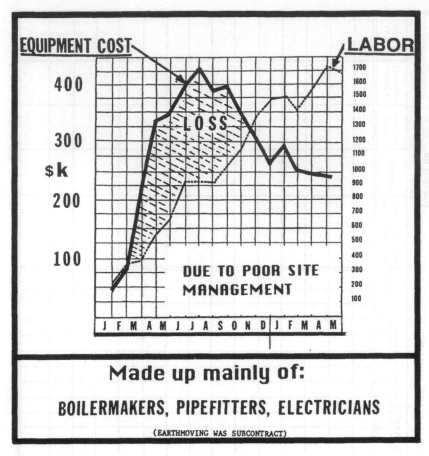

Figure 9.63   Analysis of equipment versus labor.

task, because equipment utilization is variable and dependent on numerous factors, including:

- quality of supervision,
- size of job site/number of workers,
- scheduling efficiency,
- maintenance capability, and
- weather/site conditions.

3.   *Use of Welding Machines*

Some categories of construction equipment, such as welding machines, have a direct relationship with labor. Such categories are easily tracked if such tracking is appropriate and required. Figure 9.64 is an actual case, showing a numerical relationship problem between welding machines and welders. As this is a direct relationship between workers and machines, the oversupply of welding machines was detected and remedial action taken by removing some of the machines from the job site. This technique can be used when assessing the schedule for other

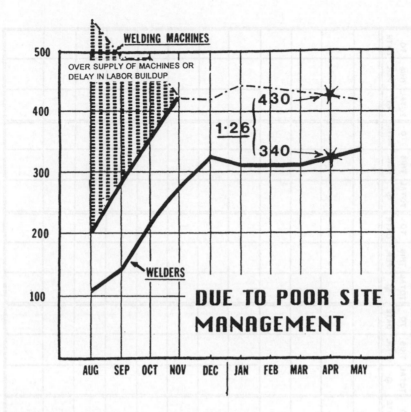

Figure 9.64 Welding machine utilization analysis.

construction equipment as well. Without additional data, the appropriate relationship of welding machines to welders could be 430/340 = 1.26.

As in the previous case, this shows additional cost due to poor site management. In both cases the use of these cost control techniques identified the problem and enabled remedial action to take place, resulting in mitigation of the cost problem.

Z. Piping Control

Piping erection is usually a major part of construction, so having an effective control system is essential. Figure 9.65 illustrates a typical control sheet that can be used to record installed piping quantities. The unit of measurement is usually feet of pipe. Five columns are shown for different pipe sizes but for the same piece of work. This is necessary because many piping drawings can have several pipe diameters for the same line or piping system. As work is completed, drawings should be marked up, and a weekly tally of installed quantities should be maintained. This control sheet shows the unit rate (direct hours per foot, etc.), total estimated quantities (Q), and quantities this period and to date. Quantities can be easily converted to earned hours.

Figure 9.65 Piping control sheet.

X. KEY PROJECT CONTROL TECHNIQUES FOR CONSTRUCTION—SUBCONTRACT

A. Introduction to Subcontract Control

This section is based on the assumption that a general contractor is responsible for all subcontract work and, therefore, is directly handling the program discussed next. Much of the reporting is the direct responsibility of the subcontractors, with some expert analysis being carried out by the general contractor. If there is no general contractor, then the owner is responsible for the analysis.

It must be recognized that many subcontractors do not use experienced project control personnel and often do not operate with detailed control systems. The key to success, therefore, is to develop a simple, practical method of control and to require that subcontractors include adequate personnel costs in their bids to use the system.

Effective subcontract control is based on the following essentials:

- good contractual documents and agreements,
- an adequate system for documenting changes and amendments,
- an acceptable scheduling system (critical path method or bar chart),
- an effective progress measurement system,
- an effective cost trending and forecasting system, and
- an adequate performance measurement system.

Most of these elements should be identified in the bid documentation.

B. Control Specifications (Large Subcontracts)

This section assumes that work will be done in several or many areas.

Immediately after the contract is awarded, the subcontractor should be required to carry out the functions, prepare the reports, and monitor progress as follows:

- quantities (on unit price subcontracts)—report monthly:

  - report installed quantities by geographical area,
  - report installed quantities by total subcontract,
  - predict final quantities;

- hours—report weekly: total hours by total subcontract;
- craft labor—report daily: daily workforce report;
- scheduling—report monthly:

  - overall critical path method or milestone schedule bar chart,
  - progress curves (area and total),
  - craft labor curves (area and total), and
  - quantity progress curves for designated work categories.

These reports and progress updates should be issued five days after the cutoff date. Progress curves can be developed on a financial basis and/or a weighted craft hour basis.

One of the objectives of this specification is to have subcontractors do the reporting and to avoid situations where the general contractor and/or the owner has to carry out this function. Subcontractors should be required to state in their

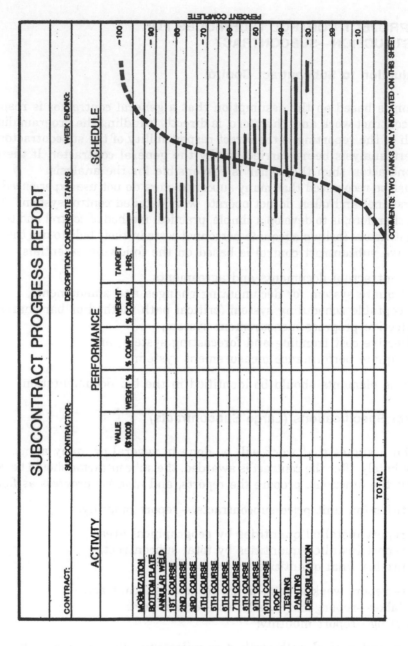

Figure 9.66 Typical subcontract progress report.

proposals that they will meet the criteria of this specification or that any objections will be stated in writing.

C. Subcontract Progress Report

Figure 9.66 breaks down the major operations, on a financial weighted basis, of a tankage subcontract. Individual activities are scheduled, and an overall planned completion curve is drawn based on the weighting. Physical completion of the activities is measured, and target jobhours are recorded. The addition of actual and planned labor levels completes the status picture.

D. Subcontract Status Report

This report is usually prepared by the general contractor from subcontractor basic information. Figure 9.60 shows a typical status report of actual progress, performance, and staffing against planned or historical data. This type of report applies equally well to a total project or to a subcontract. It is an excellent visual tool for correlating the status of the three variables and evaluating the requirements for a specific completion date or, alternatively, a likely completion date based on the current trend.

The anticipated performance profile (in dollars) is based on historical data; as shown in Figure 9.60, early poor results are due to initial learning experience and work on temporary facilities, but performance quickly improves as the proportion of direct work increases. The curve will then flatten out due to the weight of work accomplished and it is normal for the performance/productivity to then gradually reduce, due to the low value of the final punchlist and checkout work. See the performance evaluation method (discussed on p. 299) for details of applying this technique.

E. Quantity Progress Report

The quantity progress report is an efficient visual tool for scope trending and evaluating schedule performance, and can be used for many categories of work. Another technique would be to correlate the craft direct hours to the quantity and evaluate performance on a unit-hour basis. This also provides historical data for estimating purposes.

Figure 9.67 shows the quantities of mass excavation and fill for a large grassroots site. The original quantities are scheduled, and actual field installation is plotted. It is interesting to note that in this example the scope quantities increase by 20–30%. Additionally, the actual progress shows a schedule slip of nine months.

F. Measurement

As stated previously, the key element in progress measurement is quantities. This is particularly true with unit-price subcontracts. Lump-sum subcontracts also need to be evaluated for progress on a quantity basis. Field measurement can be greatly facilitated when quantity takeoffs are prepared for construction drawings. It is then a relatively simple matter in the field, as the work is

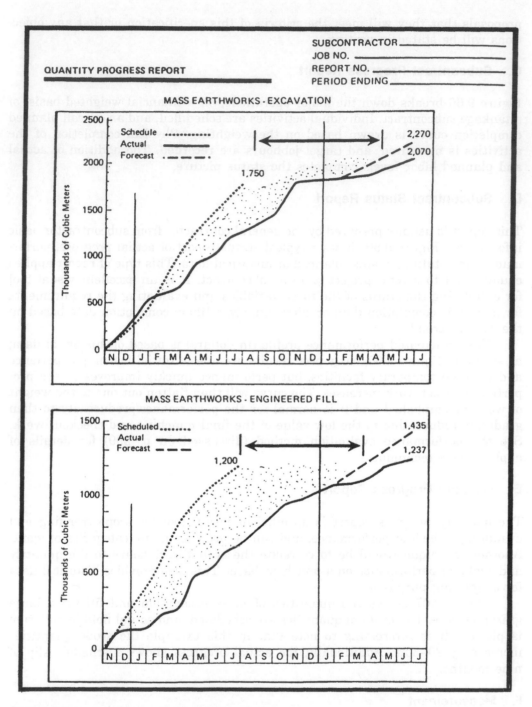

Figure 9.67 Typical quantity progress report.

completed, to use these quantities for payment and measurement of progress, together with field changes as construction proceeds.

It is important that field measurement of work in progress be recorded properly on marked-up drawings or on quantity lists so the financial billings to subcontractors can be adequately checked. It is necessary that the subcontractor's billing be checked by field construction personnel to ensure that billing reflects work completed and is for the correct facility.

At subcontract award, or before the start of construction, mutual discussions with the subcontractor on methods of progress measurement and payment should take place. At these meetings, agreements should be reached for the weighted breakdown of lump-sum subcontracts and agreements for measuring partial completion of unit prices. The frequency of reporting and the method of measurement is a worthwhile item for agreement at this early date.

G. Cost Control—Overall

The first step in cost control is evaluating the contractual documents and the contract agreement. The evaluator should look for contractual anomalies, pricing discrepancies, and conditions that might lead to future cost risk. Careful consideration should be given to schedule commitments, warehousing agreements, laydown and material handling requirements, and whether previously stated commitments can be maintained during execution of the subcontract. It is particularly important that a review of associated interface work by others, job-site areas, free-issue material supplies, and services be made, since these items often provide the source for major claims.

A log should be maintained of all engineering and contractual changes that have taken place in the agreement. All changes and appropriate cost trends of such changes should be recorded. These potential cost deviations should be estimated as definitively as possible for use in future negotiations with the subcontractor.

The following sections discuss procedures and provide examples of recommended cost control techniques.

H. Subcontract Cost Control and Forecast

Major subcontracts should be evaluated individually and tracked from the original contract price to the final cost forecast. Figure 9.68 shows a form that a general contractor can use to document such tracking. Costs are broken down into material and labor components. The control sheet shows the price of the original agreement, accumulates current experience, assesses outstanding work, and forecasts a final predicted total cost. The form is vertically divided into two parts, the top portion showing the physical scope of the work, and the bottom portion covering claims. This enables progress measurements to be made based on a financial assessment of the labor scope of the work.

Separating material and labor of lump-sum bids allows higher-quality estimating evaluations to be performed. It also enables future assessment of labor claims and conditions that do not affect material prices. The definition of scope, type of work, current conditions, and contractual agreement will largely deter-

			SUBCONTRACT COST REPORT		PROJECT	
CONTRACT FOR :					JOB NO.	
CONTRACT NO.					REPORT NO.	
CONTRACTOR					DATE	
ORIGINAL CONTRACT PRICE (LABOR & MATERIAL)		CURRENCY				

ITEM	LABOR		%	MATERIAL		REMARKS
	PREDICTED TOTAL COST	COST TO DATE		PREDICTED TOTAL COST	TO DATE COST	
1 MAIN CONTRACT						
2 WORK OUTSIDE SCOPE OF CONTRACT						
(a) MAJOR CHANGES						
(b) ADDITIONAL BID ITEMS						
(c)						
3 OPEN COST WORK						
4 OTHER (SPECIFY)						
(a)						
(b)						
DIRECT WORK S/T						
5 CLAIMS (SPECIFY)						
(a)						
(b)						
(c)						
TOTAL						

Figure 9.68 Subcontract cost report form.

mine the makeup of the items listed on the form. Claims generally will fall into one of several categories:

- change in original scope of work,
- schedule delays caused by others,
- drawing and material delays,
- interference by others, and
- changes in site conditions or site regulations.

Some subcontractors tend to greatly exaggerate adverse conditions and submit inflated claims. Consequently, it is essential that such items as daily logs, schedules, and work programs be maintained for all major subcontracts.

I. Subcontract Growth Allowances

On unit-price subcontracts (those developed with a minimum of engineering definition), factors of 20% each for scope increases and claims should be added to the original contract price (labor only). Factors for lump-sum contracts would be 5% each.

J. Subcontract Low Bids

By industry definition (custom and practice), a low bid is a bid that is under your estimate by 20% or more, assuming your estimate is a good estimate. If an evaluation indicates a low bid, a further allowance should be made for covering the probability of future poor performance, financial difficulties and/or default by the subcontractor. A low bid can result from ignorance and/or a subcontractor buying the job. Subcontractors sometimes turn in a survival bid to attempt to break into a new market or to try to block competition. Perhaps they have found deficiencies in the contract conditions which they can turn to their financial advantage, but they first have to get the contract with a low bid. Such situations can result in low bids that can, in turn, lead to serious cost and schedule consequences for an entire project if the subcontractor gets into financial difficulties and/or performs badly.

The opinion is often expressed that a subcontractor in financial trouble will maximize its field effort to finish early and get out. Yet the opposite is usually true: a subcontractor in financial trouble will generally reduce its field effort to a minimum in an attempt to improve its unit costs. This results in low levels of construction equipment and supervision and often leads to schedule extensions and poor quality. Apart from specialty subcontracts, most categories of work require subcontractors to stay until the end of the project. Thus it is rarely possible for major civil and mechanical subcontractors to finish early and get out.

1. Performance Evaluation (Dollars)

This section describes a powerful method of assessing a subcontractor's financial performance, which can then lead to a determination of future work performance and associated problems. Where there is a poor financial return, the risk of poor performance, contract default, or claims is high. Thus, a subcontractor's financial problem can, very quickly, become the general contractor's or owner's problem.

Performance evaluation is a technique for measuring and monitoring a subcontractor. Poor subcontractor financial performance increases the risk for potential schedule slippage, poor quality, claims and, ultimately, contract default. When this performance evaluation technique is used properly, it provides an early warning of potential problems and then allows time for developing alternative solutions.

2. *Performance Evaluation Method:*

1. $$\text{Performance factor (labor)} = \frac{\text{Contract billings}}{\text{Hours expended}} = \text{Dollars per hour}$$

2. Assess the subcontractor's operation cost by building up field costs, equipment costs, overhead, etc., onto the base labor cost, dollars per hour. (This does not include profit).
3. Ensure that billings truly represent work accomplished.
4. Ensure that the labor hours report is accurate.
5. If the subcontract has a material supply, evaluate it for additional profit on material to add to the profit-loss of the labor element. Check for biased bidding.

If the billing rate is significantly lower than the estimated rate, and if it remains so as work proceeds, then the risk of adverse action is high. This can lead to schedule extensions and claims. This is a relatively simple procedure. It requires up-to-date billings, an estimate of the subcontractor's all-in labor cost, accurate recording of labor hours expended, and an assessment of profit on material supply and biased bidding.

K. Subcontract Quantities (Unit Price Contracts)

On major subcontracts, tracking quantities from initial takeoffs through intermediate takeoffs to final field takeoffs is absolutely essential. Figure 9.69 is a report which illustrates such a tracking technique. The listing should show all bid items, and separate sheets should be used for labor and material. The contract column shows original contract quantities and the forecast column shows the current assessment of quantities. The difference between the contract and forecast columns is entered in the differential column. The deviation can be trended. Quantities installed to date, actual labor hours, and status are also shown.

L. Subcontract Performance Report

Figure 9.70 shows a report of a simple tabulation of financial performance per labor hour of each subcontract. In conjunction with the performance curve shown on the subcontractor's status report, this data can provide a good basis for overall cost prediction. Individual performance curves can be developed for critical subcontracts.

M. Subcontract Forecast Summary/Report

All subcontracts should be listed, in total, on the progress report shown in Figure 9.71. The report summarizes current cost and forecasted final values and identifies scope, claims, and potential trends. The figures are taken from predictions shown on the subcontract cost report (Fig. 9.68) which is a technique recommended for major subcontracts. Should this report not be used for small subcontracts, then allowances should be changed, as appropriate, to reflect small contracts. It

			CONTRACT		FORECAST		VARIANCE		TO DATE		
ITEM	UNIT	UNIT RATE	QUANT.	COST	QUANT.	COST	QUANT.	COST	QUANT.	COST	%

SUBCONTRACT QUANTITY REPORT

PROJECT
JOB NO.
REPORT NO.
DATE

Figure 9.69 Subcontract quantity report form.

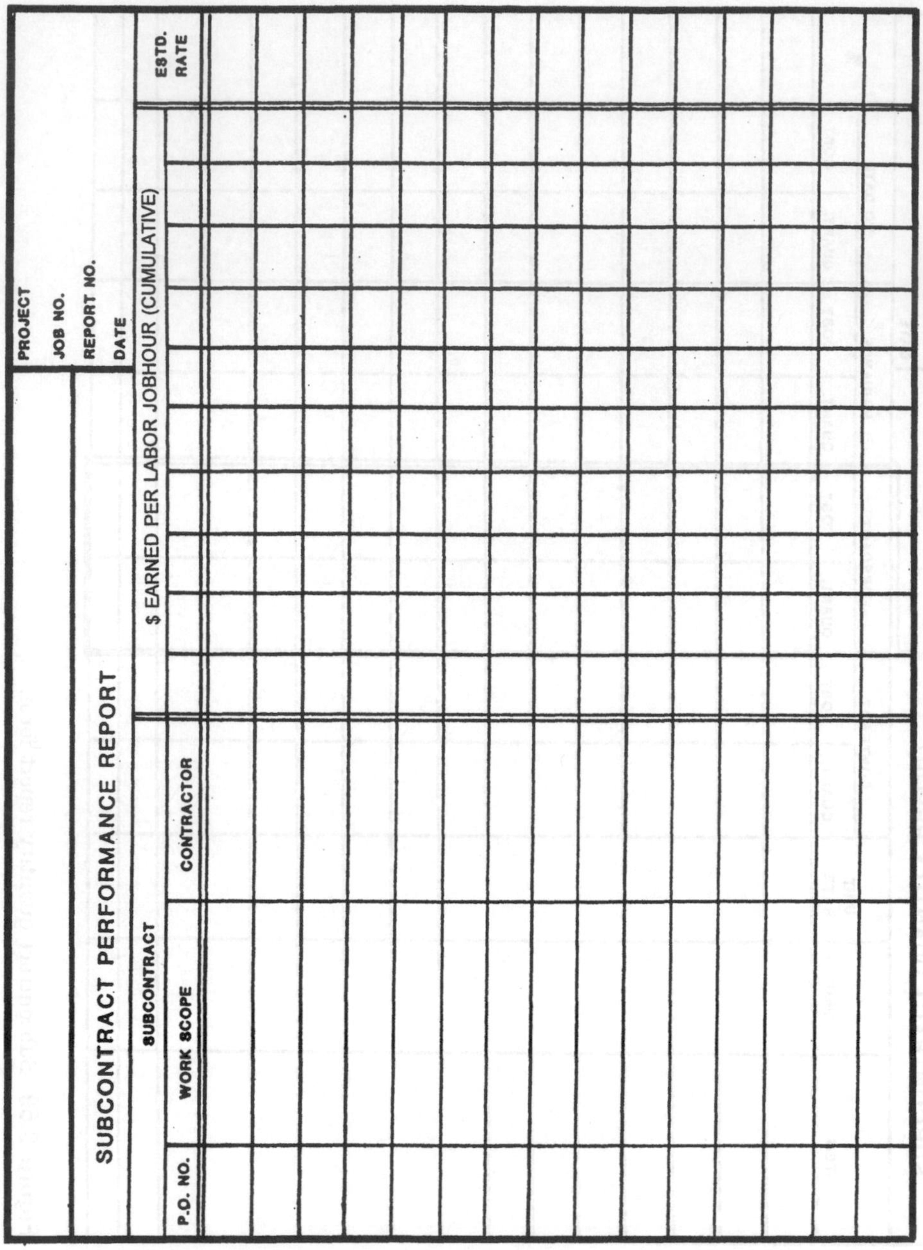

SUBCONTRACT PERFORMANCE REPORT

PROJECT
JOB NO.
REPORT NO.
DATE

P.O. NO.	SUBCONTRACT		$ EARNED PER LABOR JOBHOUR (CUMULATIVE)										ESTD. RATE
	WORK SCOPE	CONTRACTOR											

Figure 9.70 Subcontract performance report form.

CONTRACT NO.	CONTRACTOR	WORK SCOPE	ORIGINAL CONTRACT VALUE	BUDGET	CHANGES & EXTRAS	CLAIMS	FORECAST	COMMD.	EXPEND.

SUBCONTRACT SUMMARY REPORT

PROJECT
JOB NO.
REPORT NO.
DATE

Figure 9.71 Subcontract summary report form.

is recommended that assessments of changes/extras and claims be made at contract award.

N.   Field Changes and Extra Work

A procedure should be developed for efficiently evaluating, reporting, and estimating the field changes and extra work that occur during construction. The objective of this procedure is to identify all changes from approved drawings, to evaluate impacts on the cost and schedule, and to authorize the work.

Often, field changes and extra work are required on an urgent basis; however, if an efficient estimating and authorization procedure is in place, it will not delay the work. Too often, changes are requested on a crash basis where little thought has been given to the need or cost impact of the change.

During the punchlist and checkout phase at the end of construction work, changes can be required to meet operability and safety standards. These changes should be carefully assessed, as operations/maintenance staff are sometimes overly critical of the design and can be guilty of "gold plating."

Field contracts administration personnel have primary responsibility for maintaining subcontractor performance so as to ensure that previously accepted contractual commitments are met. However, contractual commitments in such areas as scope, quality, schedule, and cost often change, and an effective procedure will allow work to proceed in a timely, efficient, and orderly manner. All such changes should be identified as field changes and should be initiated, approved, and recorded in a way that will minimize the extra costs and loss of time associated with such changes.

XI.   SUMMARY

The most powerful and effective techniques for monitoring and controlling projects are effective:

- front-end planning, and
- trending programs.

Such techniques require real business skills, analytical ability and efficient project control techniques. If we get it right at the front end, we have a chance of success, though it is not guaranteed. If we do not get it right, then we have no chance of success. This is then followed with the constant evaluation of changing circumstances and their impact on cost and schedule.

This chapter illustrates a comprehensive and wide range of analytical-trending techniques and skills that, when properly applied, will lead to success and result in the desirable condition of *no surprises*.

Front-end planning is covered separately in Chapter 5.

10

Change Control and Risk Analysis

I. INTRODUCTION

Changes in the project cost and schedule baselines are as much political problems as technical problems. Even with high-quality trend analysis, as illustrated in Chapter 9, changing the baseline requires the understanding, support, and commitment of all parties to the project, especially the client or customer. This support and commitment can only be achieved when both project manager and project control analysts have competence, credibility, and effective communication channels with project parties. It is recommended that when current baselines are no longer realistic, revised baseline targets should be developed. There can be endless discussions about the definition of "realistic," but the setting of programs to unrealistic or impossible baselines is to be avoided, since it leads to reduced management credibility and poor project morale.

II. INCREASING THE COST BASELINE

Common industry practice is the company requirement that projects be completed within +10% of the approved funding. In practice, this means project cost forecasts that are over the 10% limit require a supplemental funding request. Such requests then require all appropriate justification and full backup analysis to support this position.

When quality trending is first developed on a potential basis, it is vital that the project manager, when increasing the forecast, strike the right balance between cost forecast accuracy and the need for management to know about significant cost problems. If the trend meeting is correctly implemented, all key project parties will be constantly updated about all project problems and variations, so a formal cost overrun situation would then be more routine than alarming. However, the actual numbers or forecasts, developed from potential trending, can still cause major political problems and result in loss of confidence, low project morale, and defensive engineering. This would be especially true if attempts were made to improperly assign blame for such problems. Consequently, carefully analyzing the cost forecast and carefully selecting the timing of publication are critical.

III. INCREASING THE SCHEDULE BASELINE

Industry practice and company policy are not as consistent with schedule performance as with cost performance. Very few companies have a specific *schedule quality* that is a companion to the funding requirement of contingency (probability and accuracy). A *funding schedule* of 80% probability is recommended. When such a schedule is slipping, and trend analysis shows that appropriate acceleration options have low probability, then the schedule baseline should be revised. It is, of course, possible that a major cost problem has a companion schedule problem. Trying to force a program to impossible completion requirements is of little value and often results in added costs with no schedule advantage.

IV. SCOPE INCREASES

Major industry studies have shown that the majority of cost increases and schedule slippages have been directly due to scope increases (after funding approval). Poor early definition, lack of design control, poor project interfaces, and conflicting objectives are major contributing factors to this situation. It is recommended that true scope increases, as distinguished from normal design development, should be processes for additional funds and schedule adjustment. With a lump-sum contract liability, this would be normal in a contractor operation. With owner operation, however, it is common for senior management to require that scope increases be covered with the project contingency; such poor practices are common and should be eliminated.

V. SCOPE REDUCTION

In most instances, scope reduction occurs when cost trending at the early project stage shows a cost overrun situation. Such reductions are effective during early project development, and it is essential that when the project scope and funding have been reduced to acceptable levels, the reductions are not then allowed to be reinstated. It is a common practice that operations/maintenance personnel will agree to the reductions and then, as design proceeds, will attempt to slip in the previously agreed reductions. True engineering enhancements that result in substantial economic advantage should always be properly considered, but preferential engineering and gold plating should be strictly controlled.

VI. MANAGEMENT FINANCIAL RESERVE (DOLLARS)

This reserve is not contingency, even though senior management mistakenly classifies these funds as such. A rating or risk analysis program should properly develop the appropriate contingency. Management reserve is an added allowance, controlled by management, to cover risks on projects that would have major impact on the total company financial strength if cost overruns occur.

VII. RISK ANALYSIS (PRINCIPLES, PROCEDURES, AND PROGRAMS)

Risk analysis is a tool or method for quantifying uncertainties and their inherent risk. It is a formalized structured approach defining the uncertainties and assessing the probability of risk associated with each uncertain item and/or event. Performing a risk analysis allows the project manager to qualify and quantify the sensitivity of risk to the major facets of a project and to include appropriate cost and time allowances as necessary.

The tools of risk analysis vary from intuition or "gut feeling" and judgment, to simple manual models, to computerized simulation models. Regardless of the tools used, it is usually very effective to obtain a broad review of the risks by available third parties. Other project leaders and involved groups (e.g., construction, purchasing, and facility planning) can provide valuable insights from a different perspective regarding the distribution of risk factors and elements.

A. Risk and Business Skills

The identification and proper management of risk is a vital ingredient of successful project execution. Many studies that have been performed conclude that today's project managers have inadequate business skills, resulting in poor decision-making capability and inadequate risk management.

B. Decision-Making Process

The decision-making process consists of certainty, risk, and uncertainty.

- *Certainty* only exists when the exact conditions and circumstances can be specified during the period of time covered by the decision. This is rare in the project business.
- *Risk* occurs when it is possible to specify a degree of probability for a number of likely outcomes. For example, in oil and gas exploration, there is an 80% chance that the well will be dry, 15% chance that the well will be medium sized, and 5% chance that the well will be large. It is common for probabilities to be made with historical data or, in its absence, by personal experience.
- *Uncertainty* is present when it is not possible to specify the relative likelihood of any outcome. This occurs in areas such as research and development, where there is no historical data available, or when the task is outside the experience of company personnel.

Most individuals operate in an area of some certainty, but where many decisions have an element of risk. Throughout a person's life, the element of risk is a common occurrence, and many decisions require taking some risk. With a family, the risks involve such items as the purchase of a house, education of the children, potential job prospects, etc. These are not considered big risks because there is a known history: we are not venturing outside our normal experience.

Yet consider the following unknown situations: accepting a job in a country where the culture is unknown and taking our family with us; taking up race car driving; starting a business; getting married. These are all situations that are usually outside our experience; they can be dangerous and harmful. Thus they need to be treated differently from our usual and normal risks.

In industry, risk normally involves personnel safety and the financial balance sheet. From the financial point of view, many decisions are of a potentially disastrous nature in that they could affect both the project and the future profitability of the company. For example, foreign investments, oil and gas exploration, design and new technology considerations, methods of construction, and contracting of major work all fall within this category. Because of their magnitude, some form of risk analysis becomes essential.

All projects have degrees of risks. The key to successful project management is not to wish them away, be frightened by them, or be too optimistic about them. Rather, they must be recognized and identified. Then the problem and its potential impact on the project can be quantified, and a course of action for dealing with the problem if it arises can be determined. It is unlikely that all risks will be identified by any one form of decision analysis. Further, the process of decision analysis does not minimize the likelihood of the risk occurring. Its purpose is to properly identify, quantify, and assess the cost of the risk occurring. This leads to good decisions.

C. Decision Analysis

Decision analysis methods provide a comprehensive way to evaluate and compare the degree of risk and uncertainty associated with each investment choice. The objective is to let the decision maker know the likely outcome of any decision. It must be understood that decision analysis does not eradicate risk, nor does it substitute for management judgment: it merely enables alternatives to be compared and informed decisions to be made.

Decision analysis involves:

- defining the possible outcomes for each alternative,
- evaluating the profit or loss of each alternative,
- estimating the occurrence probability of each alternative, and
- using this information to calculate the weighted average of each alternative.

This process should be carried out in a systematic way because it leads to other benefits as well: it forces a more detailed review of the project, it highlights sensitivities, and it tends to eliminate emotion and prejudice. Among the more popular decision evaluation techniques are *expected monetary value*, *breakeven*, and *Monte Carlo simulation*.

D. Expected Monetary Value

Expected monetary value (EMV) considers a number of alternative solutions to a problem, calculates the worth of each alternative, and applies an estimated probability factor to each alternative. The sum of all the probabilities at any given item is equal to either unity or 100%. The EMV is the product of multiplying the probability by the worth.

A simple illustration of this method would be spinning a coin and betting on the result. In theory, every time the coin is spun, the chance of it coming up heads is 50%, and over sufficient time the spinning relationship works out to 50:50. If a person won $10 for each time it came up heads and lost $8 for each time it came up tails, should the bet be taken? Using EMV sets the value for purposes of decision-making:

Outcome	Probability	Worth	EMV
Heads	0.5	+10.0	+5.0
Tails	0.5	– 8.0	–4.0
	1.0		+1.0

The EMV shows that accepting the gamble leads to winning, on average, $1.00 as opposed to earning nothing by not betting. With acceptance of the gamble, the expectation is to win $10.00, but analysis shows an average of $1.00. Where does this figure come from? The bet is either a win of $10.00 or a loss of $8.00; however, the probability of each event is 0.5, so the EMV of each is half the worth. The net EMV is the sum of all EMVs. This is another cornerstone of EMV analysis: it never gives the exact answer to any one question, but offers the following basic rule:

If the decision-maker bases all decisions on the highest possible EMV, the total gains from all decisions will be greater than selecting any alternative action.

Another aspect of EMV analysis is the magnitude of the risk for considerations of investment cost and cash flow. For example, since it is not known how long the 50:50 probability will take, the question arises whether the investor can afford the time it will take for the odds to even out. Consider the stakes being raised:

$100.00 (win) to $80.00 (loss)
$1,000.00 (win) to $800.00 (loss)
$10,000.00 (win) to $8,000.00 (loss)

There may come a time when an individual or a company can no longer afford a very good business gamble because of the high investment level and because the return will not occur for several/many years. The decision rule for EMV choices is to select the alternative that gives the largest positive EMV. Any number of alternatives can be selected, so long as the probability for all alternatives adds up to 1 or 100%. The worth should always be of a monetary nature.

E. Decision Trees

Decision trees graphically and logically show decision and chance options in the form of a logic diagram, as shown in Figure 10.1. The diagram shows four options, each with its own costs (net present value, NPV) and probability, and the final EMV. The diagram shows two types of nodes—the square and the circle. The square denotes a decision node: work or no work. The circle denotes a chance

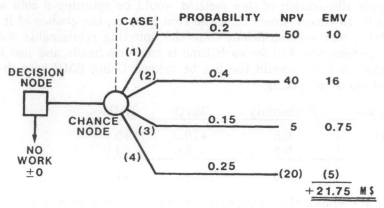

Figure 10.1 Decision tree.

node: a decision was made, and this is the possible result when the risks of the various options are assessed.

Decisions tree analysis is one of the few instances of working backward to solve a problem. The process begins by calculating the NPV for any given condition, and then working from the final outcome to the commencement of the work, applying the appropriate factors. Totalling the individual EMVs gives a positive $21.75 million, indicating the work decision to be favorable, depending on further analysis of investment funds and cash flow considerations.

Combining EMV analysis with the decision tree method results in a powerful and graphical program that is a favorite tool of senior management. The most crucial aspect of the process is the probability factor: if it is not realistic, then the analysis will be useless and misleading to the decision maker.

F. Developing the Probability

Probability is a complex subject. It is essential that whatever probability is developed, it should be the best available. The following methods will help in obtaining good probability evaluations:

*   using past success ratios (make sure that the same factors apply),
*   meeting and discussing with the "experts" to gain an understanding of the method they have used in assigning a probability range,
*   reviewing other companies' success rates, and
*   making use of mathematical models if they are available and relevant.

It must be remembered that any number obtained by any method is not a scientific certainty. There are no absolutes in risk analysis.

G. Breakeven

A planning tool developed at the beginning of the century, the *breakeven chart*, plots the relationship between cost, revenue, and profit. Figure 10.2 shows two scales, both expressed in dollars and with an identical value. The horizontal scale represents sales volume (revenue), and the vertical scale represents expenses.

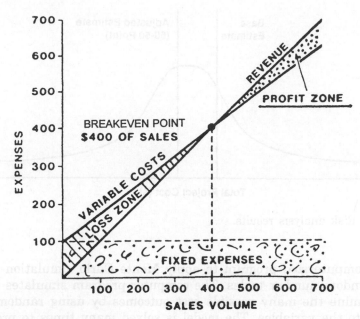

Figure 10.2 Breakeven chart.

The first element to plot is fixed cost, such as rent, which is shown as a horizontal expenditure and which does not change with sales. Then variable costs can be plotted over the fixed costs. Again, if there is no deviation in the proportion of variable costs to sales, this would be shown as a 45° line. Finally, the revenue line is drawn. Where the revenue line crosses the combined expenses line, the breakeven point occurs. This is the point of zero profit, or where losses stop and profits begin.

Breakeven is not a pure form of risk analysis but is more a form of economic analysis. It also lends itself very well to management decisions, since the normal output of the breakeven method is a graph.

VIII. RISK ANALYSIS OF AN ESTIMATE

In applying a formal risk analysis technique to an established base estimate (an estimate without contingency), the following steps are recommended:

1. Divide the estimate into groups that reflect its major disciplines (such as major equipment, bulk materials, labor directs/indirects, design costs, etc.).
2. List the variable elements that affect each item identified in Step 1 (such as major equipment pricing, piping quantities, craft productivity, etc.).
3. Estimate a range of values for each variable element identified in Step 2, and, within that range, its likelihood of occurrence (e.g., 1 chance in 4, 1 chance in 10). Essentially, a probability curve is being constructed for each variable.
4. Develop a model that assigns each identified and assessed variable to the estimate disciplines identified in Step 1.

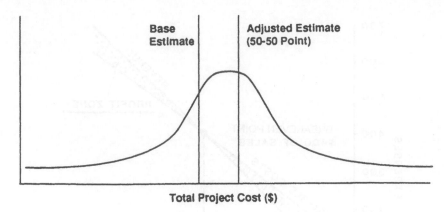

Figure 10.3   Risk analysis results.

5.   Use a computerized program to apply Monte Carlo simulation techniques using random number tables. The computer program simulates the project to determine the many possible cost outcomes by using random sampling applied to the variables. The model is solved many times to provide a set of output samples large enough that its statistical properties approximate the desired solution set.

The results of the risk analysis run described in Step 5 is shown in Figure 10.3.

Computerized risk analysis models generally go through between 500 and 1000 iterations to generate the total project cost. The plot of these total cost calculations is shown as a probability distribution (i.e., the number of times the result was observed). In Figure 10.3, the distribution shows that the base estimate has more risk of overrun than of underrun. In other words, more values fall to the right of the bare estimate line. The adjusted estimate is the base estimate plus contingency, since there is now an equal number of values falling to the left as to the right of the adjusted estimate line. A more detailed description of both manual and computerized risk analysis techniques is contained in the following sections. Range estimating, a sophisticated application of Monte Carlo techniques and probability analysis to prediction of probability of estimate overrun and underrun, is discussed in Chapter 11.

A.   Manual Technique and Example

The risk analysis method discussed in this section represents a simplified manual technique combining several of the steps discussed earlier. Because the approach is simplified (when compared with the Monte Carlo technique), the results obtained are inflated, especially at the extremes of the probability distribution. Whereas the Monte Carlo technique uses a random number generator, the manual method simply adds the probabilities of underrun and overrun of the several variables; thus there is no randomness in the calculation. Nevertheless, the assessment of the base estimate value as to whether it lies to the left or right of the adjusted estimate (the 50:50 point) is a valid conclusion.

Manual risk analysis involves the following steps:

1. The base estimate is divided into components, each representing a discrete piece of the estimate. Basically, the estimate is broken down in the same way it was built up. Each component is assigned its dollar value.

Major equipment (labor and materials)	$1000
Piping labor	2000
Electrical labor	500
Piping materials	500
Electrical materials	200
All other labor	200
All other materials	100
Design costs	500
Total	$5000

2. Each component is evaluated as to the probability of overrun/underrun at the 90/10 (90/10 meaning a 90% chance of overrun and a 10% chance of underrun), 75/15, 25/72, and 10/90 points. At each probability point, the most likely dollar value is recorded.

Estimate component	Probability of actual value being (over or under) amount shown value				Base estimate value
	(90/10)	(75/25)	(25/75)	(10/90)	
Major equipment	$ 800	$ 900	$1200	$1500	$1000
Piping labor	1500	1750	2500	3000	2000
Electrical labor	300	400	600	900	500
Piping materials	300	400	750	1000	500
Electrical materials	100	150	250	300	200
All other labor	100	150	250	300	200
All other materials	50	50	150	200	100
Design costs	450	450	800	1000	500
Total	$3500	$4250	$6500	$8200	$5000

3. The sum of the dollar values at each probability point is summed and plotted as shown in Figure 10.4.
4. The intersection of the plotted values at the 50:50 point yields the contingency required to produce an estimate with an equal chance of overflow. In the above example, the 50% point in approximately 5250, which yields a contingency of 250 (i.e., 5250 − 5000 = 250).

Technically, this manual technique is not entirely correct, in that it assumes all of the elements to be 100% dependent, allowing for the various (90/10, 75/25, etc.) components to be summed. Since this is not correct, the outer bounds will tend to be skewed, providing a smaller overall contingency. However, the error is small.

B. Monte Carlo Analysis

Monte Carlo simulation replaces the randomness and dependency of the manual method. Monte Carlo risk analysis begins with constructing a model made up of

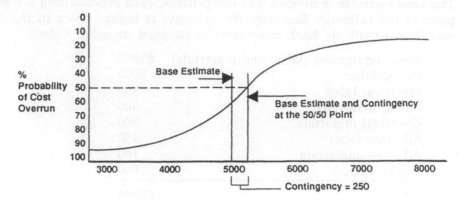

Figure 10.4   Risk analysis probability curve.

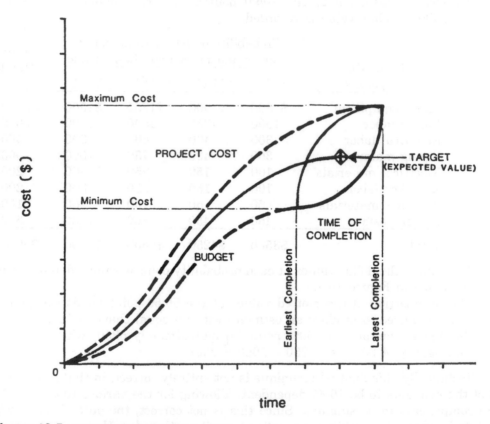

Figure 10.5   Cost/schedule risk analysis.

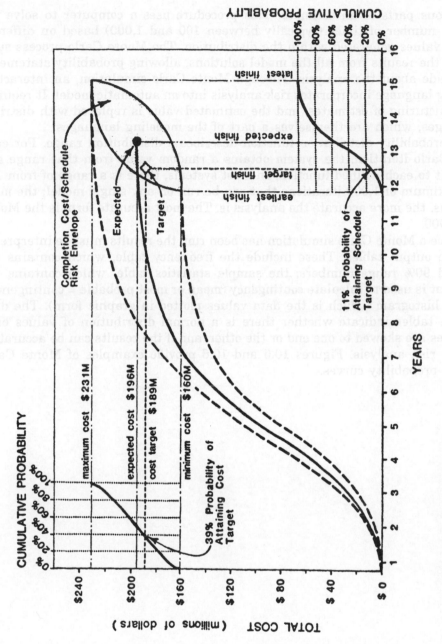

Figure 10.6 Cost/schedule performance graph.

the various parts of an estimate. This procedure uses a computer to solve the model a number of times (typically between 100 and 1,000) based on different random values generated within the distribution. The Monte Carlo process summarizes the results from all the model solutions, allowing probability statements to be made about the analysis. Through Monte Carlo simulation, an interactive modeling language incorporates risk analysis into an automatic model. It requires no restructuring of estimates, and the estimated value is replaced with distribution ranges, which are themselves a part of the modeling languages.

A probability distribution function describes a distribution range. For each Monte Carlo iteration, the system obtains a random value from that range and applies it to each defined function. In most systems, there is a range of from 100 to a maximum of 1,000 iterations that can be run on any single model; the more iterations, the more accurate the analysis is. The most accurate form is the Monte Carlo 1000.

Once a Monte Carlo simulation has been run, the results must be interpreted from the output tables. These include the frequency table, which contains the 10% and 90% range numbers; the sample statistics table, which contains the mean that is used to calculate contingency (mean or most probable = contingency); and the histogram (which is the data values plotted in graphic form). The data on these tables indicate whether there is a normal distribution of values or if the values are skewed to one end or the other, and if the results can be accurately used for risk analysis. Figures 10.5 and 10.6 provide examples of Monte Carlo analysis probability curves.

11

Range Estimating*

I. THE TRUTH ABOUT RANGE ESTIMATING

Many organizations are just beginning to learn that sophisticated software packages do not necessarily create sophisticated users. People often understand a package's functions but not the underlying decision technology. This is particularly true with Monte Carlo Simulation (MCS). Range estimating was developed to correct the deficiencies as well as to capitalize on the strengths of MCS in order to bring the benefits of risk analysis to nonstatisticians—without compromising the validity of the technique.

Since it was first coined in 1970, the term *range estimating* has been widely misinterpreted. To some, range estimating includes any attempt to measure uncertainty with ranges rather than single-point numbers (i.e., estimating ranges). To others, range estimating is MCS itself. Both interpretations are wrong.

In 1970, range estimating was defined as a combination of the good part of MCS (the sampling process), an active application of Pareto's law (defined later) and, most importantly, powerful heuristics which make the entire process easy to apply. A computer and range estimating software are needed—it cannot be applied manually.

The key differences between range estimating and basic MCS are:

- range estimating's simplified inputs, as ranges and probability factors are used rather than traditional probability density functions (PDFs) and their parameters; and,
- range estimating's ability to identify, quantify, and rank critical elements according to their contributions to bottom line risk and opportunity.

---

*By Michael W. Curran and Kevin M. Curran, Decision Sciences Corporation, St. Louis, Missouri. Copyright 1995 by Decision Sciences Corporation. All rights reserved. Reprinted by permission.

Monte Carlo Simulation (MCS) Puts User at Risk

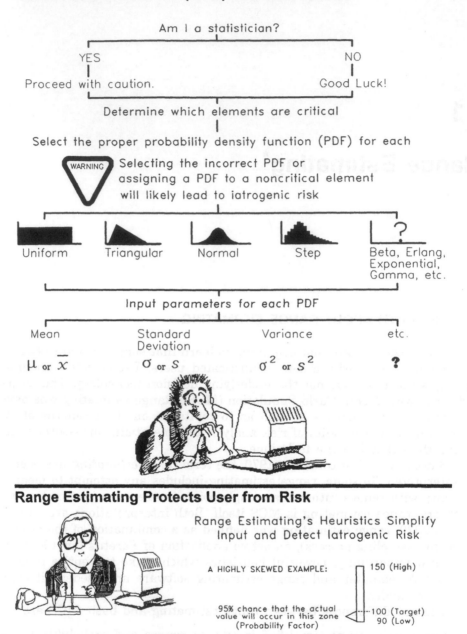

Figure 11.1   The differences between MCS and range estimating.

These differences are illustrated in Figure 11.1.

II. RISK, OPPORTUNITY, AND UNCERTAINTY

In economics, both negative and positive characteristics are ascribed to risk—high risks promise high rewards. However, this contradicts common usage of the term "risk." For example, most dictionaries (Webster's and others) define risk as the possibility or probability of loss, harm, or injury. In other words, there is nothing good at all about risk.

Range estimating uses practical (i.e., common) definitions. Several professional organizations are also moving away from the economist's definitions and toward the dictionary definitions:

- *Risk*: an undesirable potential outcome and/or its probability of occurrence;
- *Opportunity*: a desirable potential outcome and/or its probability of occurrence;
- *Uncertainty*: all potential outcomes (i.e., uncertainty is the parent of both risk and opportunity).

III. NUMBERCLATURE

Everyone knows it is virtually impossible to predict the bottom line exactly. What then does the decision maker strive to do? Manage uncertainty, that's what. But uncertainty can only be managed if it is measured and monitored!

The proper tool must be selected when measuring. The spreadsheet is a poor tool for measuring uncertainty. To select the proper tool, one must first understand the nature of the number to be measured. The are three categories of numbers: *knowns*, *known-unknowns*, and *unknown-unknowns*.

A. Knowns

A *known* is a number whose value is certain. The number of degrees in a circle is an example of a known. There is no doubt that this value is 360. Since a known has no associated uncertainty, a single-point number is used to measure it.

B. Known-Unknowns

A *known-unknown* (UNK) is a number whose value is uncertain—a number whose value is known to be unknown. The exact number of inches of rain which will fall worldwide in the year 2017 is an UNK. Many UNKs contain insignificant amounts of uncertainty and therefore pose no problems for the decision maker. A small number of UNKs, however, contain significant uncertainty. These few UNKs can cause plans to go awry, sometimes dramatically. Thus the need for risk analysis.

For the majority of UNKs—those with little uncertainty—nothing more complex than simple arithmetic is needed. However, simple arithmetic can prove disastrous if it is applied to the small number of UNKs with high degrees of uncertainty. Clearly, these need to be measured in a different way—with a range!

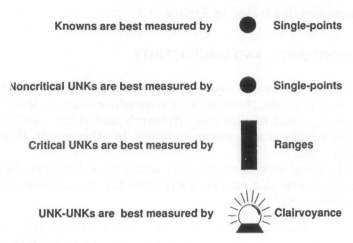

Figure 11.2   Types of numbers.

C.   Unknown-Unknowns

An *unknown-unknown* (UNK-UNK) is a number (or entity or event) whose existence cannot be imagined. An UNK-UNK cannot be predicted unless, perhaps, with a crystal ball. Once the UNK-UNK's existence is known, it is no longer an UNK-UNK. By definition, it is then either a known or an UNK. It is therefore impossible to give an example of an UNK-UNK.

The UNK-UNK is not the major culprit in planning. True, an UNK-UNK can wreak havoc at the bottom line. But history shows this does not happen that frequently. Far more often, plans fail because of UNKs—just a few UNKs with high leverage move in the wrong direction and push the bottom line over the edge (Fig. 11.2).

IV.   PARETO'S LAW

Assume a decision maker has prepared a cost estimate for a project with hundreds or thousands of cost elements and a bottom line target of $1,000,000. In that estimate there is an item of equipment for $145,000. Due to a contractual arrangement with the equipment manufacturer, the decision maker is assured that the actual cost will be exactly $145,000. This cost element is a known—it contains no uncertainty. If there is a cost overrun on this project, it won't be due to this element!

In this same project, 800 meters of material must be installed. The estimated labor cost for this is $22 per meter, and the decision maker's experience indicates it could go as low as $15 or as high as $29 per meter. In other words, the potential variability of this labor cost translates into a potential variability of $5,600 (plus or minus) at the bottom line.

The crucial issue is an element's variability, not its magnitude. Since a range is used to measure the uncertainty in UNKs, and since most of the elements in a plan are UNKs, does that mean that most of the elements in the plan require ranges? No!

Not all elements have the potential for changing the bottom line by a significant amount. In fact, just as few elements will account for the largest percentage of the bottom line's magnitude, it is also true that few elements will account for the largest percentage of the bottom line's variability. This phenomenon is called the *law of the significant few and insignificant many*, the *80/20 rule*, or *Pareto's law.*

Vilfredo Pareto, a sociologist/economist, discovered that most of the wealth in a nation is concentrated in a small percentage of the population. This same phenomenon occurs in planning. A few elements account for the vast majority of the bottom line's potential variability. Typically, these elements number between 10 and 20—regardless of the bottom line's magnitude or the nature of the decision process. This has been demonstrated innumerable times in decisions up to $12 billion.

Those few UNKs which can cause substantial changes in the bottom line are called *critical elements*. (All other UNKs and knowns within the plan are called *noncritical elements*.) Unfortunately, a majority of critical elements behave adversely with predictable regularity and thus spoil plans. That's why even good plans often produce poor decisions and poor outcomes. That's also why the variability of each critical element must be measured by a range.

V. CRITICAL ELEMENTS

Extreme care must be taken not to apply ranges to too many, or the wrong, elements. If significantly more than 20 elements are ranged, it is highly likely that an understatement of true risk will result. Understating true risk is a risk in itself. In simplest terms, the risk analyst adds risk by understating true risk. Such analyst-induced risk is called *iatrogenic risk* and is a frequently encountered pitfall for many MCS users. Range estimating actively guards against iatrogenic risk.

A. Critical Versus Noncritical

Each UNK contributes to bottom line uncertainty. But how much change to the bottom line must an UNK be capable of generating to qualify it as a critical element? Seven years of research during the late 1960s and early 1970s resulted in the development of a highly effective and easy-to-apply critical variance matrix for identifying critical elements. The use of this critical variance matrix depends upon the type of bottom line (expense or profit) and the type of planning process (detailed or conceptual).

- *Expense versus profit.* Determining the type of bottom line is a simple matter. If increasingly larger values of the bottom line are undesirable, it is an expense type bottom line. If increasingly larger values of the bottom line are desirable, it is a profit type bottom line.
- *Detailed versus conceptual.* If a plan measures all that it can be reasonably expected to measure, it is a detailed plan. Otherwise, it's conceptual.

Each percentage in the critical variance matrix (Fig. 11.3) defines the amount of change to the bottom line which an UNK must be capable of generating in order for it to qualify as a critical element. This hurdle, whether stated as a percentage

BOTTOM LINE	DETAILED	CONCEPTUAL
EXPENSE (a)	0.2%	0.5%
PROFIT (b)	2.0%	5.0%

(a) increasingly larger values are undesirable

(b) increasingly larger values are desirable

Figure 11.3 Critical variance matrix for identifying critical elements.

shown in the matrix or expressed in units of the bottom line, is called the *critical variance of the bottom line*.

While it may appear that use of the critical variance matrix will uncover many critical elements, such is not the case. Typically, there are 10–20 critical elements—regardless of the type of plan or the magnitude of its bottom line! There are exceptions, of course, but they are infrequent and nonviolent; once in a great while there may be as many as 25 or so critical elements.

The critical variances in the matrix have been successfully applied in thousands of decisions since 1972. Therefore, use of the values contained in the matrix is strongly recommended. The critical variances in the matrix are minima. They should never be made smaller since this will incorrectly classify noncritical elements as critical and will likely introduce iatrogenic risk.

Occasionally, there may be a desire to increase the critical variances in the matrix to reduce the number of critical elements to fewer than 10–20. This is not likely to introduce significant error if the critical variances are increased to no more than twice their values shown in the matrix; however, this is not a guarantee. The best advice is to adhere to the original critical variances in the matrix. (When the bottom line target is zero or near zero, the critical variance matrix guidelines for identifying critical elements do not apply. Good judgment must be used to locate the typically 10–20 critical elements.)

The critical variance, in bottom line units, is determined by multiplying the bottom line target by the critical variance percentage in the matrix. For example, a *conceptual* plan having a bottom line *expense* target of $1,000,000 has a bottom line critical variance of $5,000 (found by multiplying $1,000,000 by the 0.5% critical variance in the matrix). In this example, a critical element is:

- one which can vary enough—either higher or lower—to cause the bottom line to vary (higher or lower) by $5,000 or more, and
- one which is not composed of any other element that can do the same.

B. Finding the Critical Elements

Critical elements can be located quickly and reliably by applying the *pyramid of criticality* method (Fig. 11.4). All of the elements in the plan are imagined in the form of a pyramid with the apex representing the bottom line target. Critical elements are found by a downward search of the pyramid, starting at the apex. In effect, the search is a series of successive questions and answers predicated on the critical variance matrix.

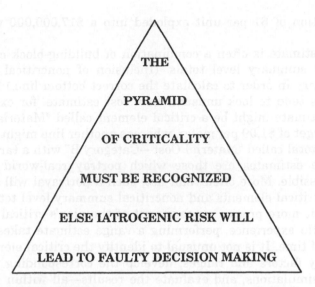

THE

PYRAMID

OF CRITICALITY

MUST BE RECOGNIZED

ELSE IATROGENIC RISK WILL

LEAD TO FAULTY DECISION MAKING

Figure 11.4 Pyramid of criticality.

Continuing with the $1,000,000 conceptual expense plan previously cited, the first question is asked at the apex of the pyramid: Is it possible that the bottom line itself could vary—either up or down—by at least $5,000? If the answer is negative, the search is terminated—the uncertainty is minimal, there are no critical elements, and a cost risk analysis is not warranted.

On the other hand, if the answer is positive, the downward search is resumed at the next lower level in the pyramid—where the components of the $1,000,000 total expense repose. Assuming the plan is an engineering cost estimate, the total expense might be composed of $600,000 in total labor, $300,000 in total equipment, and the remaining $100,000 in total materials. At that next lower level, the question is asked: Is it possible that the $600,000 in total labor could vary—either up or down—such that the bottom line could vary (up or down) by at least $5,000? If the answer is negative, the downward search in the labor portion of the pyramid is terminated; for all practical purposes, no single labor element can vary the bottom line up or down by $5,000 or more if the total labor figure cannot do so. If the answer is affirmative, the downward search is resumed at the next lower level in the pyramid—where the components of the $600,000 total labor reside.

If, during the downward search, a lower level containing no critical elements is found, that means that the element on the level immediately above is the critical element. If the downward search reaches an element having no supporting information below it, then that element is a critical element. In other words, a critical element is an element which can change the bottom line by an amount equal to or greater than the bottom line's critical variance but no number in support of that element can do so.

It must be noted that, to be classified as critical, an element need not itself vary by the bottom line's critical variance. Rather, it must have the capability of varying the bottom line by that amount. Escalation, inflation, tax rates, and other components of the plan may further compound the effect. (In one actual case, a potential variation of $1 per unit exploded into a $17,000,000 variation at the bottom line!)

A range estimate is often a combination of building-block critical elements and noncritical summary level totals. (Inclusion of noncritical summary level totals is necessary in order to calculate the correct bottom line.) For this reason, range estimates tend to look unusual. In a cost estimate, for example, one line of the range estimate might be a critical element called "Material—Tubing Unit Cost" with a target of $1.09 per meter, whereas another line might be a noncritical summary level total called "Material Cost—Category B" with a target of $326,490. The best range estimates are those which portray real-world uncertainty as faithfully as possible. More often than not, such a portrayal will be a mixture of building-block critical elements and noncritical summary level totals. Such range estimates permit more precise management of the plan's critical elements.

With a little experience, performing a range estimate takes a surprisingly small amount of time. It is not unusual to identify the critical elements, establish their probability factors and ranges, develop the corresponding model, perform 1,000 or more simulations, and evaluate the results—all within a few hours.

VI. RANGE ESTIMATING INPUTS

Once the critical elements have been identified, the variability (uncertainty) of each must be quantified. If one were using MCS, each critical element would have to be specified in terms of a PDF (e.g., uniform, triangular, normal, beta, etc.) along with its parameters (e.g., mean, standard deviation, shape variables, etc.). Using range estimating, all such tedium is eliminated without impairing the quality of decision making. For each critical element, the user supplies the target, the probability factor, and the range.

A. Target

In the traditional plan, each critical element has a single-point value. It may be called an *estimate*, *budget*, *plan*, *forecast*, etc. In range estimating, it is called the *target*.

B. Probability Factor

The probability factor, expressed as a percentage between 0% and 100%, is the probability that the actual value of the critical element will materialize between its target and lowest value. For a critical expense element, it is the probability that its actual value will materialize in the favorable portion of its range—at or below its target. For a critical profit element, it is the probability that its actual value will materialize in the unfavorable portion of its range. In other words, the probability factor is always interpreted as the probability that the critical element's actual value will fall at or below its target, regardless of whether it is an expense or profit type of critical element. The probability factor is a measure of the degree of optimism or pessimism and is expressed as an increment of 5%.

Probability factors of 0% and 100% are valid. Such a value indicates that the target is located on a boundary of the range or that it lies outside the range. A target located on one of the two boundaries is rare; it is highly unlikely that the critical element's lowest or highest value will be identical to its target. On the other hand, a target which lies outside the range is not unusual. This frequently occurs in cost-to-complete estimating. For example, a target established earlier may become infeasible. As time progresses, the range may shift completely above a critical expense element's target.

C. Range

A range of possible values is specified for each critical element in the plan by specifying the lowest and highest values the critical element can assume. These lowest and highest values are set so far apart that there is greater than a 98% probability that the actual value of the critical element will materialize within the resulting range. Specifically, the "lowest" value is set so low that there is less than 1 chance in 100 that the actual value will be any lower; similarly, the "highest" value is set so high that there is less than a 1% probability that the actual value will be any higher. If there is substantial uncertainty about the actual value of the critical element, its range will be quite broad. A lesser degree of uncertainty will be reflected as a narrower range for the critical element.

The lowest and highest values are independent of the probability factor. It is quite possible that a given critical element could have a fairly small difference between its target and lowest value and yet have a high probability of its actual value materializing in that narrow part of the range. Examples of this often occur in expense elements where it is not unusual to have a very small chance of the actual exceeding the target but, if it does, the amount by which it can exceed it is very large.

Some people have difficulty with the idea of supplying a range; some even claim that the range is nothing more than a lot of guesswork. But that's precisely why the range is valuable in decision making; it involves a lot of educated guessing by qualified people. On the other hand, the single-point value involves only a little guessing—so little that it can lead to serious errors in decision making. There is nothing wrong with guessing; Nobel prizes have been awarded for shrewd guessing!

The number of people involved in preparation of the range can vary from one to several dozen. The collective effort of a group of knowledgeable people—a group devoid of dominant personalities capable of introducing bias into the process—tends to produce the best range. Such a group is likely to steer clear of the chief pitfall—developing a range which is too narrow! The varied perceptions of a group's members translate into a wider range, one more apt to capture all of the possible values of the critical element. However, if only one person is qualified to specify the range, that is better than no one doing it!

The need to make the range wide enough to capture over 98% of all possible values cannot be over emphasized. People tend to make the range far too narrow, particularly on that side of the target which represents adverse performance! Feeling comfortable about a range which is too narrow is a mistake which is easy to make. Besides using the group approach, this problem can be minimized by challenging questions. For example, if the proposed lowest value is $500, the challenging question could be, "Can it possibly go as low as $490?" Such challenging questions often produce ranges which are much wider than they otherwise would be.

A wide range should not be interpreted as a lack of expertise. Experienced planners are not omniscient. They cannot consistently predict actual values of critical elements with high degrees of accuracy. In fact, it can be argued that their most important contribution in decision making is their understanding of what they don't know as well as what they know. A wider, more realistic, range reflects a professional judgment and not a lack of knowledge on the part of the professionals who developed it.

In the world of cost estimating, the range is nothing more than basic contingency for the element. For a cost element, that part of the range above the target is the positive contingency for that element whereas that part below the target is its negative contingency. Range estimating combines the ranges (basic contingencies) of all such elements to determine the proper basic contingency for the bottom line.

If the traditional cost estimate includes a separate line item for basic contingency, that amount must not be included in the bottom line target of the range estimate. Otherwise, basic contingency will be accounted for twice and thus will produce an inappropriately higher bottom line.

D.   Frozen Values

There are two types of noncritical elements: the known and that sort of known-unknown (UNK) which cannot cause the bottom line to vary, either up or down, by at least the bottom line's critical variance. Since such elements have little or no variability (uncertainty), they are not ranged. Each is frozen at its target, unless there is a likelihood that its actual value will be different than its target. In that case, the noncritical element is frozen at that other value.

Although noncritical elements are frozen rather than ranged, more often than not they are frozen in groups rather than as individual noncritical elements. The reason for this stems from the manner in which critical elements are identified—in the downward search through the pyramid of criticality. In the process, a group of numerous noncritical elements will often be collectively accounted for in a single balance or summary level total figure.

VII.   SIMULATION

During simulation a random value is selected from the range for each critical element, and these random values are combined with the frozen values of the noncritical elements to determine the value of the bottom line. Each such operation is a single simulation of what the real world holds in the way of bottom line results. There are many synonyms for the word *simulation*, including *iteration*, *trial*, *sample*, and *scenario* (the most popular). Many scenarios are evaluated in order to obtain a reliable bottom line profile.

Ideally, when a risk analysis is performed, a truly random number process should drive the results. However, probabilistic computer simulation typically makes use of what is called a pseudorandom number generator (RNG). A properly designed RNG will produce numbers which, when subjected to stringent statistical tests, would be capable of convincing a statistician that such numbers were likely the result of a truly random process. Most RNGs require a seed before they can produce their pseudorandom numbers (i.e., the seed number is a primer); the RNG produces the same sequence of random numbers when the same seed number is used.

It is useful to know that the results obtained in 1,000 simulations will, in fact, be identical to the results obtained from a second set of 1,000 simulations—as long as all input values are the same, the same number of elements are simulated in the same order, the same number of simulations performed, and the same seed number used. Although changing the seed would indeed produce different bottom line results, those results would not be statistically different than the first set of results. In range estimating, the same seed number is always used to ensure that differences in results are 100% attributable to differences in inputs—not statistical noise!

Range estimating never allows selection of a value outside of the specified range—this is contrary to some MCS applications and PDFs. Second, range estimating honors the probability factor in the selection process. For example, if the probability factor is 25% for a given critical element, there will be 3 chances in 4 that the value selected will be greater than the critical element's target. Third, range estimating's heuristics tend to select numbers closer to the target rather than farther away (if the target is feasible, of course)—the chance that the actual value will materialize near either end of the range is less than the chance it will materialize elsewhere in the range.

Whether basic MCS or a hybrid technique such as range estimating, elements in the simulation are assumed independent of one another. If this assumption contradicts reality, steps must be taken to include any significant interdependencies in the simulation. For example, if reinforcing bar (rebar) is directly related to concrete, the dependent variable (rebar) should be expressed as a percentage of the independent variable (concrete). The application of more sophisticated formulae to relate interdependent elements will likely produce little more than increased frustration and confusion in communicating the results to decision makers.

Most range estimates require no more than 500–800 simulations to reach stability. However, to preclude nonproductive discussions between planners and decision makers about sampling theory and to further ensure stability, a total of 1,000 simulations is highly recommended. If the results fail to stabilize after 5,000 simulations, it is extremely likely that there is an inordinate and inappropriate amount of detail in the range estimate, and the results are almost certainly afflicted with iatrogenic risk.

VIII. FIVE KEY QUESTIONS

The purpose of range estimating is to provide answers to five key questions related to the bottom line:

1. "What's the chance of coming out of this 'smelling like a rose'?" In other words, "What's the probability of success (or failure)?"
2. "If things go wrong, how bad can it get?" In other words, "What's the exposure in this decision?"
3. "What can be done now to improve the outcome?" In other words, "Which controllable critical elements contribute the highest amounts of risk and/or

opportunity at the bottom line and what can be done about them today to increase the probability of success and/or reduce the exposure in this decision?"

In the case of expense bottom lines, range estimating also provides answers to these two questions:

4. "How much contingency is needed to cover unacceptable bottom line risk?" In other words, "How much contingency must be added to (or subtracted from) the bottom line in order to achieve a desired level of confidence of not experiencing a cost overrun?"
5. "Where should the contingency go?" In other words, "How can the total contingency be distributed back into the critical elements based on each critical element's risk and opportunity contributions?"

A. What is the Probability of Success?

The 1,000 simulations are performed and the resulting 1,000 bottom lines retained for analysis. The number of bottom lines which are at least as favorable as the target bottom line is expressed as a percentage of the total number of simulations. In 1,000 simulations, for example, if there are 220 bottom lines at least as favorable as the target bottom line, range estimating reports a 22% probability of success, as shown in Figure 11.5[*]. If the bottom line is an expense, that means a 78% probability of a cost overrun.

B. How Bad Can Failure Be?

The exposure in the decision is determined by comparing the bottom line's target with the bottom line's worst value. But, what does "worst" mean? Does it mean the absolutely farthest out bottom line value the critical elements could generate—a theoretical worst, so to speak? Or does worst mean a bottom line not that far out but highly improbable nonetheless—a practical worst, for lack of a better term?

A theoretical worst bottom line value is calculated by setting each critical element at the most unfavorable value in its range, combining those values with the frozen values for all the noncritical elements and then determining the bottom line. What chance is there that this worst case scenario will occur in the real world? Consider a decision in which there are 10 independent critical elements, each with 1 chance in 100 that the most unfavorable value in its range will materialize. In such a case, the probability of the theoretical worst case bottom line scenario occurring is 1 chance in one hundred quintillion (100,000,000,000,000,000,000).

Obviously, the chance is vanishingly small that the actual bottom line will approach the theoretical worst case bottom line. In other words, the theoretical worst case bottom line (and best case bottom line for that matter) is so far out that it is useless for decision making. For this reason, the practical worst case bottom line is used to define the exposure.

---

[*] Figures 11.5 through 11.7 were created with the use of the range estimating program for personal computers (REP/PC). REP/PC is a product of Decision Sciences Corporation, Box 28848, St. Louis, MO 63123; phone: 314-739-2662; fax: 314-536-1001.

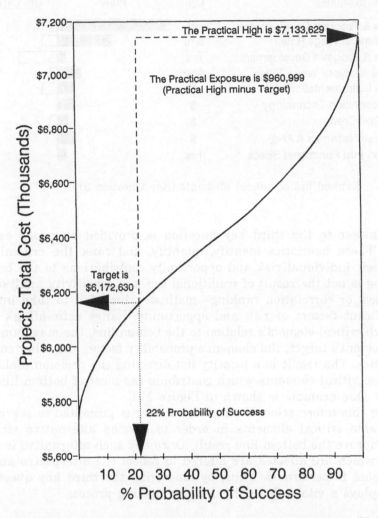

The Practical High is $7,133,629

The Practical Exposure is $960,999
(Practical High minus Target)

Target is
$6,172,630

22% Probability of Success

Figure 11.5   Probability of success (Key Question 1) and practical exposure (Key Question 2).

In the case of an expense type of bottom line, the practical exposure is the difference between the target bottom line expense and the highest bottom line value found in simulation—the practical worst case bottom line. In the case of a profit type of bottom line, the practical exposure is the difference between the target bottom line profit and the lowest bottom line value found in simulation.

C.   What Can Be Improved?

Although it can provide answers to the first, second, and fourth of the five key questions, MCS cannot answer the third or the fifth question. In other words, MCS can determine if there is a problem and how large it could be, but it cannot determine the steps needed to be taken today to improve the chance of success and/or reduce the exposure in the decision.

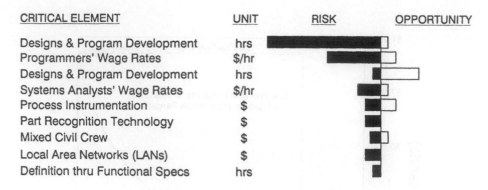

CRITICAL ELEMENT	UNIT	RISK	OPPORTUNITY
Designs & Program Development	hrs		
Programmers' Wage Rates	$/hr		
Designs & Program Development	hrs		
Systems Analysts' Wage Rates	$/hr		
Process Instrumentation	$		
Part Recognition Technology	$		
Mixed Civil Crew	$		
Local Area Networks (LANs)	$		
Definition thru Functional Specs	hrs		

Figure 11.6   Ranked list of critical elements (Key Question 3).

The answer to the third key question is provided by range estimating's heuristics. These heuristics identify, quantify, and rank the critical elements based on their individual risk and opportunity contributions to the bottom line. This ranking is not the result of traditional types of sensitivity analyses, regression analyses, or correlation ranking—methods which fail to take into account some significant factors of risk and opportunity. Range estimating's heuristics consider each critical element's relation to the bottom line, the maximum deviation from the element's target, the element's probability factor, and other contributing characteristics. The result is a priority list directing the decision maker's attention to those critical elements which contribute the most of bottom line risk and opportunity. One example is shown in Figure 11.6.

Having this information, the decision maker is prompted to search this list for controllable critical elements in order to develop alternative strategies or tactics to improve the bottom line result. Once any such alternative is identified, the ranges which are affected are altered to reflect the alternative and another range estimate is performed. Thus, by answering the third key question range estimating plays a vital role in the decision making process.

D.   How Much Contingency Is Needed?

Very often, decision makers wish to reduce the chance of a cost overrun by adding a contingency to the target bottom line. If 1,000 simulations are performed and the 1,000 bottom line results placed into an array in order of magnitude, the array can be imagined as a ladder with 1,000 steps. The lowest bottom line found in simulation occupies the first step and the highest bottom line found in simulation occupies the 1,000th step.

The range estimating user supplies a desired level of confidence in not having a cost overrun. Range estimating first locates the bottom line value in the ladder of results required to attain that goal. If 80% is the desired level of confidence, for example, the required bottom line value is found at step 800 in the ladder—20% (200) of the simulated bottom lines were greater than this value. In other words, there is an 80% chance the actual will materialize at or below the bottom line value found at step 800. The amount of total contingency is the difference between the located bottom line value and the target bottom line. Although uncommon, negative contingencies, where the target bottom line is

Target Estimate	Confidence of No Overrun	Required Confidence	Needs This Contingency	Target With Contingency
$6,172,630	22%	80%	$330,954	$6,503,584

ALLOCATION OF CONTINGENCY

Critical Element	Percent	Add This	Unit
Designs & Program Development	54.3	5,067.44	hrs
Programmers' Wage Rates	19.7	1.67	$/hr
Designs & Program Development	0.0	0	hrs
Systems Analysts' Wage Rates	9.2	1.07	$/hr
Process Instrumentation	0.0	0	$
Part Recognition Technology	6.3	1,677.47	$
Mixed Civil Crew	2.8	1,192.88	$
Local Area Networks (LANs)	5.6	1,192.86	$
Definition thru Functional Specs	2.1	16.10	hrs

Figure 11.7   Contingency required (Key Question 4) and allocation (Key Question 5).

reduced rather than increased, are possible, especially in overly pessimistic estimating and aggressive competitive bidding.

E. Where Should Contingency Be Applied?

Range estimating suggests the portion of the total contingency each critical element should receive, thus distributing the total contingency back into the critical elements (Fig. 11.7). Range estimating bases its suggestions on the answers provided by the third question (the prioritized list of critical elements) and the fourth question (the total contingency).

The portion of the bottom line's contingency which each critical element receives is based on that critical element's net risk contribution (i.e., the difference of its risk contribution and its opportunity contribution). In other words, if a critical element contributes three times as much net risk as another, the first critical element receives three times as much contingency as the second. To further facilitate this allocation, range estimating determines the necessary adjustment to each critical element's target value to reflect its suggested amount of contingency. This information is crucial in contingency management and line-item bidding decisions.

IX. RANGE ESTIMATING ADVANTAGE

MCS poses a number of serious implementation problems, especially for the nonstatistician. Range estimating delivers the power of MCS without the statistical baggage and major pitfalls (e.g., iatrogenic risk). The user of range estimating need not understand classical probability theory and statistics in order to achieve excellent results.

X.   BIBLIOGRAPHY

Anonymous, Range Estimating Provides an Edge in Running the Race Against Risk, *Engineering News-Record*, 198(25):94–99 (1977).

Anonymous, Range Estimating Gains Support, *Engineering News-Record*, 213(12):106 (1984).

K. M. Curran, and W. P. Rowland, Range Estimating in Value Engineering, *Transactions of the American Association of Cost Engineers*, pp. G.3.1–G.3.5, 1991.

K. M. Curran, Range Estimating: User-Friendly Risk Analysis, *Proceedings, First Congress on Computing in Civil Engineering*, American Society of Civil Engineers, Washington, D.C., 1994.

M. W. Curran, Range Estimating: Conquering the Cost Overrun, *DE Technology*, 49(3):10–14 (1988).

M. W. Curran, Range Estimating, *Cost Engineering*, 31(3):18–26 (1989).

M. W. Curran, Range Estimating Reduces Iatrogenic Risk, *Transactions, American Association of Cost Engineers*, pp. K.3.1–K.3.3, 1990.

L. Lewis, Range Estimating—Managing Uncertainty, *AACE (American Association of Cost Engineers) Bulletin*, Nov.-Dec.:205–207 (1977).

L. Lewis, Range Estimating, *Advanced Management Journal*, 45(Spring) (1981).

M. R. Weaver, Improved Estimating Practices using Range Estimating Decision Technology, M.S.C.E. thesis, University of Maryland, Baltimore, MD, 1989.

H. J. Welker, Range Estimating: An Owner's Experience, *Constructor*, 59(8):20–22 (1977).

H. J. Welker, Range Estimating as a Project Management Tool, *Engineering and Contract Record*, June:26–27 (1978).

12

Contracting—Claims and Extras

I. THE REALITY OF CHANGE

A. High Probability of Change and Claims

It is extremely difficult to properly and fully define the scope of work, technical quality, risk, and liability in any contract. This leads to a high probability that extras and claims will occur. Further, the inherent dynamics of any project on which scope and costs are estimated means that those estimates are always at risk as the realities of the work become known. Technical changes, material price variations, and construction conditions are rarely fixed and stable elements. In order to pass on some of these risks to the contractor, owners enter into lump-sum contracts, where many of the risks are included in the pricing agreements. In turn, contractors must properly estimate the cost of these risks and include them in their pricing proposal.

B. Change and Trending

During execution of projects, work realities can result in changes. An efficient trending program is therefore essential to detecting out-of-scope trends and changes that will result in claims. Thus cost engineers must ensure that communication channels with all design and project engineers constantly provide an accurate assessment of the developing work scope. General design specifications and equipment specifications should be monitored for conflict and gold plating. All changes to such items as engineering specifications, scope, procurement, contracting programs, and construction plans should be recorded as they occur or are being considered. Changes to the project execution plan (whether contractual, environmental, regulatory, or schedule-oriented) should be included as well.

The following list, while not all-inclusive, illustrates common changes, some of which often result in disputes as to whether or not they are scope changes:

- all scope changes;
- design changes/development, such as:

    - in early design phase,
    - by owner engineers,
    - by contractor engineers,
    - in design change log,
    - in cost/schedule consciousness, and
    - by gold-plating;

- scope reduction programs;
- project/contractual conditions, such as:

    - in plant operations,
    - in breach of contract, and
    - in site conditions;

- execution plan changes, such as:

    - in priority of project objectives, and
    - in schedule acceleration.

C. Definition of Claim and Extra

By simple definition, a claim or extra occurs because an item of cost was not covered in the contractor's estimate and the item can be demonstrated to be outside the contractor's work scope and/or terms and conditions.

D. Quality Estimating

The key to reducing claims and extras is quality estimating: the ability to properly identify all scopes, responsibilities, liabilities, and unknowns (contingency) and to then properly put a cost/price on these items. Quality estimating for lump-sum bidding is absolutely essential.

E. Litigation—Estimates As Cost Baseline

In claims disputes that result in litigation, the detail and quality of the estimates often form the *cost baseline* from which to measure damages caused by project changes. It is therefore highly recommended that lessons learned from all litigation/estimate analyses should be fed back into the estimating process to ensure that system weaknesses and technical or data errors are eliminated.

II. CLAIMSMANSHIP

Owners say that contractors devote too much effort and personnel resources to pursuing additional revenue rather than to efficiently executing the work. Contractors justify this activity by responding that owners place unfair and unpriceable risks on them, especially in a buyer's market. Spurious and exaggerated claims and disputes are wasteful and irritating, often diverting productive engineering and construction personnel from creative working functions to the chores or arguing, arbitrating, and litigating. This activity is often referred to as *claimsmanship*.

Managing claims and changes is certainly at the heart of effective contract administration. From the contractor's viewpoint, it means ensuring that claims and their associated costs are properly and fully identified. From the owner's viewpoint, it means to prevent or reduce contractor claims and their associated costs through efficient oversight and quality management.

III. RECORDS—DOCUMENTATION

The key to effectively meeting the objectives of both the contractor and the owner is records. There can be no substitute for accurate documentation, which should start at the precontract stage and continue to the final certificate of contract completion. Logs, diaries, minutes of meetings, memoranda, letters, notices, telexes, faxes, reports, and other such items should be current, efficiently collated, and filed for easy access and retrieval. Developing an effective filing system, which usually is included as part of the project coordination procedure, is a demanding task. Special notices and attention must be given to changes, delays, extra work, and all matters that appear to be out of the ordinary and that are therefore potential items for a future claim.

IV. KNOWLEDGE OF CONTRACT LAW FUNDAMENTALS

A. Common Lack of Knowledge

It has been the author's experience that many projects are executed where the project and construction personnel have little or no understanding of the contract conditions. This experience applies to both owner and contractor personnel, but especially to owner personnel. In large contracting and owner companies, it is common for contracts to be developed and negotiated by a contracts department. At contract award, the contracts are then handed to the project team for execution. The lack of continuity and the project team's lack of knowledge about front-end activities are significant failures of the contracting process that can directly result in claims of negligence and breach of contract situations.

B. Key Fundamentals of Contract Law

The following list, while not all-inclusive, outlines the more important legal relationships, liabilities, and risks inherent in the United States, United Kingdom, and Napoleonic legal systems.

General Contractor liability. Of the three parties—owner, architect/engineer, and general contractor (owner, A/E, and GC)—the GC is legally considered to be the expert. As such, the GC is legally required to foresee every and all reasonable situation that can occur to the work from start to finish. If the GC fails to foresee, it is liable for the consequences of its failure.

Noninterference, owner liability. The risk of liability assumed by the GC means that the owner, the A/E, and others are not allowed to interfere with the work operations of the GC. This applies to all contracting arrangements. Acceptable interference may be allowed, as outlined in a set of agreed-to general conditions. The epitome of interference is the equal partnership relationship clause that is appropriate for reimbursable contracts but not acceptable for lump-sum contracts.

Offer and acceptance. The offer and acceptance theoretically leads to a binding agreement, although situations can exist (described later), that negate the binding.

Consideration. A binding agreement requires an element of a bargain. It does not have to be a good bargain (e.g., the price can be low), but free agreements are not binding.

Good faith. All parties to the negotiations and the resulting contract must act with good faith (i.e., responsibly, properly, and fully).

No legal defects. There can be no legal defects, either in the agreement itself or in the process of reaching the agreement. A defect may be, for example, lack of good faith, fraud, misrepresentation, economic duress, criminal acts, drunkenness, or mental illness. Within the United States, differences in state legal systems may exist, causing an act or condition to be legal in one state but illegal in another.

Agency. The parties must be duly authorized agents of the parties to the contract.

Capacity. The parties must have agency and must be acting within the limits of their authority. They must be authorized by company policy to act and to sign on behalf of the company.

Verbal agreements. Verbal agreements are binding if there is agency and capacity. If disputes arise and the agreement has only two parties, a court or an arbitrator must decide who, if anyone, if telling the truth. Failure to be truthful can result in the case being dismissed. Thus it is always wise to reduce verbal agreements to writing.

Nonperformance. Nonperformance by any party can result in contract recision by the performing party. There can also be damages awarded for failure to perform.

Certainty. Certainty is one of the most powerful requirements for a binding contract. The law requires that the purchase or owner make certain that the other party fully understands the scope of the work and its associated risks/liabilities. If the purchaser fails to make certain by using quality contract documents and full discussions/clarifications if necessary, there is then a high risk of recision, since the contract can then be considered nonbinding. Lack of certainty is one of the most common disputes with regard to questions of a binding contract.

Common usage of language. This is an important facet of certainty and of apportioning liability in a dispute.

Bid errors. Bid errors, either in price or in contract conditions, are binding. If, however, an owner has express and direct knowledge of an error, it is illegal for the owner to enter into an agreement without disclosing this knowledge. Experience, skill, suspicion, or evaluations by the owner are not considered to be knowledge. Further, the owner or purchaser is not required to find errors.

Meeting of the minds. Whether there is a meeting of the minds is a decision made by the court when considering whether a contract is binding. If the court decides that all requirements for binding are present, then the decision will be that a meeting of the minds took place. If the requirements were not met, the opposite would be true. Meeting of the minds generally requires certainty, consideration, capacity, and no legal defects.

Intent of the parties. Under United States and Napoleonic law, the intent of the parties is equal to the written terms and conditions. In the United Kingdom, intent has a much lower value and, on occasion, has no value at all.

V. UNDERSTANDING/AVOIDING BREACH OF CONTRACT CONDITIONS

Generally, a breach of contract can only occur when the following three conditions are present:

- the item must be contrary to the contract terms and conditions;
- the item must be contrary to the intent of the parties; and
- the injured party must actually sustain additional or excess costs; if there are no additional costs, there is no breach even though the action might have been contrary to the contract terms.

In most cases, breaches of contract result in claims rather than legal action. The project manager must, therefore, ensure that neither the project manager nor any member of the project team creates a situation that breaches the contract and leads to a claim.

The following breach categories should be carefully evaluated and understood by all contracts officers, construction managers, and project managers to ensure the associated liability and risk are properly covered in the contract terms and conditions. The list represents the claims and damages items or situations that are most commonly experienced on engineering and construction projects.

Acceleration, constructive. The contractor actually incurs additional or excess costs because of accelerating the schedule when the need for schedule acceleration is not due to the contractor's poor performance.

Access denied. Every contract allows the contractor an implied right to enter, leave, and occupy the work site as required to reasonably perform the work.

Certainty, lack of. Genuine misunderstanding of scope, uncertainty of meaning, or lack of understanding/agreement based on intent and logical interpretation of the contract terms and conditions.

Damages, punitive. Damages assessed only to serve as punishment to a party that maliciously or fraudulently causes a breach of contract. Also serves as a deterrent to others. Actual malice must be proven for successful recovery.

Damages, special. Those costs that follow directly from the breach itself but which result directly and as a consequence of the breach. Recovery is granted if such damages were foreseeable by the parties at the time the contract was executed.

Delay, compensable.* Delay caused by actions or omissions of the owner and which entitle a GC to an adjustment in price and/or an extension of time.

Delay, concurrent[*]. Two or more delays in the same time frame such that the owner and GC are each responsible for delay in completing the work. If the court can properly allocate the costs to each party's delay, the recovery is then apportioned to each party. Recovery is denied and the case dismissed if each party's delay costs cannot be properly determined.

Delay, excusable[*]. A delay that is caused by forces beyond the control of and without the fault of negligence of the GC.

Delay, inexcusable. Delay that is within the GC's control and which thus bars a GC's request for an adjustment in price or time.

Differing site conditions. Some of the significant construction contract risks/unknowns that can be experienced by a new GC are the site and working conditions. Nevertheless, careful study of the contract documents, a site visit with a careful evaluation of existing conditions, and the application of good construction experience to the conditions and job requirements, should greatly reduce the risks and unknowns.

Most contract conditions require GCs to assume the risks of site conditions, particularly where the work is in an operating plant. Under some conditions, excessive standby time due to delays in work permits might be allowed on fixed-price contracts. Extreme conditions, such as those found in the offshore industry, might also be considered excusable.

As can be seen, a claim based on differing site conditions could be justified (for example, an unusual ground condition that could not be foreseen on a grassroots site project). On new offshore platform projects, detailed investigation and research is usually undertaken, so such claims rarely occur. Nevertheless, conditions on existing offshore platforms change constantly due to weather and operating requirements, and claims based on differing site conditions may still arise.

Frustration of purpose / commercial frustration[**]. This excuses a contractor's nonperformance whenever the purposes or objectives of the contract have been so frustrated that performance by the parties no longer has value to them. This situation applies regardless of whether actual performance is easier or more difficult than the parties had expected. (See *Impracticability*.)

---

[*] These categories, particularly "Delay, compensable," form the major portion of total claims presented by contractors for schedule delay. Therefore, these categories should be fully understood and all claims carefully evaluated as to merit and applicability. More important, owner-to-GC and GC-to-subcontractor activities should be reviewed constantly to prevent situations that may lead to claims and damages.

Most contracts specify that a contractor/subcontractor is excused from liability for delays due to changes in the work force majeure, major labor problems, and owner negligence/misconduct. However, if a delay results from a foreseeable situation (such as a labor dispute) and action by the GC could have prevented the delay, then the delay is inexcusable. The following are typical owner actions/negligence that can lead to compensable delays: failure to meet promised material supply dates; quality of material is out of specification; failure to meet promised drawing issue dates; delay in engineering reviews and approvals, work permit delays; failure to notify of changing site conditions, lack of site access; poor overall management/coordination of multiple, prime contractors; and scope changes of +10% of contract price.

It should be recognized that even though the completion date is achieved, a GC/subcontractor can still successfully pursue a claim for owner negligence when an owner delays the work from earlier completion. A GC can justifiably claim that its price was based on a completion earlier than the contract completion.

Impact or disruption costs. The additional costs resulting from changes in the work over and above the direct costs of the original work. The entire basis for recovery of an impact or ripple claim is that the effect of changes can and does create costs beyond those attributable to the changes themselves. Impact costs may be viewed in two ways: 1) those arising out of each change viewed individually, or 2) those that result from the cumulative impact of changes. These costs often result in a GC claim and is generally referred to as a *disruption claim.*

Impossibility[**]. The inability to meet contract requirements because the event is physically incapable of occurring. This is otherwise known as actual impossibility and is a common law doctrine accepted by all United States courts.

Impracticability[**]. The inability to perform because of extreme and unreasonable difficulty, expense, injury, or loss involved. This is sometimes called practical impossibility. It is a matter of difficulty of performance, as opposed to frustration, which is a matter of the value of the performance.

Interference. Conduct that interrupts the normal flow of operations and impedes performance. Every contract contains the implied condition that neither party will do anything to interfere with progress of the other party or parties. Actions by owners often lead to GC claims of owner interference or wilful misconduct.

Interference, contract/intentional/malicious. These changes of interference refer to actions of third parties that impact on the contractual relationship or bidding process of an owner and GC. The GC will claim that the party is acting outside its responsibility and is, therefore, interfering with the contract or the prospective contract. This typically occurs when an A/E or a consultant offers advice or proposes actions in an owner/GC dispute. If the third party has no clear responsibility or contractual right to involvement, then the GC's claim of interference may stand up in court. This type of interference is also known as *prospective advantage.*

Malpractice. This is usually a charge brought against an A/E for incorrect engineering work or when an engineer moves out of the design responsibility into contracting and construction execution matters (assuming the engineer has no project or construction management responsibility). It is difficult for GCs to successfully prosecute this charge, since engineers have a protected position in most legal systems.

Material difference. As used on changed conditions situations, a deviation from that expected or expressed in the contract to such a degree that it can provide the basis for recovery under the changed conditions clause of the contract.

Misrepresentation. This occurs mostly during the contract proposal stage, when any of the parties can misrepresent either their capability or the scope of the work (e.g., differing site conditions). Misrepresentation can result in a claim for legal defect or fraud, which can in turn lead to voiding the contract.

[**] These three conditions form the legal doctrine that excuses a GC for nonperformance. However, the courts require that the GC clearly demonstrate the extreme nature of the condition. Mere delay or circumstances that are very difficult or very hard are not usually sufficient for a claim of this nature to be successful.

"No Damage" clause. A contract provision limiting the owner's liability in the event of *certain specified delays*, regardless of fault, provided the owner is not guilty of interference or willful misconduct. The important issue here is that the certain specified delays must have been foreseeable at the time the contract was executed.

Pricing, forward (prospective). An equitable adjustment made to the contract before the work is actually performed. Forward pricing is based on estimates rather than actual cost figures. This method is now being used in claims evaluation and agreement.

Pricing, retrospective (actual). An adjustment priced after the work has been completed. Actual costs of the changed work are used to determine the amount of the change. This method is used when prior agreement on the cost is impossible to achieve.

Reasonableness standard. Costs considered reasonable if they are generally accepted in the industry and do not exceed the amount incurred by a prudent contractor. Two factors on which a determination of reasonableness is based are: 1) recognition of the costs as ordinary and necessary, and 2) restraints imposed by law, contract terms, or sound business practices.

Repudiation. Annulling or abrogating a previously made agreement.

Recision of contract. Annulling or abrogating a previously made agreement.

Responsiveness. The conformity of a bid to the requirements stated in the invitation to bid. In the public sector, lack of responsiveness is used to eliminate bids and/or bidders.

Substantial completion. The point where progress on the work can allow the project to be turned over the owner but where some punchlist items may remain incomplete.

Termination of contract, convenience. A contract clause that provides the owner with the right to terminate the contract, in whole or in part, irrespective of the liability of the GC. In exchange for this privilege, the owner is usually obligated to compensate the GC for the reasonable costs connected with the contract termination.

Termination of contract, default. A permanent suspension of the GC's work by the owner due to some material breach by the GC in failing to carry out its obligations.

Time extension. A lengthening of the contract period prompted by negotiations between the parties or the occurrence of events triggering a time extension under the terms of the contract.

Time of the essence/liability. A requirement that the work be completed within the time limits contained in the contract, with failure to do so considered a breach for which the injured party is entitled to damages.

Truth in negotiations. A term used to refer to United States Public Law 87–653, whose purpose is to require contractors to submit accurate, complete, and current costs or pricing data.

Waiver. The intentional or voluntary relinquishment of a contractual right.

Warranty, express. A promise explicitly written in the contract that a proposition of fact is true.

Warranty, implied. A promise not explicitly written in the contract but which is nevertheless recognized by the courts based on the action of the parties to the agreement.

VI. CLAIMS MITIGATION AND REDUCTION—ESSENTIAL ELEMENTS

The successful contracts administrator has a combination of personal skill and several favorable project conditions:

A. Personal Skills

- project experience in engineering and construction,
- contractual knowledge,
- business skills,
- contractual knowledge,
- cost and schedule analytical ability, and
- people and interpersonal capability.

B. Correct Contracting Arrangement

The selected contracting arrangement must properly represent the scope definition and project objectives, and must be fully subjected to the process of certainty before the contract is awarded.

C. Quality Contract Conditions

Drafting contractual terms and conditions requires great skill and experience to ensure that all risks, responsibilities, liabilities, and guarantees are properly defined and correctly assigned to all parties to the contract.

D. Effective Contract Monitoring Program

An accurate assessment of project status at all times is essential to contracting success. Depending on the form of the contract, this covers cost and schedule state, resources requirements, financial performance evaluation, and, above all, trending analysis. The cost and schedule analytical skills outlined above are necessary if this is to be done effectively. Any trends indicating poor GC financial performance often result in increased claims and extras activities.

E. Effective Negotiating Skills of the Parties

If a claim is substantiated, the final cost agreement can depend on the negotiating skills of the parties.

F. Stable Project and Site Conditions

A force majeure condition can create changes and can lead to claims at any time. Even the highest degree of skill cannot forecast the possibility of force majeure conditions.

G. No Contractual Surprises

One of the most significant factors of good contract administration is that no surprises occur. An effective program will be able to constantly monitor and evaluate the performance of the GC so that the status/forecast of scope, changes, costs, progress, and claims will always be known. This requires good working relationships within the project team and with the contractors so that all trends can be promptly identified and appropriate action initiated.

H. Establishing Good People/Contractual Relationships

Developing good working relationships should be a significant objective of the project team. Nevertheless, a good relationship can be adversely affected by aggressive negotiating and tough contract terms and conditions. Even so, a good relationship can be developed if it is based on mutual respect, cooperation, trust, and the recognition by each party that other parties may have differing objectives.

With reimbursable projects, it is sometimes difficult to establish this good relationship. The owner should ensure that the contract documents contain language referring to the *equal partner* relationship between the parties. In practice, this means the owner will have a resident project team in the GC's office so that owner and GC personnel can work closely together. This also allows the owner to monitor the GC's work. To this end, the proposal bid package should contain a requirement that the GC will provide office accommodations and services for the owner's project team at the location where the work is being executed.

On lump-sum contracts, a good relationship can be established, but it must be one *at arm's length* to allow the GC its legal right to noninterference. Nonetheless, the contracting documents should specify the owner's project control and reporting requirements, particularly in respect to progress measurement of the work and schedule/status information. On cost matters, the working relationship is centered around the contract changes provisions of the contract agreement. Low bids should be evaluated very carefully, since an award of a deliberately low bid can lead to an aggressive claims program by the GC, which can obviously jeopardize a good people/contractual relationship.

Whether the contract is reimbursable or lump-sum, an adversarial or *legal-only relationship* should be avoided if at all possible.

I. Effective Communication Skills

Effective communication skills are needed by all parties if good people/contractual relationships and successful negotiations are to be achieved. This is generally accomplished by having the following skills:

*   understanding basic human behavior so as to relate to an individual's hopes, needs, and objectives;
*   writing well and developing the skills of good structure, style, and discipline in the use of words, since verbosity or a poor vocabulary and lead to poor communication;
*   reading critically; balancing speed with retention and understanding;
*   speaking effectively and paying careful attention to the proper use of hands, eyes, voice, and presence; and
*   listening thoughtfully and bridging the communication gap by focusing on the need to develop a good listener at the other end of the speaking chain.

A person who communicates well both orally and in writing should be interested in others, be sensitive to the feelings and emotions of others, and understand their perspectives and needs (although these need not be accepted). A good communicator is both a persuader and a motivator. Similarly, a skilled negotiator can hide all personal biases and emotional characteristics, even under the most extreme conditions.

13

Managing Small, Shutdown, Retrofit, or Outage Projects

I. INTRODUCTION

There is similarity and overlapping in the execution programs for maintenance projects and small capital projects, with all operating plans having maintenance and plant engineering groups. As a result, company policy and practice must differentiate between maintenance and plant engineering projects. Not only size, but other criteria, such as engineering content and project criticality, can designate the category. Projects that are very small in size (having a range of between $1,000 and $100,000) are usually classified as maintenance projects. Those between $100,000 and $10 million are often classified as *multiple small project programs*. Figure 13.1 shows a project breakdown, by project size, which is typical of large operating and contracting companies. Four categories are shown—small, intermediate, large, and mega. Hours are listed for engineering, home office, and construction direct labor. These hours are generated using the illustrated historical breakdown. The hours relate to the cost of a project case in each category, as listed.

This chapter and this book do not address the handling of very small projects, which are usually included in a maintenance planning program. However, many of the techniques covered here can be considered for the top end of maintenance projects and critical, small-value projects where timing is of the utmost importance.

II. THE GENERAL PROBLEM (PERCEPTION AND ORGANIZATION)

In large operating companies, there is often a wide gap in the skills, experience, and management effort between the *large project* programs and the *small project* programs. To some degree, contractors have a similar situation, since they often assign less experienced personnel to a client's small projects. While numerous com-

Project Size Engineering: 10% TIC @ $45/hr Home Office: 3% TIC @ $45/hr Construction: 21% TIC @ $22/hr (direct) TIC = Total Installed Cost	Jobhours, 1000s			Months Duration
	Engineering	Home Office	Construction	
1. Small: $100,000 to $10 million ($5 million case)	11	3.3	48	18
2. Intermediate: $10 million to $100 million ($50 million case)	110	33	480	23
3. Large $100 million to $500 million ($250 million case)	550	165	2400	33
4. Mega: Above $500 million ($1000 million case)	2200	660	9600	36 Execution phase to mechanical completion.

Figure 13.1 Typical EPC projects—full scope contractor execution.

panies have experienced engineering groups that use sophisticated project management programs on large projects, a significant number of them do not appear to recognize that many of the same skills and programs are also needed on the smaller projects being handled by their plant project engineering departments. It is possible that this gap exists for two reasons: first, plant management is largely made up of production people and not project people, and second, the capital project budget forms only a small part of the plant operating budget. So, a lack of understanding of the project business often leads to a poor commitment by management to support an effective project management program. Many times there is also a failure to recognize the very real economic benefits that a quality program can bring.

Beyond the lack of management commitment to the modern project approach, however, there is often a lack of skilled personnel assigned to the smaller project. Too often, plant operations and maintenance personnel are assigned to the project engineering group without the experience of project work and without having been through adequate training in effective project management procedures.

III. SMALL DOES NOT MEAN EASY

A common fallacy is that small projects are simple and easy to manage. Project plant engineers faced with the not uncommon situation of a lack of personnel may have to personally handle many aspects of estimating, scheduling, cost control, contracting, construction, and project management. In other words, they need to be more versatile than their counterparts on large projects, where many of these tasks are handled by separate, individual experts.

IV. THE ORGANIZATIONAL ANSWER (PARTLY)

First, training and educating plant managers in project management will resolve many of the problems outlined in the preceding paragraphs. Career development and/or rotational assignments through plant operations, maintenance, and project engineering can also reduce and eliminate these problems.

Second, the project engineering group can be divided into "small" and "large" project departments. This division, however, can create a secondary problem: quality of assignment. Project professionals may believe that their careers may be damaged if they remain in the small project department too long. They may think that the large project environment more quickly expands their skills, experience, and (above all) management recognition. It is therefore recommended that careful career planning be developed for all personnel as they proceed through the two departments.

V. PROJECT DURATION

Many small projects have schedules of between 3 and 6 months, which not only allows minimal time for program development but for reaction to and resolution of problems as well. It is essential, therefore, to have a project control program that can easily, quickly, and accurately provide information and performance data.

A computerized reporting program is usually the answer, but developing a new computer program requires careful planning. Thus the recommended method is to first develop a manual program and to then computerize it with a proven software package.

VI. PHYSICAL MEASUREMENT SYSTEMS

On larger projects, the most effective method of project control is the proven program of physical measurement/quantities/earned budget/actual hours/productivity/progress. Another program is the cost/time/resource (CTR) approach, which integrates cost and scheduling and can provide equally good results.

Nevertheless, both programs require time to develop and are the most expensive of all methods for effectively managing and executing a project. It is, therefore, most probable that such programs are not suitable for managing small projects.

VII. JOBHOUR CONTROL

Jobhour control programs are probably best suited to small projects, but it must be recognized that jobhour methods have significant weaknesses, particularly in measuring progress. These inherent weaknesses can be properly handled with an adequate system of checks and balances, based on good personnel experience, adequate historical data, and with reference to the scheduling program.

VIII. PERSONNEL SKILLS AND TEAM BUILDING

Another successful program for application on small projects is to transform the engineering department into a project team. It requires ensuring that the interfaces between all the supporting groups (engineering, procurement, construction, project, contractors, and others) are unified into a cohesive group where all projects, at all development stages, are considered equally. This, of course, is one of the major objectives of total quality management (TQM). There is no question that a unified team can make up for lack of sophisticated project techniques, be technically effective, and be cost effective in managing small projects.

In the construction industry, however, matrix interface problems with TQM are common. When the project team approach is used, there is often a lack of clear-cut responsibilities, an inadequate listing of personnel duties and objectives, and poor company policies. These tend to increase matrix interface problems. The problem is *who does what and when*, which is typified by the lack of clear-cut responsibilities of engineering, procurement, and construction personnel. Questions such as the following arise:

- Who is responsible for reporting and forecasting costs, progress, and schedule completion of engineering and construction?
- Who is responsible for developing and managing the schedule?
- Who is responsible for material purchasing and forecasting?
- Who is responsible for material deliveries?
- Who is responsible for quality assurance?

In addition to clear functional and organizational responsibilities, it is vital that good communication channels be developed. Due to short reaction times on many small projects, fast and accurate communications are essential.

Projects are assigned on a multiple basis to project engineers and managers. In many cases, a small project service group (estimating, cost control, scheduling) then provides the specialist functions to each project engineer. Since service groups are often understaffed, priorities for their services have to be determined. Where service groups do not exist, the project engineers provide the services themselves.

Project engineers must be versatile, flexible, and have good, all-around experience, since they manage many projects at the same time. An ongoing training program is therefore essential to upgrading both the person and the program.

IX. A STANDARD PROJECT MANAGEMENT/CONTROL PROGRAM

An historical, five–year review of a company's annual small project programs generally will show a typical pattern of project type and size. Following this analysis, standard programs can be developed, based on the simple premise that increasing project size will require additional control and reporting. The key to a cost-effective program of this type is identifying the major and critical items of work, and then concentrating on controlling/reporting only these items.

As each project is approved and funded by management, it is recommended that a project execution plan form part of the supporting documentation for

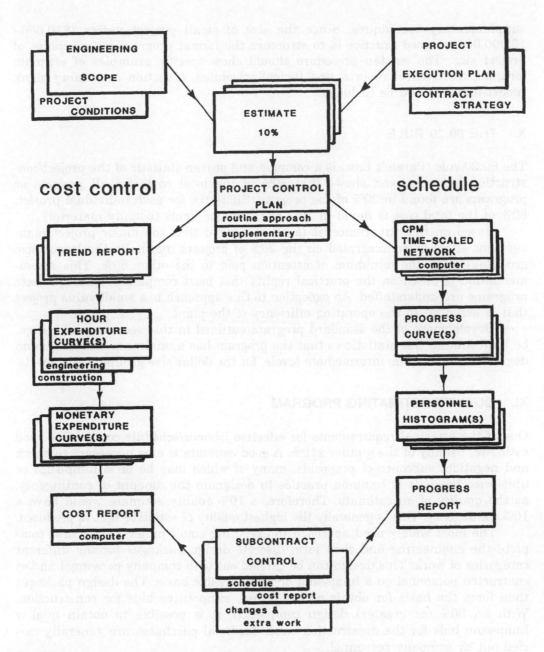

cost control

schedule

Figure 13.2   Project control of multiple, small projects.

project approval. This execution plan would outline the project management/control program or plan to be used. A project control flow chart is shown in Figure 13.2. The program is indicated as the "Project Control Plan."

In addition, a range of supplementary techniques should be developed, which in turn may be used for the larger small projects or for those projects that have critical needs and/or unique features. The standard program should be formalized into a written document (project manual) that contains guidelines for using the supplementary techniques. Since the size of small project varies ($10,000–10,000,000), a good practice is to structure the formal program by groupings of project size. The written procedure should show specific examples of estimate formats, cost control reports, and typical schedules. A section containing blank report forms should be included.

X. THE 80:20 RULE

The 80:20 rule (Pareto's Law) is a common and proven statistic of the project/construction business, and shows that 80% of the total costs for all projects or programs are found in 20% of the projects. Similarly, for each individual project, 80% of the total cost is found in 20% of the major items (usually material).

Based on this experience, it is recommended that the major project management effort be concentrated on the 20% of projects (normally the larger size grouping), with the minimum of attention paid to the other 80%. This recommendation is based on the practical reality that most companies' small projects programs are understaffed. An exception to this approach is a small value project that is essential to the operating efficiency of the plant.

Development of the standard program outlined in this section would, then, be built around this statistic so that the program has a minimum and maximum degree of control, plus intermediate levels, for the dollar size grouping of projects.

XI. QUALITY ESTIMATING PROGRAM

One of the principal requirements for effective jobhour/schedule control is a good estimate, usually of the quality ±10%. A good estimate is also necessary to check and negotiate subcontract proposals, many of which may be on a lump-sum or unit-price basis. It is common practice to designate the amount of contingency, as the *quality* of an estimate. Therefore, a 10% quality estimate would have a 10% contingency. This is generally the highest quality of estimate that is produced.

The most widely used approach to executing small projects is to first complete the engineering and then form discrete design packages for the different categories of work. This design can be carried out with company personnel and/or contracted personnel on a lump-sum or reimbursable basis. The design packages then form the basis for obtaining lump-sum, competitive bids for construction. With an 80% (or greater) design completion, it is possible to obtain quality lump-sum bids for the construction work. Material purchases are generally carried out by company personnel.

The project control program described in the section that follows illustrates the *maximum approach* of the standard program.

XII. PROJECT CONTROL PLAN (MAXIMUM)

The use of computers is strongly advocated, but care must be exercised to ensure that the software programs are user friendly, simple to run, and require a minimum of personnel resources. Figure 13.2 shows the major techniques for project control on the top end of a small projects program (i.e., the maximum approach).

A. Project Control Plan

The project control plan shows the routine program, including any supplementary techniques that are required. It is part of the project funding and approval documentation.

B. Trending Program

A formal trending report should be issued, showing all deviations (costed out) to the reporting cutoff date. The report should be accurate and timely. Responsibility for the report and the inputs from all groups and departments should be clearly established. The evaluation of potential trends should be emphasized. The trend report is then a key source for the project cost report and for all decision making and cost forecasts.

C. Expenditure Curves (Hours)

Figure 13.3 represents a control for a contractor's total home office budget, in cumulative and incremental hours. This includes engineering and all other support groups. The effectiveness of this technique is directly dependent on the accuracy of the planned curve. The curve can be developed from schedule analysis of the component groups and/or historical data. The example contained in Figure 13.3 shows a consistent overexpenditure of hours from the beginning. This need not result in a budget overrun, since the work might be ahead of schedule. A schedule status evaluation and analysis of the component parts should determine *ahead of schedule* or *overrun of hours*. In most cases, overrun is the more likely alternative, which is what the forecast shown in the figure indicates.

D. Monetary Expenditure Curves

Figure 13.4 covers the overall project budget, showing actual cumulative curves for commitments and expenditures. This case was estimated at $8 million, but the high value of commitments at an early stage was matched by an increasing forecast to $10 million in year one, and $12 million in year two. The changes to the control or current budget, barely averaging $1 million, did not properly reflect the rapidly increasing commitment curve. This indicates a project with a poor estimate or a case with many changes, where additional budget is not being allowed. Nevertheless, the trending and forecasting program is properly showing the true costs.

E. Project Cost Report

The project cost report is usually produced on a monthly basis using a computerized program. Committed costs and expenditures are provided by the accounting group. Budgets and forecasts are the responsibility of the project engineering, with assistance from the project services group and/or the engineering, procurement, and construction departments. Figure 13.5 illustrates a typical project cost report format. The keys of this report are an accurate forecast, identification of variances, and reporting on time.

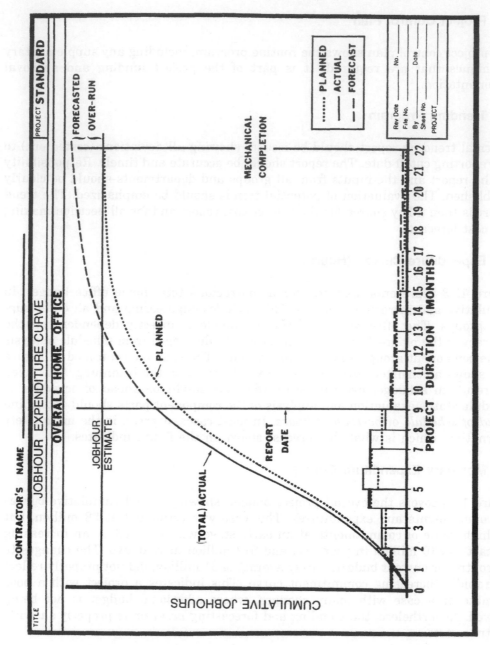

Figure 13.3 Jobhour expenditure curve.

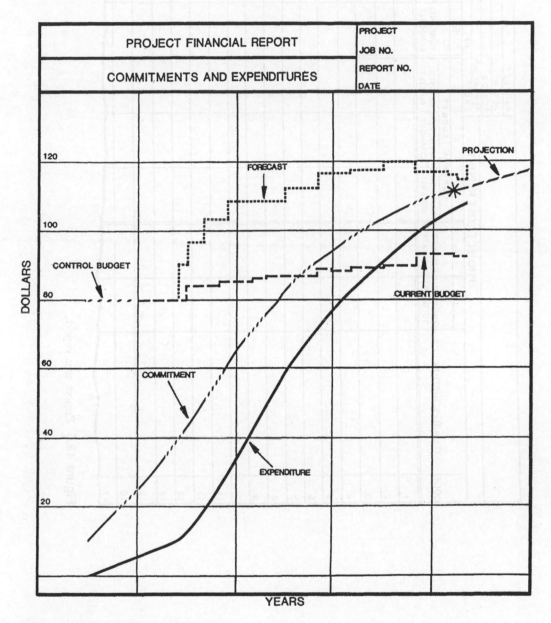

Figure 13.4 Project financial report.

CODE	ITEM/DESCRIPTION	COMMITTED TO DATE	EXPENDED TO DATE	INITIAL (AFE) BUDGET	APPROVED CHANGES	REVISED BUDGET	CURRENT FORECAST	VARIANCE (6) - (5)
		1	2	3	4	5	6	7
				CONTROL BUDGET				
1								
2								
3								
4								
5								
6								
7								
8								
9								
10								
11								
12								
13								
14								
15								
16								
17								

PROJECT COST REPORT

PROJECT
JOB N°
PERIOD
REPORT N°
SHEET   OF

Figure 13.5 Project cost report.

F. Timely Cost Reporting

An important part of cost control (commitment, expenditure, accruals, estimated final cost, escalation, and contingency) is collecting and timely reporting of comittments and expenditures. This information is needed to track the status and progress of cost objectives and also to provide a base for developing cost predictions. The project manager must ensure that commitments and expenditures are reported accurately and in a timely manner. Delayed payments or lost invoices can cause poor cost predictions. Many control techniques are based on unit costs, which can be greatly distorted if the cost-to-date is inaccurately reported or if the reports are submitted late.

G. Critical Path Method Schedule

Figure 13.6 contains an example of a time-scaled critical path method (CPM) network for a $1 million project. Only the construction phase is shown. Sixteen activities effectively show the critical path and overall construction duration. Another good feature of this diagram is its illustration of the calculation routines (hours and crew size) for the critical activities.

H. Progress and Personnel Curves

Figure 13.7 shows a combined set of progress and personnel curves. These are cumulative curves with planned and actual profiles. The CPM computer program could be used for progress measurement, with subjective evaluation of the progress of individual activities. The effectiveness of this program is dependent on the accuracy of the estimated hours and the level of detail of the activities.

Progress and personnel curves can be part of the computer program, but again, dependent on the level of detail. For effective computer application, there needs to be sufficient activities to properly represent the total work and total estimate. This requires a balance between increasing detail and the increasing cost of the program.

The example shown in Figure 13.7 indicates an increasing personnel level versus the plan, and also indicates a failure to make the planned progress. This situation would be due to increasing scope of work and/or poor productivity. A detailed evaluation would be required to pinpoint the cause.

XIII. SUBCONTRACT CONTROL

A. A Reasonable and Simplified Approach

It must be recognized that small subcontractors do not use experienced project control personnel and do not operate with detailed control systems. The key to success, therefore, is to develop a simple, practical method of control and to require that subcontractors include adequate costs in their bids to use the system. Effective subcontract control is based on several essentials:

• good contractual documents and agreements;
• an adequate system for documenting changes, amendments, and claims;
• an acceptable scheduling system (usually a bar chart);

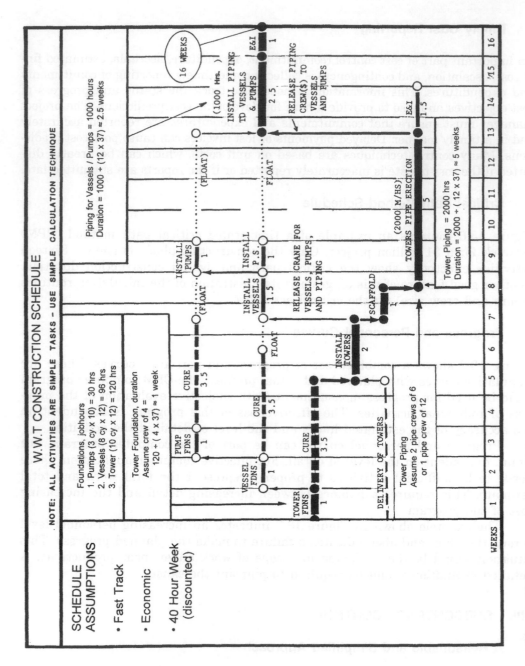

Figure 13.6 Construction schedule.

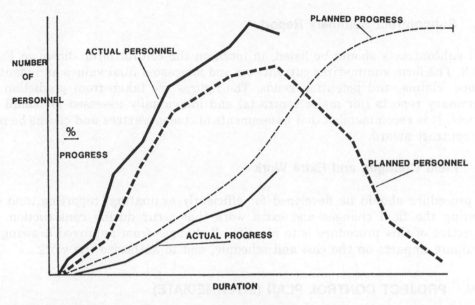

Figure 13.7 Progress and personnel curves (typical).

- a realistic progress measurement system; and
- an effective cost trending and forecasting system.

B.   Low Bids—Risks

A low bid can result from ignorance and/or a subcontractor buying the job. Subcontractors sometimes turn in a survival bid, attempting to break into a new market or to block competition, or perhaps they find deficiencies in the contract conditions. Such situations can result in low bids that can, in turn, lead to serious cost and schedule consequences for an entire project if the subcontractor gets into financial difficulties.

C.   Claims

Subcontractor claims require detailed analysis and in most cases are a combination of:
- change in scope,
- schedule delays caused by the owner and/or the main contractor,
- drawing and material delays,
- interference by others, and
- changes in site conditions or site regulations.

The subcontractors may greatly exaggerate adverse conditions and submit inflated claims. Consequently, it is necessary that daily logs, plans, work schedules, contractual situations, etc., are maintained by project/site staff for all major subcontracts. It is essential that project/site staff respond promptly to claim submissions. It is also important to consider all related claims together, or there may be common influences and consequential impacts.

D. Subcontract Summary Report

All subcontracts should be listed, in total, on the control form shown in Figure 13.8. The form summarizes current cost and forecasted final values and identifies scope, claims, and potential trends. The figures are taken from prediction cost summary reports (for major contracts) and individually assessed for small contracts. It is recommended that assessments of changes/extras and claims be made at contract award.

E. Field Changes and Extra Work

A procedure should be developed for efficiently evaluating, reporting, and estimating the field changes and extra work that occur during construction. The objective of this procedure is to identify all changes from approved drawings, to evaluate impacts on the cost and schedule, and to authorize the work.

XIV. PROJECT CONTROL PLAN (INTERMEDIATE)

An intermediate approach would eliminate much of the detail of the maximum approach, concentrate on major cost items, and rely on the skills of the project engineer to properly manage the project. A simple bar chart would provide the schedule control.

A. Key Cost Items Control/Report

This method concentrates on high-cost items (usually material and construction contracts) and evaluates the budget versus the commitments for these items on a continuous (monthly) basis. An initial review of the estimate should reveal the high-cost items, which should be identified on the report. The report should be issued monthly to show the cost situation of these items. The forecasts will probably be developed by the project manager.

B. Schedule Control—Bar Chart

Figure 13.9 illustrates a simple but adequate format for the complete project. The major phases of the work are shown with individual bars, and starts and finishes are indicated against a calendar time frame. Key milestones are identified with separate symbols. Status updating takes place monthly, and a vertical cursor is drawn at the reporting date to show the status of each bar. As milestones are achieved, the respective symbol is blacked in.

XV. PROJECT CONTROL PLAN (MINIMUM)

This is sometimes known as the zero approach. Again, based on the assumption that the 80:20 rule applies, this approach is applied to a major portion of the many small projects.

After preparation of the *authorization for expenditure* (AFE) estimate and associated schedule bar chart, the major effort is to review and approve all

CONTRACT NO.	CONTRACTOR	WORK SCOPE	BUDGET	ORIGINAL CONTRACT VALUE	CHANGES & EXTRAS	CLAIMS	FORECAST	COMM'D.	EXPEND.

SUBCONTRACT SUMMARY REPORT

PROJECT
JOB NO.
REPORT NO.
DATE

Figure 13.8 Subcontract summary report.

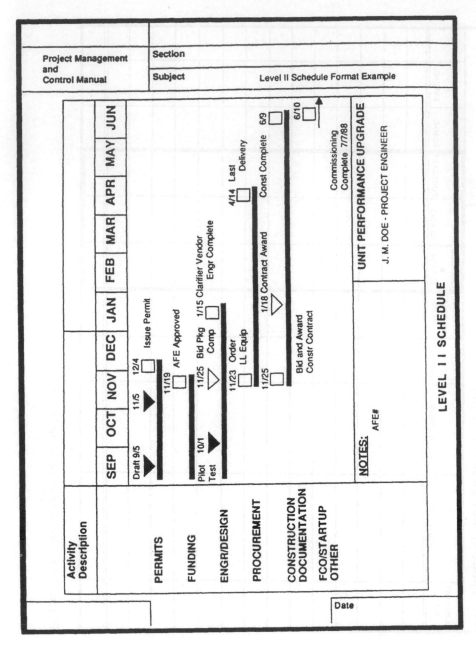

Figure 13.9 Level II schedule.

commitments. There is no periodic trending and forecasting except at the 80% committed point (total project). At this point, the project engineer will review the progress to date and carry out an *estimate to finish*. This plus the commitments to date comprise a forecast of the final costs. The schedules forms part of this estimate.

Since cost reporting/accounting is usually done on a monthly basis, it is possible that the 80% point may be considerably greater from one month to the next, indicating the presence of an overrun situation. This is a risk that must be accepted by management with the varying level approach.

XVI. MAJOR SHUTDOWNS, RETROFITS, AND OUTAGE PROJECTS

As previously outlined, this chapter does not cover maintenance planning. This section emphasizes major shutdowns of 1 or more weeks, and on retrofits with schedules of 3 months and greater.

A. Key Considerations

These projects are very difficult to estimate and, as a direct result of work/scope uncertainty, can be difficult to plan, schedule, and manage, especially as they increase in size. Successful execution of *shutdowns, retrofits, and outage projects* (SROs) requires a thorough investigation months in advance of the work to be performed. As construction is often the major cost center, this is particularly true of all construction work. In addition to problems of scope definition, a further key consideration is the type of management for these projects. Due to lack of project experience, many plant managers assign the management responsibility to an inexperienced engineer or maintenance supervisor.

The project and construction management of SROs, especially as they get larger, is one of the most difficult and demanding of project assignments. It requires intensive, creative, and careful evaluation to develop a quality project management program and execution plan. Contracting considerations require detailed analysis to ensure that both the project and the company receive the best business arrangement.

B. Major Activities—Front-End Planning

As soon as plant management has determined the need and timing for a SRO and has completed the basic engineering that evaluates the need, the information should be transmitted to the project engineering or construction group with a formal request to execute the work. At this early stage, it is vital that plant management allows sufficient time to properly develop the work scope to execute the project. Poor plant management planning and/or lack of project understanding can lead to inadequate time to properly plan and execute a SRO.

As soon as the notice or request for a SRO is received, the following major activities need careful evaluation and immediate development:

- project objectives,
- preliminary execution plant,
- work scope,

- project management program,
- contract program, and
- key estimating/planning factors.

C. Project Objectives and Program Planning

It is important that all prospective architect/engineers (A/Es) and subcontractors have experience at the plant and also with the specific categories of work to be performed. In most cases, the local subcontractors have a limited engineering design capability and so it would be normal to bring in an outside designer. It is strongly recommended that the design for major SROs be carried out at the site. This will increase the engineering cost, but the benefits of immediate access to site layouts and conditions outweigh the extra cost.

The current trend is for owners to carry out basic design and act as their own project manager; for an A/E to handle detailed engineering; and for construction to be divided between the company maintenance department for specialized work, with individual local subcontractors and/or a prime contractor handling all or part, or solely acting as a construction manager.

For most projects, economic considerations are a top priority, but with SROs, time is usually of the highest importance. The priority of project objectives is usually time first, then quality, then cost.

It is essential that the plant be brought back on stream according to plan, or earlier if possible. Sometimes quality, and certainly cost, will be sacrificed in order to meet the operational date requirement. The estimated cost can increase greatly when unplanned schedule acceleration is required to meet the planned operational date. Typical *acceleration costs* are to be found in labor overstaffing, overtime, shift work, oversupply of construction equipment, inefficient/out-of-sequence work, extra supervision, etc. One of the main reasons for extra costs/ schedule acceleration is a poor estimate of the work scope or work scope increases and changes by plant management during execution of the SROs.

D. Preliminary Execution Plan

As soon as the SRO date has been set or a plant request is received, a *preliminary execution plan* (PEP) should be developed. This plan should be developed in conjunction with plant management to ensure correct timing, design basis, availability of plant personnel; coordination with operations personnel; and identification of regulatory requirements and the contracting program for the planned division of the work.

A preliminary schedule similar to the one shown in Figure 13.10 should be developed in conjunction with the PEP. In fact, the schedule could actually be the PEP. At its very earliest stage, the PEP should then be approved by plant management to ensure their full commitment to the proposed program. Securing such a full commitment means that plant management will allocate adequate plant personnel, time, and resources to the program.

A major decision that must be made during the early stage is delegating responsibility for managing, contracting, and inspecting the construction work. Most companies and plant organizations have separate maintenance and engineering departments, either of which can be responsible for the construction work.

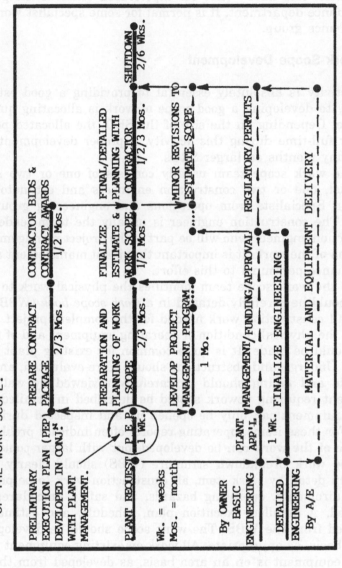

Figure 13.10 Shutdown/turnaround preliminary schedule/program.

Plant management can make a serious mistake by delegating the construction responsibility of larger SROs to maintenance supervision. Maintenance personnel can be effective with small, hourly, or 1–to–2–day SROs, but larger SROs are usually beyond their experience and capability. An exception to this general observation can exist where an experienced construction group is part of the maintenance department. It is normal for some specialist work to be given to the maintenance group.

E. Work Scope Development

This activity is absolutely essential to providing a good estimate of the work. The key to developing a good scope of work is allocating qualified personnel to the take. Depending on the size of the SRO, the allocated personnel or team is part- or full-time during this activity. A proper development of work scope can take many months on larger SROs.

The work scope team usually consists of one or two senior maintenance personnel, one or two construction engineers and estimators, and a schedule engineer. Specialists from operations and engineering groups are used as required. The construction engineer is usually the team leader, as it is only the construction engineer who will be part of the project management program during execution of the work. It is important that plant management allocate experienced maintenance personnel to this effort.

As the work scope team identifies the physical work to be carried out, the work should be carefully detailed in a *work scope book* (WSB), which essentially covers all construction work needed for the complete project. Detailed investigations of the physical condition of operating equipment and of the appropriate site area should be made. It is quite common for existing plant drawings to be out of date. Underground obstructions should be re-evaluated, and new routing runs for cable and piping should be carefully reviewed by walking the route. All equipment requiring rework should be described in detailed inspection reports. Some equipment can only be opened up and inspected during the actual shutdown. For these items, operating reports often indicate problems, from which an estimate of the work can be developed, but with larger contingencies allowed.

The work breakdown structure (WBS) should clearly and accurately describe, in detail by work item, all construction work to be performed. Work permit requirements, operating hazards, and safety procedures should be clearly identified. A detailed execution plan, schedule, and estimate should then be developed from the WBS. The work scope should be developed on an area and system basis, as appropriate. All work on existing equipment and the installation of new equipment is on an area basis, as developed from the equipment list (a major source document). Major plant equipment (both process and utility) have specific reference numbers by which are used to identify them in the WSB, estimate, and schedule. Changes to existing systems (piping, electrical, instrumentation, etc.) and the installation of new systems can be on an area and system basis.

F. Project Management Program

This program consists of:

- project organization,
- project control/reporting,
- schedule(s), and
- worker resource histograms.

As shown on the PEP (schedule) contained in Figure 13.10, the project management program should be developed within one month after approval of the PEP.

G. Project Organization (Owner)

Structure of the owner's organization will depend on whether or not the actual construction work is contracted out. As previously stated, the assumption of this program is that this is the case. Again the organization will depend on the contracting arrangement with (and the responsibility of) the contractor. A full responsibility, fixed price contract based on a well-defined scope-of-work, let to a competent contractor, requires a reduced project organization. Conversely, a reimbursable or unit rate contract (which is usually the case), requires a full project organization. In such instances, the project management responsibility and liability is with the owner. Depending on the contract terms and conditions, a fixed price contract may still place the liability for schedule with the owner. In summary, the project organization is directly affected buy the terms and conditions of the contract agreement.

As engineering and purchasing are the responsibility of the owner and/or the A/E, the project organization will require allocation of specialist engineering, purchasing, contracting, construction, estimating, cost control, and scheduling resources. The personnel allocation is a combination of part- and full-time workers. During the actual shutdown period, the construction staff greatly increases from the staff level during the preparatory period.

For larger projects, the project manager should be selected carefully. Good, all-around experience in the plant, engineering, contracting, planning and scheduling, and construction are desirable. The personal attributes of dedication, leadership, creativity, and a calm nature are essential. For SROs, a project manager needs to be a workaholic. During the potential 24–hour workdays of an SRO, the project manager has to be mentally prepared for any and all things to occur. Crisis management, risk analysis, and decision-making skills are essential capabilities.

Figure 13.11 shows an organization chart for a $6 million SRO. The chart outlines the organizations for the owner, a separate A/E responsibility for detailed engineering (15,000 hours), and construction being handled by one subcontractor (50,000 hours) for the three-week, two-shift shutdown program. The construction organization for the earlier five-month preparatory program is not shown.

H. Project Control/Reporting

Generally, this consists of two distinct phases. First comes design, procurement, and construction preparatory work. All construction work that is not constrained by the actual shutdown should have been completed before shutdown. This earlier construction work is often scheduled on a daily or weekly basis, and the engineering/procurement on a weekly or bi-weekly basis. Second, the actual construction shutdown work should be scheduled on a hourly basis, with work being performed

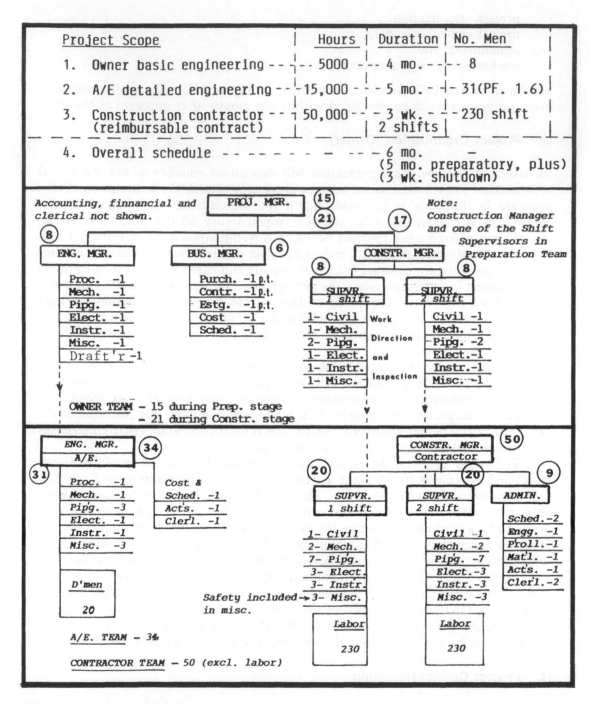

Figure 13.11 Turnaround/shutdown typical project organization.

on a two-shift basis. Planning and scheduling is the main project control program, with cost reporting and analysis being carried out as an accounting function.

It is common for work to be executed on a work order system, with CPM-type schedules and planning boards showing plan and progress of the work. In most cases, jobhours identify scope of the work and are also used to measure progress. This requires accurate jobhour estimates for effectively planning resources and measuring progress. All work orders are individually estimated, and the field accounting system allocates and collates actual jobhours to each work order. Prompt analysis of deviations of actual versus estimated hours is necessary for an effective project control program. A field computer system is necessary to provide timely and accurate commitments, expenditures, and jobhour data.

Scheduling of shutdown work should allow for unforeseen circumstances, with overstaffing factors and additional construction equipment forming part of the resource calculations.

Engineering and procurement activities should be generally scheduled with a simple method. However, it should be recognized that engineering design also has a retrofit nature, with field investigations providing input to actual design drawings. On major SROs, a field design office is often essential to effectively progress the engineering effort.

Apart from an overall program, it is very difficult to use detailed CPM schedules/computer systems during the shutdown period. Due to the rapid time changing of work tasks per individual shift, such changes and the required display flexibility are best accommodated with a large, manual planning board. This board is usually mounted on a wall in the project office and is the focal point of the planning and execution program. Work tasks/orders are mounted on magnetic tape and placed on the board by priority for each shift. Thus they are highly visible for all to see. As new work is identified, the work tasks/orders are mounted in a different color so that the accumulation of new work can be clearly tracked.

I. Overall CPM Schedule

An overall CPM-type schedule covering the complete project duration can be an effective tool. It is relatively easy to increase the detail of the PEP schedule (Fig. 13.10) and to work on a weekly basis for actual versus planned progress. This can be very effective in ensuring the A/E's engineering performance meets the purchasing and construction program. Figure 13.12 illustrates an overall CPM schedule for the construction phase of a $9 million shutdown.

J. Personnel Curves

These should be generated by key discipline for the A/E's engineering program and the construction shutdown work. A planned and actual progress curve for the work might be needed or useful to show overall progress against plan.

K. Contract Program

A lack of good scope definition severely limits contracting options, which are also affected by the current financial state of the marketplace. SROs, by their very nature of *cut-and-fit*, operational unknowns, schedule acceleration, and overstaff-

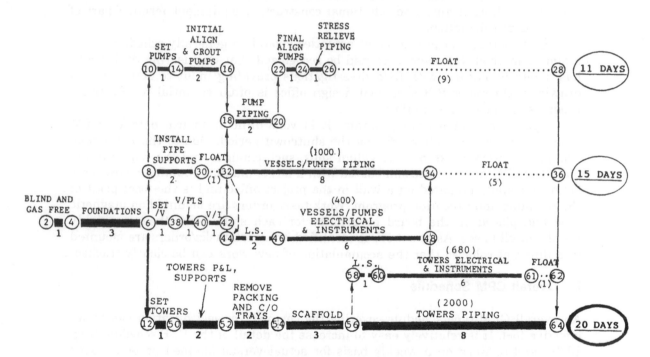

Figure 13.12 Shutdown critical path schedule sub area IV.

ing, are difficult to estimate. This difficulty has a direct impact on the contracting process. As a result, the reimbursable type of contract is widely used. With this form of contract, it is also quite common for the owner to take responsibility for management and execution. Thus the contractor acts as a labor broker, supplying labor, construction equipment, and supervision under direction of the owner's construction management. If a good scope definition can be established, and if project conditions (such as unplanned acceleration) remain firm, then a fixed price and/or unit price contract is possible.

The following are typical reimbursable contracting arrangements:

1. Reimbursable Hourly Rates

This is the simple (low contractor risk) form of contract, since the contractor quotes all-inclusive rates (including overhead and profit) for labor, supervision, and construction equipment. The rates may be on an hourly or weekly basis. Miscellaneous services and expenses are reimbursed at cost. The number of resources is specified by the owner. This is known as a *time and material* (T/M) type of contract. Purchasing is carried out at cost, with a percentage markup added to the cost to cover the contractor's administrative costs.

2. Fixed Fee/Reimbursable

This is similar to the T/M contract, except that the contractor's overhead and profit are removed from the hourly rates and included in the fixed fee. The fixed-fee/reimbursable contract has a higher risk than the T/M contract, since the contractor takes a risk on the fixed-fee portion.

The fee does not change with a change of the reimbursable hours, so the contractor will not recover its overhead or an a profit if the scope of work is underestimated. Most SROs experience an increase rather than a decrease in hours, so the contractor must evaluate this risk carefully. Many contractors will not accept the risk and will attempt to negotiate it away. This can be achieved by a formula, or percentage, arrangement that ties the fixed fee to the reimbursable hours or the total cost.

3. Adjustable Fee/Reimbursable

This arrangement is a *risk compromise* of the full, fixed-fee arrangement. The fixed fee is determined on estimated or target jobhours and then adjusted on a percentage basis for jobhour variations. There also may be additional bonus/sharing on jobhour underruns and a penalty on jobhour overruns, such as an hourly rates reduction on a percentage basis.

4. Percent Fee/Reimbursable

This arrangement is similar to the adjustable fee/reimbursable form of contract, except that the fee is determined as a percentage of the total reimbursable costs. There are usually no bonus/penalty arrangements. The problem with this arrangement is that there is an incentive for the contractor to increase the costs, which will result in a larger fee.

All reimbursable-type contracts have the general problem for the owner that poor contractor performance will increase the jobhours and costs, resulting in greater profit to the contractor. Obviously, poor performance should not be rewarded.

Overtime	Recreation facilities	Geographical area
Prefabricated assemblies	Schools	Climate
Specifications	Permanent community	Site access
Local codes	Schedule	Soil conditions
Procurement	Inflation	Earthquake factors
Origin of materials	Escalation	Site elevation
Origin of equipment	Currencies	Offshore platforms
Export packing	Financing	Environment
Construction facilities	Overseas premiun	Attitude of community
Temporary facilities	Cost of living	Political climate
Housing	Cost of traveling	General business climate
Logistics	Taxes	Prime contractor
Communications	Insurance	Joint venture
Warehousing	Legal assistance	Unionized labor
Guard service	Government agencies	Qualified labor pool
Site fabrication facilities	Letters of credit & bonds	Recruiting & training
Construction equipment	Language problems	Labor productivity
Medical facilities	Local cultures	Labor contracts
Food & catering	Sanitary facilities	Labor cost
	Experience and Imagination	

	RETROFIT / REVAMP	
Hazards - work limitations	Health factors	Security
• Workers	Contractor training	Clearance
• Equipment	Standby allowance	Permits

Figure 13.13   Pre-estimating survey.

L.   Key Estimating/Planning Factors

In addition to the difficulty of recognizing the work to be carried out, a further problem is that many companies do not possess a formal jobhour database to apply to the units of work. Developing such a database is a substantial and costly task, where a comprehensive code of accounts has to be established. It then takes several/many years of tracking of actual jobhours to ensure a reliable database. A faster, less costly and reasonable method is to purchase a jobhour estimating system (e.g., Richardson Engineering Services, R.S. Means, etc.) and adjust the data to the specific on-the-job application, based on actual work experience. As such estimating systems are generally based on new construction work and fixed conditions; the impacts of geography, weather, plan operating conditions, and SROs on labor hours/costs require individual adjustment.

When using commercial estimating programs designed for new construction work, it is recommended that factors of 1.5–2.0 be used against these jobhour databases for SROs. Judgment is required in the application of this *factor range* and generally 1.5 applies to work during plant operations and 2.0 applies during a shutdown. These factors recognize the hours necessary for unknown conditions and productivity losses for overtime, shift work, overstaffing, and schedule acceleration. Indirect labor/nonproductive hours will also increase for SROs as compared to an open, grassroots or new-site conditions. The norm for new construction is 30% and for SROs a figure of 40–50% is probable. Note that commercially published systems do not generally cover engineering design hours and costs.

As the published systems generally can only be used for the installation of new facilities, another method must be developed for the cut-and-fit, repair, demolition, retrofit, refurbish, and rearrangement-type work of the SROs. The most commonly used method is to determine the number of workers and the time required to carry out the work. This determination requires considerable practical experience, plus knowledge of the working conditions for such estimates to be accurate. The development of good jobhour estimates is essential for the effective planning and scheduling of the work.

XVII. SUMMARY

A successful multiple small projects program requires informed plant management, skilled project personnel, a varying level of control/reporting (structured and preplanned), quality estimating, and extensive early planning for SROs. With all these ingredients in place, success is not guaranteed, but it has a good chance. Figure 13.13 lists many of the items that need to be considered when a detailed estimate is being prepared.

14

Project Closeout—Lessons Learned and Historical Data

I. INTRODUCTION

The major emphasis of this chapter is on the project control function, not on project management, where assessments of all facets of project execution are outlined. In essence, at the end of a project, a detailed assessment should determine the success, or otherwise, of the following key factors:

- project objectives achieved,
- personnel skills of all key players,
- project organization (project control),
- team building,
- techniques and procedures,
- project cost consciousness (trending),
- lessons learned,
- failures,
- recommendations for change, and
- historical data.

This list is not all-inclusive and can be modified for application to specific projects. The resulting information should be included in a confidential report, since some of the issues would be sensitive and, perhaps, critical.

II. PROJECT OBJECTIVES ACHIEVED

The project control objectives, by simple definition, are completing the project on time and within the approved cost appropriation. Achieving early completion and a cost underrun may be clear evidence of superior project performance. Unfortunately, current studies clearly show that cost overruns and schedule slippage are the norm, although most companies have the policy that a cost deficiency of up to 10% equates to successful within-budget completion.

A detailed analysis of final costs versus the approved estimate will identify, at the operating code of accounts level, the individual variance as well as the total value.

III. PERSONNEL SKILLS OF KEY PLAYERS

An assessment will focus on all major parties and individuals: owner, engineer, contractor, and subcontractors. The assessment should cover key individuals directly responsible for the estimating, cost control, and scheduling operations. The report should name individuals and their positions, with assessments of their skills and experience.

IV. PROJECT ORGANIZATION—PROJECT CONTROL

This evaluation should cover the effectiveness of the total project control group, since specific individuals should be included in the preceding section.

The evaluation should include factors such as sufficient project control personnel; project relations; communication channels; work done in engineering offices and at the construction site (separate analyses); the interface between engineering and construction; the coordination of contractors at the site; and the interface with procurement.

Of great significance is the interface with the project manager. Was it direct or otherwise? Did the cost engineers consider themselves bean counters or did they feel they were strong, active contributors throughout the project? Was the cost control effort properly coordinated with the cost accounting group?

V. TEAM BUILDING

In relation to the preceding comments, was the project control group an active participant in the team building effort of the total project organization? Did it obtain the right information at the right time? Within the group (depending on the size of the project), did appropriate team-building efforts take place to ensure that all required inputs and all necessary outputs were effective?

One of the keys for successful cost and schedule control is the proper development of relationships and contacts so that a full and steady stream of information is obtained. Without this information, effective trending is impossible. Did the responsible project control personnel attend all appropriate progress and status meetings, both within the project and within the company?

VI. TECHNIQUES AND PROCEDURES

This part of the report should focus on all major techniques and procedures, listing them with appropriate descriptions of their use, application, and effectiveness. The assessment should cover the range and level of techniques and procedures as well as their quality. However, with large contractors and sophisticated owners, there is always the tendency to over-control and over-report.

Management and leadership skills, especially on large projects, are required to ensure a cost-effective program. Particular care should be given to assessments of computer programs.

VII. PROJECT COST-CONSCIOUSNESS— TRENDING

After completion and acceptance of a quality estimate, proper trending of the work is then the major requirement. As previously discussed, trending can only occur by turning data into useful and useable information.

- The cardinal principle of *no surprises* should be considered. Was it achieved, apart from force majeure items?
- Were changing scopes of work and project conditions properly identified and correct impact studies carried out in time? If so, the result would be a good potential trend report.
- Were potential claims identified well in advance of their occurrence?
- Was there a conscious effort to establish and maintain a cost-consciousness project culture?
- Was a claims prevention program established?
- Correct trending leads to accurate forecasting. Was this the case?
- Was a weekly trend meeting held?
- Did the correct individuals attend the meeting?
- Were weekly progress meeting held at the site?

Specific examples of key techniques and efforts should be highlighted, as should failures.

VIII. LESSONS LEARNED

This section should deal with positive issues, covering items that were modified and deficiencies that were covered. Such items could include personnel, organization, techniques, or procedures. The items should be of significance, so that lessons learned would then become standard operating procedures in the future. The analysis needs to be as objective as possible.

IX. FAILURES

This section, as with lessons learned, covers only items of significance. It assesses individual and/or company performance, methods, techniques, procedures, and organization, and should focus on situations that should not recur on future projects. This is a sensitive area, so care is essential in properly identifying these items, especially if some of them reflect poor performance by an individual.

X. RECOMMENDATIONS FOR CHANGE

The major issues identified in the "Lessons Learned" and "Failures" sections are further evaluated, resulting in a series of specific recommendations to change current practices, procedures, and company policies. There is always a reluctance to change, so all recommendations need to be properly evaluated for future application as well as for the impact of change on the current operation.

XI. HISTORICAL DATA

Before project startup, all project control feedback data should be properly identified. This includes both estimating/cost data and scheduling information.

Since this information can be voluminous and therefore costly to compile, care should be taken to ensure that the data is actually needed. It is normal for estimating and cost data, at the detailed code of accounts level, to be required. However, further analysis is probably required to obtain accurate costs for each equipment item and category of material. As far as it is practical and cost effective, identifying feedback requirements enables a cost accounting system and scheduling program to all these items to be reported when the system is being established.

15

Project Management Fundamentals—Key Essentials

I. INTRODUCTION

In today's difficult and challenging business environment, it is vital that the management of projects results in:

- identifying risks,
- maximizing cost savings,
- minimizing time delays, and
- improving economic return.

These results can only be achieved through:

- effective management of people,
- tough but fair project objectives,
- efficient business techniques, and
- outstanding leadership skills.

The following project management chapters cover these subjects in considerable detail. The roles, functions, and interfaces of company management, project management, engineering management, construction management, and support-service groups are explored. Overall relationships and personnel relationships, essential for successful project execution, are outlined. Throughout, the emphasis is on practical approaches and the relationships of personnel in the project team.

The emphasis is on the need for and substance of early project planning and the related scheduling of time and development of costs. The material concentrates on the need for every project leader to be an effective professional project manager—to be an organizer, a planner, a motivator, a communicator, and a businessperson.

Current studies of state-of-the-art project management methodology by the Construction Industry Institute (CII), have shown that the top category for successful project execution is front-end planning/project organization. The studies have concluded with the following simple but true premise:

> If we get it right at the front end, we have a chance of success, though not guaranteed. If we do not get it right, then we have no chance of success.

The following material, mostly flowcharts, illustrates the major factors, functions, and project phases of the *project lifecycle*. Full understanding of all these elements can ensure effective communication channels, tight schedules, low cost, and a clear path for efficient decision-making and economic actions.

But it is in these very elements where many problems develop and become impossible to solve or reduce as the project progresses. The result is more money is spent than is necessary, more time is consumed than is needed, and there is a constant recycling of options and alternatives that have been eliminated at an earlier stage. Project ignorance, lack of skill, poor cooperation, and refusal/reluctance to participate properly are commonplace. Biases, prejudices, and personal interests are also part of the equation. All of these elements are compounded by industry-wide lack of up-to-date project training.

II. DEFINITION OF A PROJECT

A project can be defined loosely as an item of work which requires planning, organizing, dedication of resources, and expenditure of funds in order to produce a concept, a product, or a plant. This chapter focuses on plant projects, all of which require design engineering, the purchase of material, and the installation of that material to the previously completed design engineering.

III. PROJECT MANAGEMENT FUNCTION

Almost all companies have personnel who are trained, skilled, and dedicated to the execution of projects for the company. The individuals who lead these efforts are called project engineers and/or project managers. Supporting these project managers are such personnel as design engineers, procurement personnel, contracts officers, estimators, cost engineers, planners, construction managers, and a variety of technical specialists.

In many cases, the type, size, and complexity of projects vary greatly and, therefore, the skills and experience of project engineers, project managers, and support personnel similarly can vary in capability.

A. Major Factors

Figure 15.1 shows the essential major factors for the successful execution of projects.

1. Cost Management

Many projects have project cost as the top objective and this, then, requires the project to be completed at, or less than, the budgeted cost. Adequate business skills are essential to meet this objective.

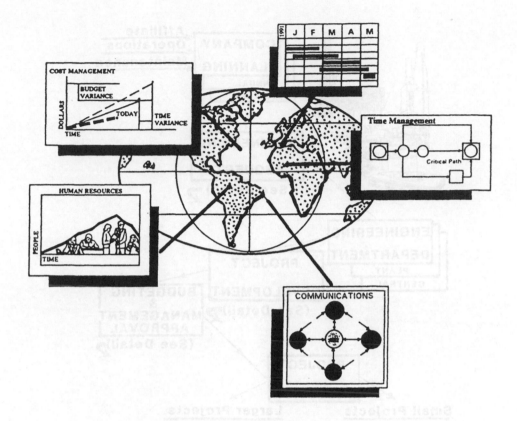

Engineering

• Procurement } **(EPC)**

• Construction

Figure 15.1 Major factors essential for successful project execution.

2. Time Management

To meet the cost objective, it is necessary to manage time efficiently. This means the predetermined schedule, upon which the cost was based, must be met and met economically. Some projects may have schedule as the top objective. In such cases, acceleration programs are planned, and it is probable that there will be corresponding cost increases to the economic-based project.

3. Human Resources

Of all the resources required for plant projects, the people resources are the most difficult to manage. Interpersonal skills and the effective motivation of people, at all levels, are essential for successful project execution. Lack of human resources, plus a corresponding lack of the correct mix of people skills, are becoming

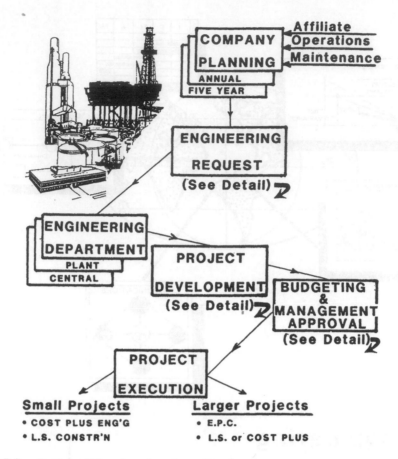

Figure 15.2   Projects lifecycle—functions flowchart.

an increasing feature of the project business. One of the most abused people resource concepts is the "lean and mean" program. The management intent is that a reduced group of people, through advanced skills, can execute as effectively as a larger group and, therefore, reduce costs by reducing staff. There is some merit in this concept but in many cases it is a device used by poor management to cut costs. If there is a significant lack of people, there is almost certain to be a corresponding inefficiency in project execution, coupled with an increase in costs.

4.   Communications

A formal and informal structure of effective communications is absolutely essential for successful project execution. In addition to weak people skills, many company organizations and cultures have poor administrative practices that also form barriers to project success. These barriers are common to all companies and are generally referred to as *matrix interface conflicts* (*MICs*). The conflicts or barriers are caused by departmental jealousies, rivalries, and failures by management to create a culture where project consciousness and esprit-de-corps are common to all personnel. The total quality management programs sweeping the industry are an attempt to solve these problems.

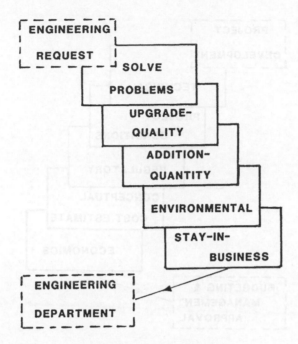

Figure 15.3 Engineering request—functions flowchart.

B. Overall Company Projects Lifecycle

Figure 15.2 shows the general steps common to all plant projects. Experience in this process, recognition of each company program's individualities, and the skills of bridging the MICs, are necessary for project success. Getting the front-end planning right is the key to success.

C. Engineering Request

Figure 15.3 illustrates the major factors that generate the capital project work. Timely and quality assessments of plant requirements are difficult to achieve but are essential for company profitability. Such assessments result in formal engineering requests for the project work.

D. Project Development

Figure 15.4 shows the major components for developing the scope of each project. Each of these components (technical, project conditions, regulatory, cost, economics) is then further defined and prioritized. It is vital that the priority be clearly established.

E. Budgeting and Management

Development of the scope in terms of risk, cost, time, and resources is followed by approval, partial approval, or rejection of each proposed project. Figure 15.5 shows

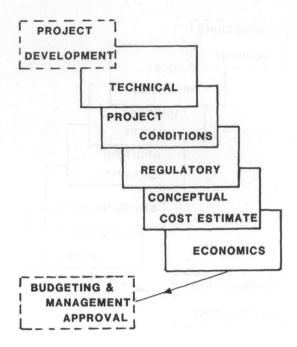

Figure 15.4   Project development—functions flowchart.

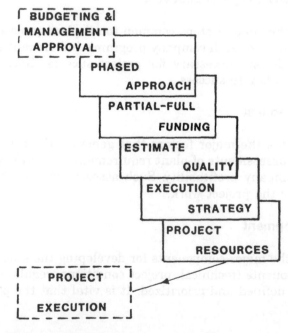

Figure 15.5   Budgeting and management—functions flowchart.

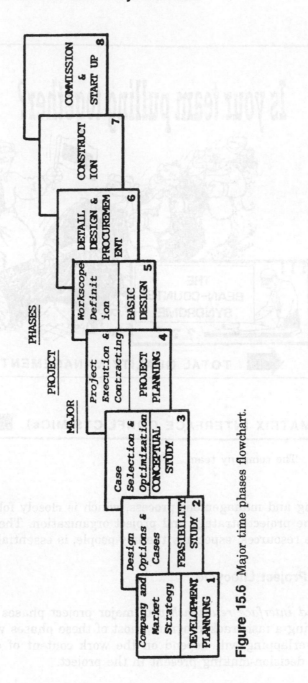

Figure 15.6 Major time phases flowchart.

Figure 15.7 The company team.

the budgeting and management process, which is closely followed by the development of the project strategy and project organization. The correct assessment of the people resources, especially the key people, is essential at this early stage.

F. Typical Project Lifecycle

The *time and interface relationship* of major project phases is shown in Figure 15.6. Assuming a fast-track program, most of these phases will overlap, and the degree of overlapping will depend on the work content of each phase and the efficiency of decision-making present in the project.

G. Project Organization (Culture)

Finally, Figure 15.7 poses the question of company personnel working as a team. Without question, this matter has become the vital issue to profitability, especially as companies downsize and reduce the core. Greater personnel efficiency and increased operational quality are essential requirements in today's difficult business environment. The *bean-counter syndrome* is highlighted. This dangerous and unacceptable practice is covered in Chapter 9.

16

Managing the Feasibility Study (Preproject Planning)

I. INTRODUCTION

The management of the first phases of a project requires outstanding communication and people skills. The task is to take multiple sets of ideas, rough notes, sketches, and somebody else's thoughts and develop the information into a recognizable design scope (with options) and to develop an associated cost and time budget. As there is often little definition to the thoughts and ideas of others, the ability to motivate others to a firm and common approach is, essentially, a people problem.

Figure 16.1 illustrates the schematic relationships of these first project phases and the major elements of a feasibility study.

II. TYPICAL FEASIBILITY PROJECT APPROACH

This discussion is based on the typical approach to feasibility studies for larger projects, where a combined owner-contractor team is executing the work, usually on a cost-reimbursable basis. The project manager is generally from the owner's engineering group, and most of the work is coordinated with the owner's internal operating groups. Due to the number, independence, and lack of project experience of these groups, the coordination of the work with them is normally far more difficult than is the working relationship between owner and contractor personnel.

III. OVERALL OBJECTIVE

Preproject planning is required to ensure that early evaluations, discussions, and overall planning properly interpret the owner's requirements and lead to ade-

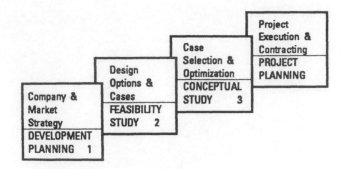

1. OVERALL OBJECTIVE

2. TYPICAL PROBLEMS

3. WORK INITIATION

4. STATEMENT OF REQUIREMENTS (SOR)

5. FEASIBILITY STUDY COSTS

6. KEY DELIVERABLES

Figure 16.1   Early project phases and major elements of a feasibility study.

quate work scopes, preliminary plans, and necessary resources. This, then, develops into basic engineering with a resulting conceptual cost estimate, schedule, and assessment of project visibility for owner review and approval.

The deliverables of the feasibility-conceptual design study are:

- technical program (preferably, a single case),
- cost estimate for funding or further development,
- associated risk analysis, and
- preliminary execution plan (schedule, project organization, etc.).

IV. TYPICAL PROBLEMS

This early work needs to be of the highest quality, as many significant problems, such as poor scope definition, inadequate planning, lack of resources, organizational and owner–contractor–A/E conflicts occur at the early stage of a project. If these typical early problems can be identified, prevented, or reduced, then the subsequent execution of detailed engineering, procurement, contracting, and construction is greatly enhanced. In other words, the project manager should set up the project at the earliest stage, with proper planning, quality organization, skilled/experienced personnel, good owner relationships and effective communication channels. This will not necessarily guarantee success but will greatly increase the probability of success.

The following breaks down typical problems encountered by the project manager in the preproject planning stage:

- poor owner scope definition,
- failure of owner and engineering division to recognize that each party has a differing interpretation of the project requirements,
- unclear priorities and priority conflicts of project objectives,
- inadequate planning,
- lack of resources,
- organizational difficulties of the matrix,
- internal owner conflicts,
- inexperience/lack of understanding of involved parties,
- new/changing technology, and
- difficult project conditions (site, operating plant).

It is essential that these typical early problems be clearly identified, prevented, or reduced by the project manager.

V. PROJECT MANAGER AS COMMUNICATOR/MOTIVATOR

The project manager must assume the leading role in developing effective communication channels and good working relationships with all interested groups. Lack of proper liaison and poor coordination, especially in the early stages of a project, will add to the preceding problem list and, during project execution, lead to:

- lack of commitment and support,
- lack of discipline during project execution,
- poor personnel relationships,
- ineffective approval/signatory procedures, and
- inadequate owner-contractor relations.

Prior to contract agreement for the execution phase, owner communications, reporting requirements, authority levels and approval procedures should be outlined in the project coordination procedure, so that contractors can properly respond during the bidding period.

VI. OPERATIONS INTERFACE AND SCOPE DEVELOPMENT

The project manager must work very closely with operations/maintenance to ensure proper technical definition (a precise, unambiguous definition of requirements) and project scope definition leading to an approved *statement of requirements* (SOR). Thereafter, the project manager must maintain an effective interactive working relationship to gain the operator's full understanding, acceptance and commitment to the project strategy, project objectives and implementation plan, including development of the *project finance memoranda* (FMs) and *authorizations for expenditure* (AFEs).

The most cost sensitive part of virtually any project is changes to the scope. Such changes can easily arise from early failure to allocate total corporate attention to the project scope in general and scope control in particular. Scoping must not be solely confined to the producing facilities (e.g., oil/gas/chemical/food production) but must also include all major peripheral facilities, such as power, water, shipping, storage, communications/telemetry, fire protection, flares, buildings, roads, regulatory/environmental, etc.

It is also important to note that, during development of the SOR, many potential impediments are likely to arise, particularly in scoping minor facilities. It is essential to avoid all unimportant delaying factors and to allow small uncertainties to be covered by the estimating methodology and incorporated into the contingencies element of the overall cost estimate. It is probable that the project will gain from early release of an approved SOR document rather than delay to await clarification of minor areas of uncertainty.

VII. OWNER AUTHORITY AND APPROVAL

The roles of the various owner groups in authorizing, approving and decision-making must be clearly established as early as possible in the life of the project. Until the project coordination procedure has been developed, these approvals may have been determined by verbal discussions and agreements. It is good practice for the project manager to confirm such verbal agreements with an internal memorandum, copied to all involved parties.

VIII. OWNER PROJECT DISCIPLINE

Sometimes individual owner groups deviate from the agreed project and contract procedures, change priorities, develop additional scope, delay decision-making, work directly with the contractors and/or other groups, and then, after the fact, inform the project manager of these actions. Such behavior can be very damaging to the effective execution of the project and to the establishment and maintenance of firm contractual and technical control. Cost increases, schedule slips, and lower morale due to poor working relationships can rapidly develop. The project manager must therefore work to eliminate or reduce the potential for poor owner project discipline.

IX. GOVERNMENTAL AND LOCAL AUTHORITY PERMITS

These requirements need to be promptly identified and properly established at the earliest stage of the project. Good community relationships, as well as conformance to governmental regulations and permit requirements, are essential. The role of the owner and/or other parties in these activities must be clearly outlined.

It should be noted that project costs may be seriously underestimated unless full account is taken of the stated health, safety, and environmental requirements of the local, state, and national governmental authorities. The project manager must also be aware of the full practical implications for consequent redesign of, for example, licensor' packages that do not properly follow these requirements. Allowances should be made for any likely upgrading and tightening of such requirements by local and governmental authorities.

X. COMPANY SERVICE DIVISIONS—WORK INITIATION

The required inputs and relationships with service divisions inside and outside of an *engineering division* (ENG) in support of the project must be properly developed, established, and coordinated. Project work undertaken by an ENG on behalf of an owner should be initiated by the *universal service request* (USR) system, which also provides a means of monitoring and controlling the work.

A project normally commences when ENG receives a USR from an owner, along with a statement of requirements (SOR), which may be a broad outline at this stage. These requests are handled by the engineering/development group whose responsibilities include the preparation of feasibility studies and engineering proposals to establish preliminary project programs and cost estimates (conceptual technical definition). At this stage of project development an overall project strategy is prepared together with the consolidation of the SOR, as agreed by all parties. Approval and funding, via the AFE control system, will then enable the project front-end engineering design to be implemented. Before full funding, management may require further front-end engineering design to be developed to give a clearer technical definition and identification of risks.

XI. STATEMENT OF REQUIREMENTS

A preliminary SOR is prepared following the preproject stage discussions with ENG and is submitted with the USR. It may be only a broad outline if feasibility studies are required. On completion of these studies and the engineering proposals, ENG will assist the operations group in preparing a more comprehensive SOR. In subsequent stages, the number of options will have been reduced and management will have made a decision on a particular option.

The engineering/development group would normally take the lead in preparing the SOR. As a preliminary document from which the project specification will be developed, the SOR takes a broad overall view of requirements, referring where appropriate to process and plant specifications. It is important that this is a *full and complete record*, as any subsequent amendments or additions could lead to confusion, delay, and extra costs.

The SOR should include:

- introduction;
- scope (marketing requirements to be met, including production flowrates and product specification; for technical and safety projects, the SOR would include the justification requirements);
- site selection or location;
- outline design parameters and engineering requirements, including associated facilities, utilities, services, and chemicals applicable;
- technical definition appropriate to a conceptual cost estimate and a forecast of expenditure to meet the development requirements;
- timescale with any production constraints and access problems (such as seasonal weather windows);
- applicable codes and standards;
- level of quality assurance, inspection, and testing;
- external approvals certifying and statutory authorities;

- commissioning;
- safety, environmental and hazard studies; and
- project strategy requirements:
  - organization,
  - communications,
  - reporting,
  - mobilization for engineering design and construction,
  - contracts, and
  - spares policy.

XII. FEASIBILITY STUDY COSTS

One of the earliest and most critical project decisions revolves around the following key questions:

- How much front-end engineering (FEE) do we need, to select the best technology?
- How much FEE do we need for an acceptable estimate?
- How good an estimate do we need to properly assess the financial risk?
- How much should this FEE cost?
- Should we contract out this FEE?

There are no simple, or single answers to these questions. The answers very much depend on a company's technical strength, estimating skill, management bias and preference to a specific process and, ultimately, on the skill to properly balance technical viability with economic need. Time consciousness and discipline are vital as the project program is being formulated. Of equal importance is the ability of management to authorize full funding from preliminary design project cost information and market strategy.

From the project viewpoint, there is a direct correlation between the quality of technical-project information and estimate quality. However, the continuous funding for further engineering development to improve the estimating and hence lower the financial risk, has a point of zero or minimal cost return. In general, industry practice has established two approaches.

Approach 1. Develop a *minimum design package* and execute the project on a cost-reimbursable basis. This method assumes that the economic return of the shorter execution period is worth the risks that are inherent in the cost-reimbursable arrangement. With lack of adequate front-end engineering definition, it is difficult to obtain good or acceptable lump-sum bids. Many companies have reduced the *inherent reimbursable risk* by developing a strong company project management capability, to contain this risk. This minimum design package is usually referred to as a *basic design package*. However, the degree of definition can vary widely.

Approach 2. The second approach is to develop a *maximum design package* and execute the project on a lump-sum, turnkey, basis. With a better design package, a much greater identification of risk/cost is possible, and contractors will generally respond positively, subject to current/future market conditions and the size of the project. This maximum design package can be referred to as *front-end engineering design* (FEED).

There can be other considerations which can increase the time and/or add to the cost of both approaches. New technology, with a higher risk of engineering changes, would favor Approach 1. Management with a top priority of low cost risk, would favor Approach 2. Inappropriate and/or unnecessary cases would add to the cost of both approaches. Lack of coordination between operations, maintenance engineering, owner, contractor and management can add substantially to the costs of both approaches.

Figure 16.2 illustrates the cost of feasibility studies for Approach 1, where process design is categorized at 20%. Add factors for cost (30–40%) and time (2 months) are shown for Approach 2, where process design is shown as 60%.

XIII. KEY DELIVERABLES

Figure 16.3 (see p. 393) shows a typical arrangement of the major project phases, overall key deliverables, and an overall schedule for Approach 1. Figure 16.4 (see pp. 394–395) is a list of detailed key deliverables (31 items) for Approach 1, where process design is approximately 20%. Figure 16.5 (see pp. 396–397) is a similar list for Approach 2, where process design is approximately 60% (FEED).

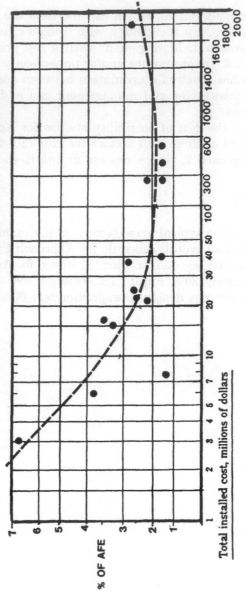

FEASIBILITY STUDY COSTS - % of Capital   AFE or Base Proj. Cost

See Figure 17.4, Key Deliverables, for Work Content with 20% Process Design

• For quality Reimbursable Bidding - For L.S. Bidding(if required), add costs as shown below.

Total installed cost, millions of dollars

% OF AFE

• Includes Contractor and Owner services.

• Confidence level at plus/minus 30% with 80% probability, as work depends on no. of processes, new technology, no. of cases and extent of "project" work(early purch. etc.)

• For L.S. Bidding, recommend Process Design at 60% and add 30-40% cost & 2 mo. time.

Figure 16.2 Feasibility study costs.

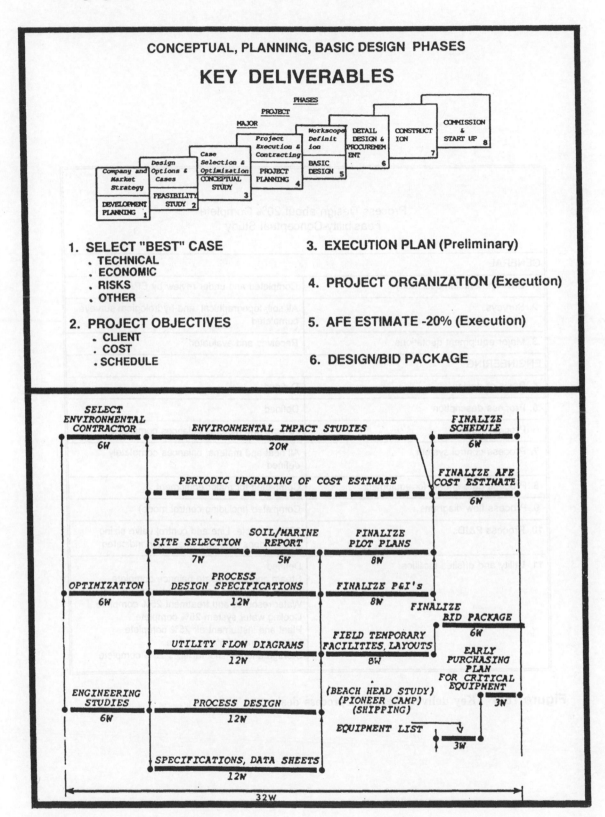

Figure 16.3　Key deliverables.

Process Design about 20% Complete
Feasibility-Conceptual Study

GENERAL	
1. Environmental impact study	Completed and under review by EPA
2. Surveys	All soil, topographical, and hydrological surveys completed
3. Major equipment quotations	Received and evaluated
ENGINEERING	
4. Basis of design	Defined
5. Process description	Defined
6. Process design	All heat and material balances completed
7. Process control system	All heat and material balances completely defined
8. Process data sheets for equipment	Completed for major equipment
9. Process flow diagrams	Completed (including control mode)
10. Process P&IDs	25% complete. Line and control valve sizing, control loops, and instrumentation indicated
11. Utility and offsites facilities	Defined Steam and condensate balance complete Steam generation and distribution complete Water resource and treatment 25% complete Cooling water system 25% complete Plant and instrument air 25% complete Effluent control 25% complete Storage and terminal facilities 25% complete

Figure 16.4   Key deliverables: 20% process design.

12. Utilities and offsites P&IDs	20% complete
13. Location and site plan	65% complete
14. Plot plans and elevations	45% complete
15. Grading, paving, and drainage plans	Preliminary drawings
16. Structural and foundation design	Design sketches; estimated quantities
17. Building requirements	Preliminary drawings
18. Equipment list	Major equipment complete
19. Mechanical data sheets	Completed for major equipment
20. Engineering specifications	Completed for major equipment Job specifications for piping, civil, electrical Instrumentation 75% complete
21. Engineering guides	Appropriate sections included
22. Above ground (A/G) and underground (U/G) piping layouts	Approximately 15% complete
23. Piping takeoff	Estimated from P&IDs and layouts
24. Line and valve sizing material and selection	45% complete
25. Instrument schedule (including control valves)	Control valves sized
26. One line electrical diagrams	Complete
27. Electrical equipment list	Preliminary
28. Electrical takeoff	Based on one line diagrams
29. Painting and insulation	Quantities factored
30. Fire protection	25% complete
31. Tie-ins	Existing facilities surveyed and measured to verify scope of tie-ins and valving

Figure 16.4 (Continued).

High Quality Design Specification for EPC Lump Sum Bidding (Equivalent to FEED) Process Design about 60% Complete	
GENERAL	
1. Environmental impact study	Completed and under review by EPA
2. Surveys	All soil, topographical, and hydrological surveys completed
3. Major equipment quotations	Received and evaluated
ENGINEERING	
4. Basis of design	Defined
5. Process description	Defined
6. Process design	All heat and material balances completed
7. Process control system	All heat and material balances completely defined
8. Process data sheets for equipment	Completed for major equipment
9. Process flow diagrams	Completed (including control mode)
10. Process P&IDs	75% complete. Line and control valve sizing, control loops, and instrumentation indicated
11. Utility and offsites facilities	Defined Steam and condensate balance complete Steam generation and distribution complete Water resource and treatment 75% complete Cooling water system 75% complete Plant and instrument air 75% complete Effluent control 75% complete Storage and terminal facilities 75% complete

Figure 16.5   Key deliverables: 60% process design.

12. Utilities and offsites P&IDs	60% complete
13. Location and site plan	Complete
14. Plot plans and elevations	Complete
15. Grading, paving, and drainage plans	Preliminary drawings
16. Structural and foundation design	Design sketches; estimated quantities
17. Building requirements	Preliminary drawings
18. Equipment list	Major equipment complete
19. Mechanical data sheets	Completed for major equipment
20. Engineering specifications	Completed for major equipment Job specifications for piping, civil, electrical Instrumentation 75% complete
21. Engineering guides	Appropriate sections included
22. Above ground (A/G) and underground (U/G) piping layouts	Approximately 50% complete
23. Piping takeoff	Estimated from P&IDs and layouts
24. Line and valve sizing material and selection	75% complete
25. Instrument schedule (including control valves)	Control valves sized
26. One line electrical diagrams	Complete
27. Electrical equipment list	Preliminary
28. Electrical takeoff	Based on one line diagrams
29. Painting and insulation	Quantities factored
30. Fire protection	75% complete
31. Tie-ins	Existing facilities surveyed and measured to verify scope of tie-ins and valving

Figure 16.5   (Continued).

17

Front-End Planning and Project Organization

I. INTRODUCTION

In today's difficult, global business environment it is vital that the management of projects results in identifying risks, maximizing cost savings, minimizing schedule delays and improving economic return. This can only be achieved with a quality program of *front-end planning and project organization* (FEPPO).

The most comprehensive study done to date of the construction industry was carried out by the Construction Industry Institute (CII). They concluded that the No. 1 category of project management methodology is FEPPO. (Note that CII refers to it as strategic project organizing). To state the principle very simply:

> If we get it right at the front end, we have a chance of success, although it is not guaranteed. If we do not get it right, then we have no chance of success.

The following are the major constituents of FEPPO:

- project organization,
- establishing objectives,
- scope definition control,
- communication and information utilization, and
- constructability planning.

Careful attention to the details of these constituents will provide a good start to any project.

II. PROJECT ORGANIZATION

It is an obvious, but not well understood fact, that it is people in single, medium, or large-size groups who design and build projects, not companies. Therefore, there must be a consistent and long-term interest in people needs, their development, and their training. When there is little interest, or the interest is not genuine, the long-term success of the company is unlikely. The entire total quality management (TQM) program is built around the needs and development of people, and there is unanimous acceptance by industry that TQM is the key to success. In essence, *develop the people and, in turn, the people will develop the profits.*

A. Should the Owner Be Its Own Project Manager?

A very fundamental consideration in today's world of company re-engineering is the question of the owner functioning as its own project manager. Too often owners arrive at an affirmative answer, through poor analysis. It is a matter of having experience with the specific project (particularly size), having adequate in-house or consulting resources (skills and numbers), and having a good project management program with controlled costs.

This is currently a major consideration with many operating companies as they downsize their operations. Many companies confuse the issue due to technical/engineering considerations. Having competent engineering personnel, they take the project management responsibility but without adequate project experience or project resources. *Engineering design competence does not necessarily translate into project capability.*

This question was a major issue during the early development of the North Sea (1972–1977) oil resources. The answer, at that time, was an emphatic "No" from all the large oil companies (as all had limited resources), except for the rare case where contractors declined to bid due to lack of capacity. This problem of limited contractor resources and inadequate technical and engineering expertise was resolved with a *project services contract* (PSC) and ultimately partnering and integrated project teams. With a PSC or reimbursable form of contract, owners can *direct* engineering.

B. Does the Organization Structure Fit the Contracting Arrangements?

Different skills and different numbers of personnel are directly related to contracting arrangements (i.e., lump-sum, reimbursable, unit price, PSC, agent, independent contractor, etc.). From an owner's perspective, reimbursable contracts can require three times as many owner personnel as a lump-sum contract and require personnel with extensive analytical skills. For a lump-sum contract, a good design package and strong project discipline (with no/little design changes) are essential. Often there is a mismatch of people resources in relation to contract arrangements. Having both the wrong contract and the wrong organization/people is a recipe for disaster. Also, a poor management application of the "lean and mean" principle will result in a serious lack of resources, leading to poor project execution and cost overruns/schedule slippage.

Figure 17.1 is a flowchart that outlines the major work activities and associated decision points.

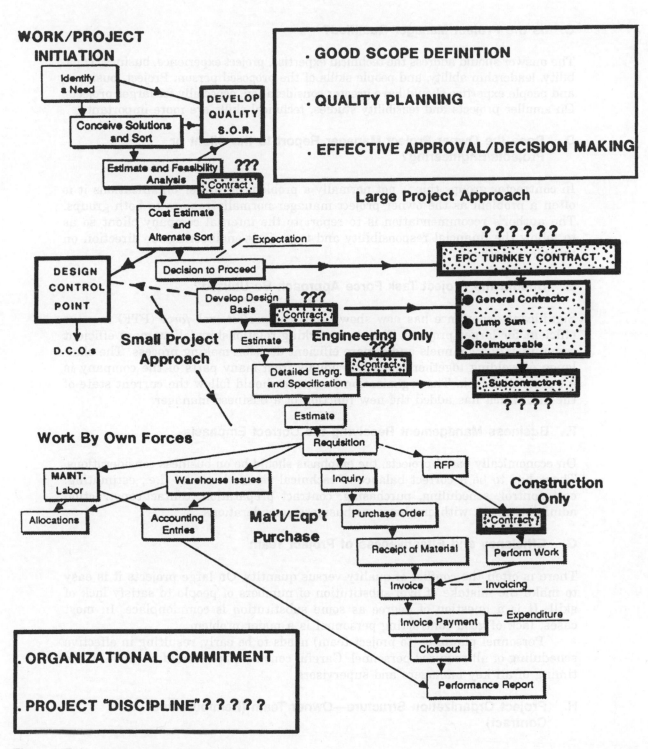

Figure 17.1 Contracting arrangements.

C.   Is the Project Manager Qualified?

The answer should address the technical expertise, project experience, business capability, leadership ability, and people skills of the proposed person. Project, business, and people expertise should have greater consideration, especially for larger projects. On smaller projects and feasibility studies, technical skills are more important.

D.   Does the Owner Project Manager Report to the Client or Projects/Engineering?

In contractor groups this is not normally a problem. In owner organizations it is often a problem as the owner project manager normally reports to both groups. The author's recommendation is to report to the internal company client so as to follow the financial responsibility and to projects/engineering for direction on technical methods.

E.   Should the Project Task Force Approach Be Utilized?

Significant experience has now shown that the *project task force* (PTF) is more efficient for larger projects. The close working relationships allow more efficient communication channels and a more efficient decision making process. The challenge of welding together many individuals from many parts of the company is a substantial task. The organization structure should follow the current state-of-the-art, which has added the new function of a business manager.

F.   Business Management Receiving the Correct Emphasis

On economically based projects, the emphasis should be on business considerations. There has to be a correct balance of technical versus business (i.e., estimating, cost control, scheduling, purchasing, contract preparation, contact/construction administration), with emphasis on business considerations.

G.   Efficiency and Effectiveness of Project Team

There is often the conflict of quality versus quantity. On large projects it is easy to make the mistake of over-substitution of numbers of people to satisfy lack of skill. It is a question of degree as some substitution is commonplace. In most cases, lack of good contracting personnel is a major problem.

Personnel planning (to project team) needs to be early, resulting in effective scheduling of all required personnel. Careful consideration should be given to the timing of all key managers and supervisors.

H.   Project Organization Structure—Owner Team (Reimbursable Contract)

On a reimbursable project, the owner should have a directing position in order to contain the risk of a contractor taking advantage of the reimbursables and manipulating day-to-day execution to enhance profitability. This risk has been well documented by owners and its practice is a key technique of large international contractors. The balancing force to contain this risk is the quality and skill of the owner's project team, both in the precontract activities and the postcontract work execution program. An *equal partner relationship* (EPR) is an essential requirement and should be built in to the agreement with the appropriate contract clause.

These precontract activities should be undertaken by the project team; they are to ensure that contractor provides competent personnel. It is a contractor practice to train new/inexperienced personnel on their client's reimbursable projects since the major cost risk is to the client.

I. Precontract Activities for Contractor Evaluation

- Evaluate the quality of contractor's program, using individual criteria for technical, project management, commercial/pricing, project control, contractual, and construction.
- If the project is of a substantial size, interview key personnel (previously nominated).
- Ensure that correct contracting arrangement/conditions are in contractor's proposal, especially the EPR clause. Assess the required liability of agent or independent contractor.
- Evaluate contractor proposal program/execution plan and key interfaces of local, corporate, and governmental entities.

J. Project Organization Charts

These charts should be dynamic, up-to-date documents, used to identify owner and contractor positions. During execution of the work they will constitute the current and future personnel plan, as generally agreed in precontract meetings. These charts need to be properly recognized and the organization clearly understood by all project team members. The use of formal job descriptions and duties is recommended.

K. Project Manager Authority

The individual must have full authority to make both design and cost decisions, with appropriate limits of authority and management reporting requirements. On reimbursable projects, the authority of the contractor project manager must be adequate so as to allow efficient day-to-day operations.

L. Project Control Function Reports Directly to Project or Business Manager

Many hold the concept that cost control should be an audit function of the project and, therefore, report to higher/senior management. The author does not support that concept as it can lead to an adversarial relationship and dilute the trust and cooperation that is absolutely essential in the cost effort in the project. There are always independent, periodic cost reviews by senior home office personnel that should be more than adequate for a management audit need.

III. ESTABLISHING OBJECTIVES

In many cases, the process of developing objectives can also assist in building team commitment and understanding. Objectives will always be a compromise among quality, cost, and schedule and are used as a guide to make decisions. These major objectives then guide development of more detailed goals, procedures, technical criteria, cost targets, and individual milestones. Ideally, a common set of objectives should guide owner, engineer, and constructor. These objectives provide the work direction to all parties and, as such, have to be compatible and acceptable. *The key to successful acceptance by all parties is a set of well-defined objectives.*

A.   Client Satisfaction

Criteria and a measurement program acceptable to the client should be developed to produce a periodic and timely report. Client satisfaction should be the single most important objective; reports, showing poor performance against this objective, should receive top management attention and immediate resolution.

B.   Scope Objective

The objective is that the technical and project scope, as identified in the approved project budget appropriation, will be achieved. A well-written, but brief, scope definition is developed for issue to all.

C.   Cost and Schedule

Overall and intermediate milestone objectives should be developed. A risk analysis program should identify the ranges of risk for both cost and schedule. Chapter 11, "Range Estimating," describes an effective way of assessing risk. The responsibility and management of contingency should be clear and precise. It is the author's judgment that project contingency should be the project manager's responsibility and not be treated as a management reserve, which it is not.

D.   Quality

Clear and unambiguous criteria should be developed and be fully acceptable to all project parties. The criteria need to be measurable so that a status/progress report can be issued on a regular basis. Quality of project operations, as well as quality of design and construction, needs to be covered.

E.   Other

Training, technology transfer, and other necessary considerations must be fully defined.

F. Prioritizing, Documenting and Communicating the Project Objectives to the Project Team

If project objectives are not properly organized and constantly maintained, then acceptance leading to commitment will be lacking. Establishing clear priorities, with each objective having its relative priority, will allow the multiple groups to work in harmony with each other. Thereafter, a constant effort (part of team building) must be made to keep the project objectives viable.

G. Effective Project Team Building

Assembling a group of individuals, especially on large projects with large companies, does not make a team. Personnel can come from different locations (worldwide) and different cultures. Time is needed for individuals to recognize and control their individualities, as appropriate, and learn to work together. Individuals usually accept project assignments hopefully, but with little or no knowledge of the individuals with whom they will be working or the work itself. It is, therefore, essential that management and project leaders (in all groups) develop a team building program and maintain it throughout the life of the project. Working togetherness, project commitment, cost consciousness, personnel satisfaction, etc., are the deliverables. The cost for this activity should be a recognized budget item.

H. Effective Community Relations—Local and/or Overseas

There is an ever-increasing opposition from local communities to process projects, due in part to the hazardous nature of many of these projects. An effective and positive public relations effort, in conjunction with direct financial investment in local matters, is necessary and essential.

IV. SCOPE DEFINITION CONTROL

This is a matter of project discipline and design control to prevent or identify scope changes that are all too common on fast track projects. A CII study reports: *Poor scope definition and loss of control of the project scope rank as the most frequent contributing factors to cost overrun.*

A. Effective Interface with Operations and Maintenance for Scope Approval

Achieving proper design input from all project parties is a formidable task. This work is usually the direct responsibility of the project engineering manager, strongly supported by the project manager. If there is no project team, then it is the responsibility of the project manager.

All parties (and especially the design decision makers) must reach consensus and full understanding, as well as approval, of the design basis. The design basis must be shared openly with all participating parties. When the design basis is sensitive or proprietary, security procedures must be established. In addition to design, the project execution plan and financial program must be part of the approval process.

B.   Defining Scope Before the Start of Detailed Engineering

The purpose of the feasibility study is to develop a well-defined scope (see Chapter 16, "Managing the Feasibility Study," for greater detail). However, the quality and extent of the early design and project work is a matter of management decision and can vary widely. A poor design package at the start of detailed engineering will result in significant change, rework, and substantial cost increase.

The major deliverable of the feasibility study is the basic design package—statement of requirements (SOR). It should consist of well-written documents that properly define the technical requirements and have sufficient depth to provide clear direction for all major design issues. The package should clearly communicate the intent to the designers and set appropriate boundaries on the project design for detailed decision making.

The scope document should cover:

- project description:

    - project justification, project objectives;
    - economic justification, if pertinent; and
    - facilities description;

- design basis and specifications:

    - process definition:

        - description of process,
        - process flow diagrams,
        - tabular heat and material balance,
        - process conditions/special conditions,
        - materials of construction, and
        - startup and shutdown requirements;

    - mechanical definition:

        - P&ID preliminary sizing and tie-ins,
        - preliminary plot plan,
        - preliminary general arrangement, and
        - preliminary equipment list;

    - instrument definition:

        - defining primary control points and purpose, and
        - defining instrument set points/low-level alarms, etc.;

    - safety system:

        - hazards analysis (hazops),
        - list of safety devices and their design criteria, and
        - interlock logic description and diagram;

- project location (elements):

    - engineering and construction productivity factors (versus database),
    - logistics reviews and delivery to/at site,
    - infrastructure requirements at site, and
    - weather concerns and impacts;

- project conditions:

    - offshore installations/suppliers,
    - prefabrication and modules,
    - operational restraints/conditions, and
    - site and access problems; and

- estimate (definition):

    - work quantities and takeoffs,
    - engineering and labor/staff hours,
    - contingency and budget limitations, and
    - risk analysis and identification.

C. Timely Decisions on Scope

Timely decisions on scope can only be achieved if there is a cohesive, dynamic trending program. Effective communication channels and working togetherness are direct contributors. Management approval process is also a factor.

D. Dynamic Design Change Control—Formal Program

Project trending and reporting systems, such as the design change order log, are essential. The weekly trend meeting and regular progress meetings provide much of the early identification of change. An effective design control program is centered around an engineering milestone, called the *design control point* (DCP). If the feasibility study is extensive the DCP could be operational at the end of the study and, thereafter, all changes would be formally documented. If the feasibility work was minimal, then the DCP would be established at the early part of the project engineering phase. The DCP is reached when the project's original scope is properly defined, agreed and approved by all parties. As this approval is reached, the project manager will inform all appropriate parties that the DCP has now been implemented. On very large projects there can be multiple DCPs. It is emphasized that the DCP is not a design freeze as viable design changes should always be an option.

Figure 17.2 illustrates the format of a design change order.

V. COMMUNICATIONS—INFORMATION UTILIZATION

The requirement is to turn data into useful and useable information. Current computers and software systems make the gathering and collation of data a relatively simple task. Thus, correctly establishing the available input data results in obtaining the required output and information. With the establishment

```
┌─────────────────────────────────────────────────────────────┐
│                                                             │
│        SCOPE CONTROL - DESIGN CHANGES                       │
│                                                             │
│        (Post Design Control Point)                          │
│                                                             │
├─────────────────────────────────────────────────────────────┤
│        Facility Engineering & Construction                  │
│             Design Change Order          DCO No. _____    │
├─────────────────────────────────────────────────────────────┤
│                                              Y    N         │
│  Project_____    Within Scope  ☐   ☐     │
│    No./AFE_____     Value Added  ☐   ☐     │
│  Location_____  Funding Revision Required ☐ ☐ │
│                                                             │
│  Originated by_____ Date_____             │
│                                                             │
│  To:          Endorsement        Comments    Signatures/Date│
│  Discipline Engr.    ☐    _____  _____   │
│                                                             │
│  Constr. Engr.       ☐    _____  _____   │
│                                                             │
│  Operations          ☐    _____  _____   │
│                                                             │
│  Other               ☐    _____  _____   │
│                                                             │
│                                                             │
│  Brief Description of Change:_____│
│                                                             │
│                                                             │
│  Estimated Cost and Schedule Changes:                       │
│  Cost ($)_____Project Forecast including this Change ($)_____ │
│                                                             │
│  Funding AFE Upper Limit ($M)_____         │
│                                                             │
│  Schedule Impact (Weeks)_____         │
│                                                             │
│                                                             │
│  Reason/Justification for Change:                           │
│                                                             │
│                                                             │
├──────────────────────────────┬──────────────────────────────┤
│  Originator                  │  Project Control             │
├──────────────────────────────┼──────────────────────────────┤
│  Project Leader              │  Approved by                 │
└──────────────────────────────┴──────────────────────────────┘
```

Figure 17.2   Scope control—design changes.

of effective communication channels, the information is then directed to the correct recipient.

A. Execution Plan—A Formal Written Program

The *execution plan* should be a dynamic document, being revised and updated as conditions/scope change, with proper/timely inputs from all parties. Commitment to the plan must then be achieved with all project parties and, especially, management.

These are the three major categories of a good execution plan:

- What is the scope of work? (discussed earlier),
- How is the work to be executed? and
- When is the work to be carried out?

1. How is the Work to be Executed?

- statement of project objectives;
- proposed division of work:

  - in-house,
  - work contracted out, and
  - development of work packages;

- contract strategy:

  - required scopes and degree of definition,
  - forms of contract, and
  - risk allocation versus cost of liability;

- detailed engineering;
- procurement program:

  - competitive,
  - domestic and international,
  - single source, and
  - plant compatibility and spares requirements;

- construction:

  - preplanning program (critical highlights),
  - prefabrication/modules, and
  - precommissioning and testing program;

- commissioning and startup;
- quality assurance—control and inspection;
- project organization (described earlier); and
- project coordination procedure (as follows).

2. When is the Work to be Carried Out?

- schedule and probability:

  - economic versus acceleration, and
  - critical path and float analysis;

- resource analysis:

  - engineering/construction availability, and
  - skills and trade union climate;

- marketing interface (limitations or constraints);
- cash flow limitations;
- access problems (weather windows, traffic limitations); and
- shutdown (retrofit program).

B. Other Requirements

- All scope-cost matters are routed through the project manager to ensure project consciousness and dynamic trending.
- A team cost culture is developed on the project and is essential for information utilization.
- An internal project charter program is a method for motivating working togetherness. It is a one- or two-page document, signed by all parties, lists all major parties and their responsibility/accountability as well as the project objectives.
- All project parties need to maintain open communication lines at all times to foster effective organization, project commitment, working togetherness, and leadership.
- A *project coordination procedure* (PCP) clearly defines communication channels to all. It includes:

  - limits of authority,
  - responsibilities of parties,
  - correspondence procedures,
  - filing and reporting codes,
  - document and action schedule (for all drawings, documents, and reports),
  - public relations procedures,
  - security and safety procedures, and
  - project close-out report.

- Weekly trend and progress meetings are a must for the project manager. They are a vital communications tool and are detailed in Chapter 9.
- The project management information system needs effective levels of detail. Information must be accurate, timely, and useful. Because unnecessary detail can easily be generated with today's computer, a vigorous screening effort is called for.

VI. CONSTRUCTABILITY PLANNING

Constructability and *construction preplanning* are often used interchangeably to describe the function of each category. Constructability is largely concerned with technology, methods of installation, and the associated cost. Preplanning is largely to do with scheduling of resources, organization, site access, and infrastructure.

The purpose of constructability is to reduce costs by considering alternative design and/or installation methods. Typical examples are steel or precast concrete for a building and, for process plants, greater prefabrication and preassembly, even modularization.

A. Early Economic Path of Construction Program

This is an evaluation of the physical sequence of construction work to produce the lowest cost. Many factors are involved, such as:

- physical site conditions and weather;
- restraints of drawings, material delivery, and schedule critical path sequences;
- economics of crew sizes and supporting resources; and
- plant operations, safety regulations, etc.

With such early planning, design or material alternatives can be considered at little or no additional cost.

B. Formal Constructability Programs

Formal constructability programs are an integral part of project execution. They ensure that the early initiative, as outlined above, is maintained to project completion.

C. Relationship of Front-End Planning and Construction Input

It is essential that capable and experienced construction personnel are assigned to the project at this early stage and that their constructability and preplanning evaluations are a proper part of project development. Sometimes, owners are not prepared to pay for this service and do not appreciate the cost benefit of this early work.

D. Construction-Driven Scheduling as the Key to the CPM Program

Construction-driven scheduling is also known as *backwards scheduling*, meaning that the project CPM schedule is structured around the construction schedule, assuming that the construction schedule has been developed on a best economic basis. Engineering and material deliveries can then be matched to the economic construction program, at no cost penalty.

VII. SUMMATION

It is again emphasized that FEPPO, as reported by CII, is the number one activity on any project, and when this work is properly executed, the dollar payout is immediate and substantial.

If a project is started with a good scope definition and good organization and all parties are committed to working togetherness, then project success is a reality.

Areas for Improvement	Current Average/Percent Utilization		
	Owner	A/E	Contractor
Constructability Planning	56	52	
Contract Incentives	37		33
Using Design Evaluations	60	60	
Implementing QA/QC Systems	76	66	66
Quality Effectiveness Analysis	62	51	54
Use of Work Force Motivation Programs	40		52
Site Training	51		43
Substance Abuse Programs	55		27
Materials Management Systems (Planning and Utilization)	63	58	62

Constructability planning is utilized at only a moderate rate on projects. This includes the use of both formal constructability programs and incorporating construction input early in project planning and design. Case studies have clearly shown large payoffs from practicing constructability planning, yet early design and construction interaction remains low.

Prepared by
The Construction Industry Institute
Strategic Planning Committee

A Special CII Publication
April 1990

Figure 17.3 Potential areas for increased utilization.

The Construction Industry Institute also reports that this program, FEPPO, is the least used of major project management methods; current research shows its utilization at an average rate 54%. Figure 17.3 is a chart of the CII study.

18

Managing Engineering—Project Control Keys/Interfaces

I. INTRODUCTION

Effective management of engineering must be equally concerned with:

- proper technology development, and
- efficient and cost-effective management of design program.

Obviously, the design has to be of sufficient quality so that the plant will start up and operate at the required performance levels. However, ultra-conservative and/or overdesign practices are common and can result in added costs. There has to be a balance between design quality and the cost of that quality, where project economic requirements have equal consideration. These issues are covered in detail in Chapter 16, "Managing The Feasibility Study," and Chapter 17, "Front-End Planning and Project Organization."

Efficiently managing the design program requires the achieving or under-running of cost and schedule budgets and the constant trending of the work to identify change and project status, leading to accurate and timely forecasts.

The project manager/engineer should take the lead in developing a quality statement of requirements (SOR) and, thereafter, motivate the design engineering group to an economic design program. Technical responsibility is divided between the design groups (discipline chief engineers) and the project manager. Ultimately, technical integrity cannot be a shared arrangement; final responsibility will generally reside with the chief engineers.

The project manager/engineer must ensure that constructability, plant operability, and maintenance considerations are properly interfaced into the early stages of design and are developed on an economic basis.

II. EARLY ENGINEERING PHASES

Engineering design generally falls into the following successive phases.

A. Conceptual/Feasibility Definition

As discussed in Chapter 16, studies are undertaken to establish the potential technical and economic viability of the various options to a plant addition or production problem. Each option is then evaluated for cost, schedule, technical risk, the required resources, and the availability of these resources.

After eliminating the unattractive options, the next objective is to produce a detailed evaluation of the attractive options in order to choose the best case. During this phase, it is normal for the nucleus of the future project team to emerge.

B. Front-End Engineering Design

With complex projects, the best case requires a great deal of design effort in order to properly ascertain the time and cost to complete the project and to develop clearer technical definition. This front-end engineering design (FEED) phase can equal 10% of the total design effort. The work is led by an owner project team and often involves the assistance of an A/E or engineering contractor.

In addition to efficiently determining the best case, another major objective is to provide a good design basis that will enable lump-sum bids to be sought for the final engineering phase. However, a lump-sum approach requires good project discipline in limiting design changes as the work proceeds. If there is a significant probability of design changes, then a reimbursable approach is recommended.

III. PROPER TECHNOLOGY DEVELOPMENT

The major development of design occurs during the early engineering phases with selection of the best case, associated process-utilities technical base, and subsequent optimization of the selected technologies. Thereafter, the technology development takes the technical base through detailed engineering, where economic engineering becomes the main issue.

A management control, measuring, and reporting system must be developed to show that the correct balance is being achieved among the three major components:

- technology requirements,
- economic considerations, and
- operational efficiency (design works).

A proper control and reporting system covers the following key considerations.

A. Design Document Accuracy and Degree of Changes

Both design document accuracy and degree of changes are measured by the cost reporting program. Cost codes should be established for tracking design rework due to design error, normally about 5% of total design engineering hours, and for field design rework due to construction error, again about 5% of total design engineering hours. It is common for this rework to be hidden or under-reported. Accurate records are essential for proper tracking. Design and scope changes are easier to track as design contractors doing the work will ensure that these are fully recorded.

B. Usability of Design Documents

Usability of design documents is a requirement of the reporting of field supervision, for example:

- Did field supervision find the issued-for-construction (IFC) drawings:
      Excellent   Good   Adequate   Poor?
- Were there significant discipline design errors? If so, give details.

C. Constructability Of Design Consistently Carried Out

Carrying out the constructability of design in a consistent manner is vital, as significant cost efficiencies can be achieved from this work. Refer to Chapter 17, "Front-End Planning," and to Chapter 20, "Managing Construction," for detailed information.

IV. DETAILED ENGINEERING/PROCUREMENT

The final part of engineering design is the development of the conceptual design, resulting from FEED, into full working drawings. This then allows procurement activities and construction planning to commence. This detailed design work may be done in-house by the owner, by an A/E, or an engineering contractor.

Detailed design packages are then developed to match construction work packages which can then be contracted on a lump-sum basis, insofar as is possible. This can only be achieved with quality construction preplanning.

V. ENGINEERING DESIGN/PROJECT MANAGEMENT INTERFACE

As the design engineering group has the primary responsibility for *design integrity*, the role of the project manager is mostly that of business management and project direction, as follows:

- working with operations/maintenance to develop a quality SOR,
- ensuring an economic design (no "gold plating"),
- managing/limiting the impact of operations/maintenance design changes,
- motivating the engineering group to work to the schedule,
- ensuring that the design/estimating interface is effective,
- ensuring that the design/procurement interface is effective,
- ensuring that the design/construction interface is effective, and
- ensuring that the design/contracting interface is effective.

VI. EFFICIENT AND COST-EFFECTIVE MANAGEMENT OF DESIGN PROGRAM

The realization of this objective is found in the following benchmark evaluations.

A. Cost of Design, Actual Versus Estimated

From the cost control system, a constant assessment is made of actuals versus budget for:

*   engineering productivity, through an earned value system;
*   hourly cost;
*   expense and other support costs;
*   design cost as percent of total cost (typically 10%);
*   total engineering hours per drawing (typically 120–150);
*   hours per P&ID (typically 400–500 for high process severity); and
*   hours per piping isometric (typically 4–5 for CAD systems).

The status of these criteria will, in turn, demonstrate the status of this benchmark. Chapter 9, "Cost and Schedule Trend Analysis," fully illustrates techniques that will determine these assessments.

B. Economy of Design—Overdesign, "Gold Plating" and Pre-Investment

Full application of this category is in the exercise of value engineering and the application of a quality assurance program. The ongoing design needs to be constantly evaluated for design characteristics that are not part of the originally approved design basis and/or are preference items.

C. Performance Against Original Schedule

As with the cost benchmark, the engineering schedule must also be constantly assessed for actual progress against plan. Overall progress should always be measured and on larger projects, individual disciplines should also be measured.
   The following are key milestones that should be assessed:

*   Were IFC drawings issued to schedule?
    If not, what was the slippage (by discipline)?
*   Did design packages meet purchase order dates?
    If not, what was the slippage (by discipline)?
*   Did design slippage cause material delivery delays?
    If so, give details.
*   Did design delay cause delay in letting contracts?
    If so, give details.

D. Ease of Start-Up

This benchmark is assessed by comparing the actual cost and schedule of commissioning against the estimated or budgeted numbers. On larger projects there is an individual group or crew for commissioning and start-up, and the effectiveness and interaction of this group with the regular construction staff can also be part of the assessment of this benchmark.

The following are key milestones that should be assessed:

- Did start-up program meet schedule?
  If not, give details.
- Did start-up costs meet budget?
  If not, give details.
- Was the assigned start-up team adequate (numbers and skills of personnel)?
  If not, give details.

VII. CORRECT BALANCE BETWEEN TECHNICAL AND PROCUREMENT CONSIDERATIONS

The cost of equipment and material can be more than 50% of the total project cost. As such, it is vital that full and proper business considerations be in place in the selection of all equipment and materials. It is very common for engineers to select on a technical preference basis or for operations/maintenance personnel to ignore capital costs and select on compatibility (existing equipment), spares availability, maintainability and operating characteristics. Both technical and business considerations are equally important, and their relative priorities should have been properly identified as part of the value engineering-quality assurance program.

VIII. ENGINEERING-PROCUREMENT CONTRACTOR/PROJECT MANAGEMENT INTERFACE

When using an engineering contractor, it is essential that the specified technical, maintenance, and operational design requirements are being met. This can be achieved by checking and approving key drawings and documentation produced by the contractor. It is also necessary to provide the contractor with backup guidance on the standards being used where such explanation will enable the design intent to be more clearly understood.

On lump-sum contracts, the project team must ensure that such explanation/guidance does not lead to scope changes and/or contractor claims for delays, due to owner interference or negligence.

IX. EFFECTIVE CONTRACTOR ENGINEERING-PROCUREMENT PERFORMANCE

An effective and/or acceptable contractor engineering/procurement performance results in the design (drawings/specifications) and purchased equipment fully meeting the project specification. In addition, the work is on time and to the approved hours and/or financial budget.

A. Lump-Sum Basis

As the contractor has most of the financial risk, the major activities of the project team (subject to contract conditions) are:

- approval of specified drawings (design quality),
- technical review of equipment/material bid packages,
- monitoring detailed contractor schedule,
- monitoring engineering progress/staffing levels,
- monitoring equipment delivery time/purchase order placement,
- ensuring design is structured to planned construction packages,
- change order review and approval of all scope changes, and
- progress payment verification.

With lump-sum contracts, the major emphasis of the project team is on quality and schedule, plus the tracking of any scope changes. It is assumed that the contractor's design basis/specification was fully reviewed and approved during the proposal evaluation program.

B. Reimbursable Basis

As the owner has the major financial risk, the contractor monitoring effort is much more detailed, and the contractor requires the following major procedures:

- detailed cost control/reporting system;
- engineering change log (weekly) maintained by design groups to show design changes from the approved for expenditure (AFE) design basis and then evaluated by the cost group for possible impact on project cost/schedule;
- detailed planning and scheduling, complete with the progress/completion curves, staffing histograms and productivity profiles;
- earned value progress/productivity measurement system (larger projects);
- hours tracking program/curves (small projects);
- purchasing/procurement plan;
- quality equipment bid summary procedure; and
- trending system/weekly trend report.

The key to an effective contractor monitoring effort is to have the contractor develop a full and effective system, subject to the contract arrangement, so that the owner project team has a review, analysis, and approval role.

Major attention should be given to:

- progress measurement,
- productivity analysis, and
- trending and forecasting.

On reimbursable projects, the daily interaction of the design group with the contractor personnel should achieve the desired quality of design/procurement.

X. TYPICAL DESIGN PACKAGE FOR CONSTRUCTION WORK PACKAGES

When required, engineering/procurement contractors structure their work into design packages to match a predetermined breakdown of construction work packages. In some cases, the engineer assists in determining the packaging of the construction work. The quality of these design packages should be sufficient for lump-sum construction contracts, subject to schedule considerations.

A typical design package includes:

- index;
- summary;
- general work scope;
- design/operating conditions;
- design regulations/standards/codes;
- construction standards (safety procedures, etc.); and
- detailed work scope, by discipline, for example:

    - list of all drawings;
    - bills of material (if unit price subcontracting);
    - equipment list (major or tagged items);
    - paint, insulation, corrosion specifications;
    - weight penalty, for offshore work;
    - material information (free issue);
    - inspection and testing;
    - documentation/certification;
    - manuals/spare parts lists; and
    - drawings (shot-down size).

XI. FULL SANCTION AND CLASS II COST ESTIMATE

With most companies, projects are executed with a phased approach, and the funding of the execution phase of a project is often referred to as *full funding*. The early development stage of a project is usually funded by the operations/production group from its own budget or from company expense.

Full sanction requires a specified quality of cost estimate so that management can evaluate the financial risks when approving/authorizing the funds. These funding or appropriation estimates generally range from 10–20%. The manufacturing industries normally require a 10% quality estimate for full sanction, whereas the petrochemical industry normally requires a 20% estimate. This lower quality estimate is accepted because of the larger size of projects, plus the need to execute these projects with the fast-track approach. The fast-track method works from a lower degree of engineering definition and, therefore, only allows a lower quality of cost estimate.

A quality feasibility study and FEED will allow a Class II (+20%) or better cost estimate to be completed for subsequent application for sanction (management approval) of full funds. To obtain a Class I (+10%) cost estimate, the engineering and procurement has to be greater than 80% complete. Work carried out in producing the feasibility study and FEED are funded by the client division

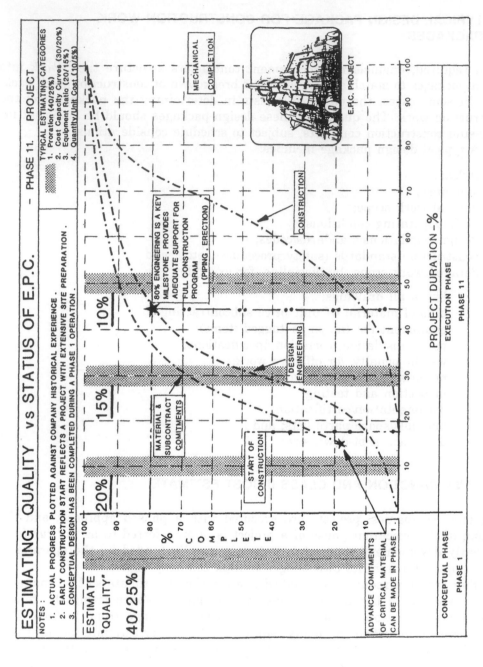

Figure 18.1 Estimating quality versus status of EPC.

from the sanction of a Class III (+30%) or better cost estimate. Figure 18.1 illustrates the historical quality of the cost estimate in relation to completion of engineering and procurement.

XII. TECHNICAL APPROVALS AND HANDLING PROCEDURES

All required technical approvals must be prompt and/or meet the required contractual timing if the work is contracted. Likewise, the associated handling procedures must be effective to prevent delays. This is to ensure that the contractor cannot accuse the owner of compensable delays and/or negligence, as these are common claims.

19

Managing Procurement—Project Control Keys/Interfaces

I. INTRODUCTION

Effective procurement means getting the *right material to the right place at the right time at the right price*. Procurement comprises the three functions of purchasing, expediting, and inspection. The work is carried out in accordance with the contract and procurement strategies developed by the project manager during the preproject stage and in compliance with corporate purchasing policies. Figure 19.1 illustrates these essentials.

II. POLICY—VALUE AND PRICE NEGOTIATING

The best procurement policy within the private sector obtains the *best value for the lowest cost*. In addition to price and delivery, best value takes into account other commercial and services considerations, as follows:

- appropriate quality of materials and equipment as per the technical specifications (i.e., reduce/eliminate extra quality);
- terms of payment, conditions of purchase, performance guarantees, quantity discounts, import duties and taxes, etc.;
- availability and cost of vendor services, startup support and spares;
- quantity sensitivity analysis of unit price tenders to evaluate total tender price;
- standardization/compatibility with existing plant and equipment; and
- potential for future price increases due to changing specifications.

There are two distinct policies in relation to price negotiations. The first is a *sealed bid/lowest competitive price*, which precludes price negotiating if there are not technical/contract condition changes. Price negotiations, then, are only

1. **Right Material,** *to*

2. **Right Place,** *at*

3. **Right Time,** *and at*

4. **Right Price**

5. **Procurement Program**
   - **■ Purchasing**
   - **■ Inspection**
   - **■ Expediting**

6. **Pre-Project Strategy & Execution Plan**

7. **Compliance with Corporate Policies**

Name some extraordinary efforts to achieve "right time."

- ELIMINATE LATE BIDS
- SINGLE SOURCE
- NEGOTIATED PURCHASE
- SELECT FOR BEST SCHEDULE, REGARDLESS OF PRICE
- OFFER PREMIUMS-INCENTIVES FOR SCHEDULE
- ADDITIONAL EXPEDITING
- "RESIDENT" ENGINEERS
- AIR FREIGHT

Figure 19.1   Essentials for managing procurement.

allowed if there are technical or contract condition changes. However, companies having this policy sometimes have a double standard as price negotiations are undertaken with attractive bids which do not meet the conditions and with all scope changes, extras, and claims. This can cause problems, as the company personnel are not skilled in price negotiations.

The second is an *open policy*, which allows price negotiations at any time, even if there are no technical/contract condition changes. This approach is sometimes referred to as *price-alone* negotiating and must be carried out on an ethical basis. For example, this would require that quoted prices are never disclosed to other parties. It is also important that price-alone negotiating be carried out with good business practice and judgment. To negotiate a price at or below the other party's costs is rarely good business and often results in poor project execution, bad quality, aggressive claimsmanship, and default of contract.

Normally, companies use competitive bidding for the placing of orders and contracts. Major bids are usually invited in two parts, commercial and technical. As appropriate, the responsibility for maintaining the confidentiality of such bid documents rests with either the contracts engineer or procurement coordinator. They will ensure that access to the documents is limited to project management, those undertaking bid analysis, and those executing the work.

III. PROCUREMENT RESPONSIBILITY

For intermediate and large projects, it is normal to establish a *project procurement coordinator* (PPC) within the project team. When working with engineering contractors, the responsibility of the owner's PPC is to ensure that the contractor's procurement effort is effective and that the owner's interests and project objectives are being realized. This provides a full service to the project manager and project team to ensure the timely, economic, and coordinated supply of all plant, equipment, and materials to the job site. If the owner is directly handling the procurement function, then the PPC will be working directly with corporate purchasing to ensure that the project objectives are being realized.

The PPC's scope of responsibility is to manage, direct, monitor, and overview all procurement and materials control activities.

For all projects it is important that the PPC or the procurement function be involved at the preproject stage for formulating the contract and procurement strategies. This ensures that the procurement systems and staffing are appropriate. In addition, the PPC/procurement function should be involved in prequalification of design and procurement contractors.

One early and important procurement/project activity is the development of a *purchasing plan* for all critical material.

IV. MATERIALS PROCUREMENT AND CONTROL ESSENTIALS

The following activities, illustrated by Figure 19.2, are essential for a quality procurement program. Such activities are necessary for effective execution and control:

- developing a purchasing plan for critical material;
- establishing a current and approved bidders list;
- using sealed bids and/or price-alone negotiating;
- preparing quality bid summaries;
- issuing material requisition and purchase orders in a timely and proper manner;
- using effective vendor drawing control;
- expediting items adequately, including material status reports;
- maintaining close liaison with planning engineer on delivery times and MSRs;
- keeping to an inspection schedule;
- keeping an efficient material receiving and inspection system;
- having adequate material documentation/control system;
- forecasting final costs of major purchase orders/contracts;
- maintaining close liaison with cost engineer on prices and bid summaries; and
- evaluating carefully spare parts for plant startup.

The most difficult, and in many respects, the most important of the above activities is forecasting the final cost of major purchases. This activity requires considerable skill in developing accurate forecasts and should be a regular part of the monthly project cost forecast. Figure 19.3 shows and emphasizes these key elements.

Essentials

1. Critical Material Purchasing Plan
2. Current Approved Bidders List
3. Formalized Price Negotiating Policy
   ■ Sealed bid and/or price alone
4. Quality Bid Summaries
5. Effective MR & PO Procedures/Timing Issue
6. Good Vendor Drawing Control
7. Coordinated Expediting & MSR's
8. Close Liaison with Planner On Delivery Times
9. Inspection Schedule
10. Efficient Material Receiving
    & Inspection System
11. Adequate Documentation/Control System
12. Periodic Forecasting of Costs of Major PO's
13. Close Liaison With Cost Engineer on Prices
14. Careful Evaluation of Spare Parts

Figure 19.2 Essentials for effective procurement execution and control.

1. Most Difficult of Purchasing Control Effort
2. Most Important of All Activities
3. Requires Considerable Analytical Skill
4. Requires Close Coordination
   With Engineering
5. Should Be Regular Monthly Cost Activity
   ■ Part of trending and cost report

See Case History

$
PLUS
INTEREST

Figure 19.3 Forecasting final cost of major purchases.

A. Contractor as Owner Purchasing Agent

With a reimbursable-type design and procurement contract, the contractor is responsible for procurement of the majority of the project materials. If the contractor acts on behalf of the owner, then the contractor is acting as the owner's agent. This relationship carries no liability to the contractor, other than normal responsibility for reasonable performance and professional competence. To protect the interests of the owner, the owner's PPC has full authority to review and approve all procurement activities carried out by the contractor. A further feature of the agent relationship is that the owner is also responsible for and liable for the contractor's actions. This is the form of contractual liability preferred by A/Es.

It should be noted that A/Es and contractors can function as an independent contractor on reimbursable forms of contract. The cost, of course, will be significantly different for these two forms of contractual liability.

B. Contractor as Independent Contractor

With a lump-sum design and procurement contract, the contractor is fully liable and responsible for the quality and cost of all the materials and subcontracts. Full liability and responsibility is termed an *independent contractor relationship*. However, quality should be carefully monitored to ensure that all materials meet specification and that checks are made on the progress, schedule, and execution of the work.

C. Owner's Project Buyer

The *project buyer* is a buyer from corporate purchasing who has been assigned the purchasing function for a specific project. The individual is resident in the corporate office. This provides the project with the facility for fast-track purchasing when the work is being done in-house.

D. Owner's Corporate Purchasing Division

Corporate services and expertise are generally utilized on small projects. Alternatively, corporate services can be required by the owner's PPC due to high workload problems or the need for specialist services. When there is no assigned project buyer, the lack of designated responsibility can result in purchasing delays and a poorly coordinated program. The impact on the project(s) can be disastrous.

Using corporate purchasing division services can result in more advantageous terms being obtained for price, delivery, conditions of purchase, terms of payment, maximum net discounts for volume and bulk purchases, etc. It also has the added advantage of enabling other owner project material surpluses to be assessed for compatibility with the project materials requirement.

E. Construction Contractor

As required, the construction contractor or a subcontractor may purchase materials, using approved purchasing procedures and be closely monitored and controlled by the construction manager. Detailed attention to specification and certification is required, particularly when discovered shortages are required urgently on site.

It is quite common for a construction contractor to provide the bulk materials that they are going to install, but it is not common for them to provide major equipment, unless they are also the engineer. Major equipment is more efficiently purchased by the party that is producing the specifications and data sheets.

V. PROCUREMENT PROCESS

Procurement normally follows a process of inquiry, bids comparison, ordering, expediting, inspection and delivery. Figure 19.4 is a schedule that shows the purchasing steps and the time, based on historical experience, that are required to carry out the work. These durations are for a normal level of activity and the durations could be reduced, with special action, for critical situations. As shown in Figure 19.4, the typical/average time for the purchasing cycle is 3–4 months (12–18 weeks) after the preparation of design documents and material specifications by the engineering group.

A. Inquiry

Inquiries should only be issued to companies with adequate and proven capability. These companies should be shown on the current bidders list. The practice of seeking bids only for budget/estimate checks is to be avoided if at all possible. If there is no serious purchasing intent, the supplier/contractor costs increase due to the cost of spurious bidding, and the business credibility of the company is impaired.

The project bidders list should be developed by the PPC, in conjunction with corporate purchasing and approved by the project manager.

The technical requirements contained within a material requisition should be as complete as possible in terms of specification, inspection documentation requirements, etc. The information should be stated in a concise and unambiguous manner to obtain quality bids from suppliers. Requisitions should be properly checked against the control estimate, prior to authorization by the project manager.

B. Bidding Strategy—Bids Comparison—Quality Evaluation

Bidding strategy has the following major considerations:

*   competitive open bids,
*   competitive negotiated bids,
*   single-source competitive (as perceived by supplier) bid,
*   single-source negotiated bid,
*   competitive sealed bids, and
*   single-source sealed bid.

As previously stated, the bidding strategy will largely depend on the company negotiating policy.

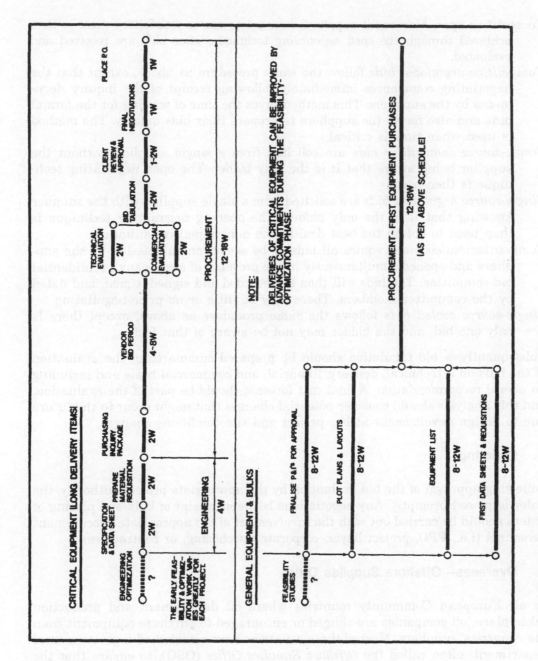

Figure 19.4 Front-end engineering and procurement.

Competitive open bids solicit supplier's best prices and a satisfactory price is then achieved through the *open negotiating* technique, after bids are received and evaluated.

Competitive negotiated bids follow the same procedure as above, except that the negotiating commences immediately following receipt of the inquiry documents by the suppliers. This method saves the time of waiting for the formal bids and also forces the suppliers to prepare their bids quickly. The method is used when time is critical.

Single-source competitive bids are solicited from a single supplier, without the supplier being aware that it is the only bidder. The open negotiating technique is then used.

Single-source negotiated bids are solicited from a single supplier, with the supplier knowing that it is the only bidder. The opening negotiating technique is then used to obtain the best deal in the noncompetitive situation.

Competitive sealed bids require all bids to be sealed on submission by the suppliers and opened simultaneously in the presence of a company's confidential bid committee. The bids will then be recorded and signed, timed, and dated by the committee members. There may be little or no price negotiating.

Single-source sealed bids follows the same procedure as above, except there is only one bid, and the bidder may not be aware of that fact.

Subsequently, a bid tabulation should be prepared summarizing the evaluation of the bids on a technical, delivery, financial, and commercial basis and including an award recommendation. A final cost forecast should be part of the evaluation, and this analysis should consider potential changes that might occur in the future due to design development and/or project and site conditions changes.

C.   Ordering

Following approval of the bid evaluation by the appropriate project authority, the order is placed promptly. Any negotiations between receipt of bids and placing of orders should be carried out with the involvement of the appropriate procurement personnel (i.e., PPC, project buyer, corporate purchasing, or contractors).

D.   Overseas—Offshore Supplies Office

In all European Community countries where oil development and production takes place, oil companies are obliged or encouraged to purchase equipment from the countries' suppliers. Most of these countries have established a governmental department, often called the *Offshore Supplies Office* (OSO), to ensure that the companies follow this policy. For example, in the United Kingdom, the government-appointed OSO applies pressure on oil companies to buy within the UK.

To comply with the OSO directive, companies are obliged to obtain OSO clearance for all bid lists and bid summaries for any purchases, with anticipated value in excess of 250,000 pounds sterling. The procedure applies whether the item is contractor- or owner-generated, and prior to inquiry or order placement. Equipment suppliers are obliged to complete, when submitting a tender, a "Statement of UK Contents" form to assist this process.

Many other countries have a similar program, and such restrictions should be recognized in their cost estimate, project schedule, and purchasing program.

E. Expediting

Expediting covers both the delivery program and the submission dates for drawings and certification. This requires a close liaison between the project planning engineer and the expeditor. Construction frequently takes over the progress chasing materials where critical site delivery problems are anticipated.

Project materials status reports (MSRs) are produced regularly by suppliers, contractors, and corporate purchasing. These are collated and updated by the PPC and distributed to appropriate members of the project team for input and action. The prime purpose of MSRs is to highlight potential problem areas so that appropriate action can be taken.

F. Inspection

The degree of inspection is determined by the quality control program in conjunction with the design specifications. Some companies use a *criticality rating* system which automatically establishes the quality/inspection requirements for the equipment and materials being purchased.

Project materials and equipment inspection will be coordinated by the PPC and the quality assurance engineer in liaison with those responsible for the procurement to ensure that the established requirements are satisfied in full. This would include any third party inspection required by statutory legal requirements and the supply of all certification and documentation.

G. Delivery

The method of delivery and delivery responsibility must be clearly defined when ordering. Equipment and goods must be properly checked by the site materials controller against the order details for correctness, completeness, and transit damage. A goods receipt notification is then issued and the goods, if not immediately required for installation, should be suitably protected before being consigned to stores.

VI. EQUIPMENT SPARES AND OPERATING DOCUMENTATION

It is essential at the inquiry stage to include the requirement for suppliers to provide equipment spares lists, operating and maintenance instructions and illustrated parts manuals. Also recommendations from the suppliers for spare parts for startup and their associated cost. The project manager, in conjunction with startup specialists, should approve the purchase of these spare parts. Failure to properly investigate the need for startup spares can be very detrimental to an orderly plant startup. On the other hand, an oversupply of spare parts can be very costly.

VII. SURPLUS MATERIALS

The lack of materials during construction can have a disruptive and costly effect on a project; it is equally important that project funds are not wasted by over-ordering. Disposal of surplus materials normally recovers only a small fraction (typically 5–10%) of the original cost of the materials. In order to minimize the amount of surplus materials, the project manager and his team need to take the following steps:

1. When possible, ensure early completion of detailed engineering and accurate materials takeoff prior to purchase.
2. When accurate materials takeoff cannot be achieved at the time materials must be ordered, take particular care to ensure the accuracy of the estimated quantities.
3. Use manufacturers' standard products and stock materials wherever possible. This approach will:

    –   reduce purchasing lead time and costs,
    –   render surplus materials attractive to a wider market, and
    –   increase the potential recovery from sales of surpluses.

4. Carefully monitor the contractor's procurement to prevent excessive over-ordering.
5. During the course of the project, undertake periodical reviews of bulk materials to identify any that are unlikely to be used or are clearly surplus. This should be done by the PPC.
6. Dispose of surplus material early to increase the chances of higher rates of cost recovery as the opportunities to sell these materials to the operations group, other company projects, or third parties are likely to be most favorable.
7. Transfer surplus materials available during and at the end of a project to other company projects at, or approaching, the purchase value, provided:

    –   they are required,
    –   they are in good condition, and
    –   necessary certification is available.

20

Managing Construction—Project Control Keys/Interfaces

I. COST-EFFECTIVE BUSINESS MANAGEMENT OF CONSTRUCTION

As construction is a major part of the project costs, it is imperative that all construction work be executed on a good business basis, as well as to required quality standards. Too often, construction managers ignore sound business practice in order to push the installation of the work. The project manager must ensure that good business management is practiced by construction management.

Cost-effective management is equally important as efficient planning and coordination of the site and craft labor.

With most *economic path of construction* (EPC) projects, construction costs represent 40% of the total project cost. Failure to manage construction properly, therefore, can be extremely costly. Many owners make the serious mistake of completely underestimating the needs and demands of an effective construction management program. Leading contractors rarely make this mistake and their construction managers are among the most senior members of their organization.

Figure 20.1 illustrates a benchmarking evaluation of cost-effective construction management, where the major categories are weighted to 100%. As item 1 of this figure can be difficult to measure, this category is broken down into greater detail, as shown in Figure 20.2. With accurate assessments of the criteria shown in Figures 20.1 and 20.2, an overall evaluation can be made of the cost effectiveness and business management of the construction team.

II. CONSTRUCTION PREPLANNING

Preplanning for construction at the early stages of a project is vital. At an early stage, detailed planning is restricted by lack of scope definition. However, there are areas where preplanning can be effective, for example:

433

	% Wt.	Actual	Wt'd %
1 Construction Manager Properly Motivated by Business Considerations	20		
2 Economic Construction Path Program Fully Functioning	20		
3 Cost of Construction vs. Estimate	35		
4 Performance Against Original Construction Schedule	25		
TOTAL	100%		

Remarks:

Figure 20.1 Benchmarking assessment: cost-effective business management of construction.

- work accessibility—operations limitations,
- traffic patterns—material staging,
- laydown areas—support services,
- rigging studies—heavy lifts,
- preassembly and modularization,
- weather window constraints,
- material selection, and
- temporary facilities.

There are, in addition, many other interdependent activities that require early attention. For instance, the payout of quality preplanning, in improved construction efficiency and productivity, is very significant.

Many companies are now developing the project schedule/execution plan backwards. This means concentrating on the construction schedule and developing at this early stage, the most economical construction program possible, usually referred to as the EPC. Engineering and material delivery requirements are then matched to the construction program. The matching must not be a force-fit, as such an action would render the program invalid by producing an impossible schedule.

On EPC projects the typical relationship between engineering and construction labor hours (direct) is the ratio of 1:6. Thus, the deliberate attention to

```
┌─────────────────────────────────────────────────────────┐
│                                                         │
│  1.  Was C.M. involved in preparation ................. │
│      of construction contract RFQ's?                    │
│  ─────────────────────────────────────────────────────  │
│  2.  Was C.M. part of all contractor .................. │
│      bid reviews and negotiations?                      │
│  ─────────────────────────────────────────────────────  │
│  3.    Was there an effective bid evaluation .......... │
│      program for major contracts, with                  │
│      appropriate criteria for selection?                │
│  ─────────────────────────────────────────────────────  │
│  4.  Was C.M. required to approve ..................... │
│      all bid recommendations?                           │
│  ─────────────────────────────────────────────────────  │
│  5.  Did C.M. ensure that Site Team fully ............. │
│      met Owner/M.C.'s contractual obligations?          │
│                                                         │
│  ─────────────────────────────────────────────────────  │
│  6.  Did construction supervision know and ............ │
│      understand general contract conditions?            │
│                                                         │
│      Remarks:                                           │
│                                                         │
│                                                         │
│                                                         │
│                                                         │
└─────────────────────────────────────────────────────────┘
```

Figure 20.2 Benchmarking assessment: construction manager motivation.

preplanning and constructability at an early design stage can reduce the ratio and save construction labor hours. The potential cost saving is substantial.

The following is not all-inclusive but is typical of major preplanning considerations:

- construction organization—personnel assignments,
- detailed layouts for temporary facilities,
- labor resource studies—training program,
- labor productivity evaluations,
- material handling—logistics studies,
- site survey—soil report—weather window limitations,
- rigging studies—onshore/offshore,
- construction permits—environmental matters,
- site preparation and early fieldwork,
- contract strategy,
- construction management procedures,
- construction planning and coordination program,
- construction project control and reporting program,
- site and safety regulations, work permits and clearances, and

- labor contracts and site agreements.

Refer to Chapter 17, "Front-End Planning and Project Organization," for more information on early planning activities.

III. OWNER/CONTRACTOR COORDINATION

When the work is contracted out, the owner project team function:

- provides overall site coordination,
- ensures an acceptable quality of construction,
- ensures that work is installed to the drawings and specifications,
- evaluates the contractor performance,
- interfaces with the operating group, and
- institutes correct handover procedures.

On small projects, the owner construction management function is often handled by the project manager. However, this dual function is rarely effective on intermediate/larger projects, as the construction manager responsibility is too demanding. It is strongly recommended that owners only assign experienced, professional construction managers to larger projects.

It is now widely accepted that today's construction managers need to be professional project managers. They need to be:

- organizers,
- planners,
- motivators,
- leaders,
- communicators, and
- business-oriented.

IV. CONSTRUCTION MANAGER/PROJECT MANAGER INTERFACE

The construction manager should be the project manager's representative onsite and in this context is directly responsible to him/her for completing the works to the drawings and specifications, on time and within budget. To reinforce this relationship, project managers are now residing at the construction site to strengthen the day-to-day business relationship with the construction manager. The day-to-day execution of the work and management of the site is the direct responsibility of the construction manager. Therefore, the function of the project manager is to direct and support the construction manager in all business matters.

The construction staff varies in composition and strength and is, essentially, determined by the contractual arrangement. A reimbursable contract requires a larger staff, particularly in the project control function so as to maintain full financial control.

Generally, the technical side is covered by various discipline engineers with ability to monitor/control their counterparts employed by the contractor(s). Similarly, schedule and cost control can be monitored/controlled with site planning and cost engineers. Quantity surveyors are also used in measuring work done and administering contracts based on measured work/unit prices, usually supported by one or more contracts officers/administrators.

V. CONSTRUCTION MANAGER/OPERATING STAFF INTERFACE

The project manager, construction manager, and staff have a sensitive and important interface with the existing or future (on grassroots sites) operating staff, particularly on matters of safety and handover. The sensitivity occurs at the precommissioning stage when the project/site team and operating staff have a joint responsibility for an efficient plant start-up. At this stage the site team has the task of full cooperation with the operating staff and, at the same time, must resist excessive changes requested by the operating staff. This, therefore, requires sensitivity and tact to meet the project cost objectives and maintain a good relationship with the operating personnel. An effective way to enhance a good working relationship is the early assignment of an operating individual to the site team to act as a permanent liaison/coordinator for all operating/plant matters.

VI. CONSTRUCTION MANAGER/REGULATING AUTHORITIES INTERFACE

A further interface takes place with external regulating authorities and bodies in the implementation of national and local regulations, work permits, environmental reports, etc.

VII. EFFECTIVE OVERALL SITE COORDINATION AND SAFETY

When the work is contracted out, the construction manager has a coordination role which, if not properly executed, can lead to serious legal implications and/or contractor claims. In this coordinating position the construction manager has to ensure satisfactory overall management of the project schedule, site safety, and labor relations. The safety and security of the site or location as a whole is his/her responsibility. Although each contractor will be responsible for the safety and security of its own working area, it is necessary to ensure that common and adequate standards are observed by all site contractors.

As the project manager's representative onsite (particularly on remote and large multicontract sites), the construction manager has an overview of all aspects of the works and their relative priorities. Thus the construction manager may directly administer those contracts covering general site services such as staffing buses, catering, medical facilities, and camp accommodations.

An owner can assign this coordinating liability to a managing contractor/construction manager, but the owner must step back and allow this representative full, total, and day-to-day management of the site.

1. Number and value of contractor claims for ....
    poor access or site coordination problems?

2. Were weekly Progress Meetings held ....
    with all key & major contractors?

3. Were weekly Trend Meetings ....
    held with Site & Project Teams?

4. Did Owner/M.C. Site Team have ....
    "official" Safety Engineer?

5. Did major contractors have ....
    Safety Engineers?

6. Was safety a problem?..

Remarks:

Figure 20.3 Benchmarking assessment: efficient site management and multiple contractor coordination.

VIII. INDUSTRIAL RELATIONS

The *industrial relations* (IR) policies/programs, especially on multicontractor sites, are usually determined by the construction department/project manager before the construction phase commences and should be reflected in the various construction contracts. The construction manager is then responsible for ensuring that the agreed IR policies are implemented by the contractors in an orderly and productive manner. The main factors to be considered are:

• specific application of national agreements regulating the terms and conditions of employment and the procedure for resolving disputes;
• provision of suitable amenities and facilities for the workforce;
• recruitment of competent and reliable labor and their induction into site safety and IR provisions;
• overall coordination and control of site IR through main or managing contractor and/or site contractors committee; and

1. Did Owner's/M.C. RFQ require a labor..
   resources chart for all major contracts?

2. Did Owner/M.C. Site Team then properly ....
   evaluate theses charts (for resource adequacy)?

3. Did lack of resources cause any ....
   significant schedule problems?

4. Were there any significant ....
   construction quality problems?

5. Was labor productivity ....
   assessed by Owner/M.C. Site Team?

6. What was level of labor absenteeism?...

Remarks:

Figure 20.4 Benchmarking assessment: correct labor resources as to skill and number workers.

- management by each contractor of its workforce with particular reference to working practices, disciplinary procedures and handling of disputes.

IX. EFFICIENT MANAGEMENT OF SITE AND CRAFT LABOR

The realization of this objective is found in the following benchmark evaluations:

- adequate organization (i.e., quality and numbers of staff),
- efficient site management and multiple contractor coordination (Fig. 20.3),
- correct labor resource planning (i.e., number and skills of workers; Fig. 20.4),
- effective work planning and scheduling at all levels (Fig. 20.5),
- efficient materials management and warehousing program,
- effective monitoring of contractor(s) performance,
- independent trending analysis/report (multiple contractors),

---

1. Was the "overall" construction planning ....
   by Owner/M.C. adequate?

---

2. Was the "planning" properly fed to ....
   the contractors?

---

3. Was contractors detailed ....
   scheduling adequate?

---

4. Was the critical path(s) and ....
   float always evident?

Remarks:

Figure 20.5   Benchmarking assessment: effective work planning and scheduling at all levels.

- properly functioning industrial relations, and
- cost/schedule/progress reporting (multiple contractors).

Figures 20.3, 20.4, 20.5 are benchmarking reports for some of the above items.

A.   Effective Monitoring of Contractor(s) Performance

The contractor monitoring/reporting requirements will vary according to the type of contract (i.e., lump-sum, measured/unit price, and reimbursable).

1.   Lump-Sum

As the contractor has most of the financial risk, the major activities of the owner construction manager/team, subject to contract conditions, are:

- overall site planning/coordination;
- inspection of construction work to assess quality;

- checking that work is to drawings and specifications;
- monitoring construction progress/staffing levels;
- punchlist, checkout, and acceptance;
- progress payment verification; and
- change order review and approval of all scope changes.

With lump-sum contracts, the emphasis of the construction manager/team is on quality, meeting the design requirements, schedule, and the tracking of scope changes and claims.

2. Measured/Unit Price

With these contracts, the owners assume the risk of quantity change, and the contractor assumes the financial risk of unit price. The major activities of the owner construction manager/team, subject to contract conditions, are:

- overall site planning/coordination;
- inspection of construction work to assess quality;
- checking that work is to drawings and specifications;
- monitoring construction progress/staffing levels;
- evaluation of quantity (bills of materials) changes/forecast;
- punchlist, checkout, and acceptance;
- quantity and payment verification (by quantity surveyors); and
- change order review and approval of all scope changes.

With measured/unit-price contracts, the emphasis of the construction manager/team is on quality, meeting the design requirements, schedule, tracking/verifying quantities and payment, and the tracking of scope changes and claims.

3. Reimbursables

As the owner has the major financial risk, the contractor monitoring effort is much more detailed and requires the following major procedures:

- detailed cost control/reporting system;
- detailed schedule/program, completion/progress curves, staffing histograms, and productivity profiles;
- physical progress/productivity measurement system (large projects);
- labor hour tracking program/curves (small projects);
- inspection and quality control system;
- purchasing plan/system;
- quality material bid summary procedure; and
- trending program/weekly trend meeting/report.

The key to an effective contractor monitoring effort is to have the contractor develop a full and effective system subject to the contract arrangement so that the owner construction manager/team has a review, analytical, and approval role. Major attention should be given to:

- inspection and quality,
- progress measurement,
- productivity analysis,
- staffing expenditure and approval, and
- trending and forecasting.

Of the many effective monitoring techniques, as listed, the single most effective technique is the weekly trend meeting and trend report.

B.   Independent Trending Analysis/Report

On large multiple contractor projects, the construction manager/team has the overall management/coordinating role. As such, a weekly trend analysis/report should be developed to cover current influences, problems, variations, changes, and claims that could have cost and schedule impacts. Early identification of potential trends can provide time to prevent harmful situations and provide valuable input for accurate forecasts.

X.   QUALITY CONTROL AND DOCUMENTATION

It is the construction manager's responsibility to monitor the quality control function onsite and to witness and approve all site testing. He/she must also ensure that all documentation referring to quality control and testing is recorded and maintained in an orderly manner.

XI.   PRECOMMISSIONING—MODULES AND PRE-ASSEMBLIES

On offshore projects, in order to minimize/reduce expensive onsite and offshore work, it is essential that a good precommissioning system be developed during the engineering stage. This system will, essentially, involve the design and construction groups in engineering a physical testing program that will adequately check out the operational state of the facility prior to load-out.

   With major pre-assemblies and modules, it is planned that the degree of completion at load-out, be in the 95–98% range. This can only be accomplished with an effective precommissioning system. It has been common for modules to be loaded out in a less complete state, resulting in additional offshore work and at a significant cost addition.

XII.   HANDOVER OF CONSTRUCTION WORKS

The three stages between the start of construction work and plant operation are:

*   construction (erection and physical testing),
*   precommissioning (functional testing and preparation for operation), and
*   start-up (introduction of feedstock).

As the project moves from construction to precommissioning, the commissioning superintendent progressively takes on overall responsibility for the site from the construction manager and the contractors as various areas and systems reach mechanical completion. Although regular work is still done by the contractor, much of the precommissioning work will be undertaken with the assistance of the operating company's staff. This provides a means of familiarizing operations staff with the plant prior to the introduction of process feedstock.

XIII. HANDOVER CERTIFICATES

To ensure that the transition from each stage is clearly defined for any particular section of the project, a series of certificates is used, typically referred to as:

* mechanical completion certificate,
* takeover certificate,
* performance certificate,
* certificate of completion, and
* final acceptance certificate.

XIV. SAFETY DURING PRECOMMISSIONING/COMMISSIONING

The potential hazards arising during the changeover from construction through precommissioning to commissioning are probably greater than at any other time during construction or subsequent fulltime operation. This reflects the transfer of responsibilities, the involvement of commissioning and operating personnel unfamiliar with the plant, and the introduction of potentially hazardous materials into previously safe areas.

It is, therefore, essential at this stage that the responsibilities of all parties for complying with safety requirements are clearly defined and understood. Furthermore, it is also essential that the permit to work system, to cover the safe use of utility systems during precommissioning activities, is taken over at the appropriate time by the operating company in order to cover all plant systems.

Index